S. P. King

6 / 11 / 08

PLUMBING
ENCYCLOPAEDIA

PLUMBING ENCYCLOPAEDIA

Fourth Edition

R.D. TRELOAR

*Curriculum Development Manager for
the Professional Training Centre
Colchester Institute*

WILEY-BLACKWELL

A John Wiley & Sons, Ltd., Publication

This edition first published 2009
First edition published in 1989 by Blackwell Publishing Ltd
Second edition published in 1996 by Blackwell Publishing Ltd
Third edition published in 2003 by Blackwell Publishing Ltd
© 1989, 1996, 2003, 2009 R.D. Treloar

Blackwell Publishing was acquired by John Wiley & Sons in February 2007. Blackwell's publishing
programme has been merged with Wiley's global Scientific, Technical, and Medical business to form
Wiley-Blackwell.

Registered office
John Wiley & Sons Ltd, The Atrium, Southern Gate, Chichester, West Sussex, PO19 8SQ,
United Kingdom

Editorial office
9600 Garsington Road, Oxford, OX4 2DQ, United Kingdom

For details of our global editorial offices, for customer services and for information about how
to apply for permission to reuse the copyright material in this book please see our website at
www.wiley.com/wiley-blackwell.

Library of Congress Cataloging-in-Publication Data

Treloar, Roy.
   Plumbing encyclopaedia/R.D Treloar. – 4th ed.
      p. cm
   ISBN 978-1-4051-6164-0 (pbk. : alk. paper) 1. Plumbing–Encyclopedias.   I. Title.

   TH6109.T74 2009
   696′.1–dc22

                                                                         2008017682

British Library cataloging-in-publication Data

A catalogue record for this book is available from the British Library.

Set in 10/12.5pt Sabon by Graphicraft Limited, Hong Kong
Printed in Singapore by C.O.S. Printers Pte Ltd

1   2009

INTRODUCTION

This book is designed to be a multi-purpose aid, presenting the definition or explanation of fundamental words and terms to the apprentice, and enabling the more experienced to retrieve information quickly when faced with an unfamiliar task or term on various aspects of their craft. The original aim of the first edition, to preserve and define terminology, that has become increasingly hard to define and explain, has been maintained. This new edition updates the text with the latest terminology. Out-of-date terms have been removed. They have been replaced by over one hundred new entries, so providing even greater coverage of the terms used today. This edition takes account of the many changes in legislation and British standards that have occurred since the publication of the last edition.

The material is presented in encyclopaedic form to assist in locating the information required. When the main entry, a word or phrase, is followed by another word or phrase in brackets the latter represents an alternative term sometimes used. Within the text there are a vast number of cross-references, shown by the use of SMALL CAPITAL LETTERS. These words serve two purposes; firstly to indicate to the reader whether another related term is to be found elsewhere in the book and secondly to assist in reading further into the subject. Sometimes a particular term has more than one meaning and in many cases an alternative has been given, but remember more alternatives may exist.

It must be emphasised that this book is not intended to be read from cover to cover, but be used for reference as and when required – ideally little and often.

In these times when more and more plastics are being used, push fit joints are suitable for mains supply and the wiped solder joint is fast becoming building history, it is clear the old craft skills are diminishing. Nevertheless the field of plumbing covers a wide spectrum. Learning to become a knowledgeable skilled worker can take considerable time yet today one is expected to become competent in the trade very quickly. One of the main requirements is to act and work safely, logically and without panic should things begin to go wrong, thus confidence is something which needs to be gained. Confidence develops with experience. So if things do not proceed according to plan or one is asked to tackle a new task, take a few minutes to prepare yourself by looking through this book for the answer. It may well be there, if not, then panic! Forward planning is essential, even for the experienced. Should your new job be zinc roofwork – prepare by browsing through a few relevant entries in this book. As a student at college, should the subject in a future technology lesson be sanitary pipework, read about the subject the day before and you will find the topic more familiar and the lecture much more absorbing due to your interest.

Happy reading and good luck in your future life.

R.D. Treloar

ABLUTION FOUNTAIN   See WASHING FOUNTAIN.

ABLUTIONARY APPLIANCES   See WASTE APPLIANCES.

ABOVE GROUND DRAINAGE
A system of pipes which conveys surplus water or liquid sewage down to the house drains. The systems must be designed so as to prevent foul air from the drain entering the building. The above ground drainage system could be designed to remove either foul water from soil and waste appliances (see SANITARY PIPEWORK) or to remove surface water from roofs (see GUTTERS and RAINWATER PIPES). See also RAINWATER CONNECTIONS TO SOIL DISCHARGE STACKS.

ABS
Abbreviation for ACRYLONITRILE BUTADIENE STYRENE.

ABSOLUTE PRESSURE
The gauge pressure plus the ATMOSPHERIC PRESSURE. The gauge pressure is created by the head of water only:

$$\text{gauge pressure} + \text{atmospheric pressure} = \text{absolute pressure}$$

ABSOLUTE ZERO   See TEMPERATURE.

ABUTMENT
The position where a roof meets a wall or surface which extends above its surface. A good example would be where a chimney passes up through the roof.

AC   See ALTERNATING CURRENT.

ACCELERATED CIRCULATION
The speeding up, for example of water circulated through a hot water system, by means of a pump.

ACCESS COVER (ACCESS CAP OR PLATE/INSPECTION CAP, EYE)
A bolted on cover used to gain access to the inside of a drain or waste pipe, see ACCESS POINT. An access cover could also be fitted to a hot storage vessel, a boiler, or a flue pipe, etc. for internal inspection.

(See illustration over.)

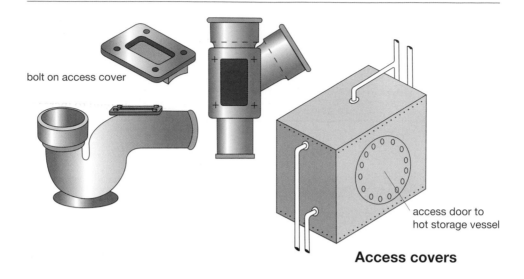

bolt on access cover

access door to
hot storage vessel

Access covers

ACCESS GULLY (INSPECTION GULLY)
A trapped gully which has a rodding point to permit easy cleaning. For example, see GULLY.

ACCESS PANEL
A removable panel set in the side of a wall or floor. Its purpose being to gain access to concealed pipework.

ACCESS POINT
A position where access can be gained for inspection and maintenance to a soil, waste or discharge pipe. All installations should be designed so that all internal pipework is within rodding distance and particular attention should be given to areas where blockages are most likely to occur, such as bends, branch junctions, and changes in gradients such as RAMPS and BACKDROPS. The top of a ventilating pipe is regarded as an access point, as is the removal of a trap. There should always be a rodding eye at the highest point of a drain and no drainage run should exceed 90 metres in length. See also ACCESS COVER.

ACETONE
A solvent used to dissolve acetylene gas to enable the acetylene to be compressed into cylinders for use when welding, etc. Acetone will dissolve and absorb about 25 times its own volume of acetylene gas at atmospheric pressure.

ACETYLENE GAS
A gas used in the oxy-acetylene welding process. Acetylene is produced by bringing calcium carbide into contact with water, when the two meet acetylene gas is given off freely and collected. In its natural state acetylene gas cannot be compressed or pressurised because it is prone to self-detonation and explosion, therefore it must first be dissolved in ACETONE to stabilise it. Acetylene needs to be pressurised in order to be stored and used in cylinders on site. When the acetylene gas has been dissolved in acetone it is often referred to as DA (dissolved acetylene). The cylinders

used are painted a maroon colour for easy identification. It must be noted that the thread into which the regulator screws is left-handed. See WELDING EQUIPMENT.

ACID RAIN
Rain that has fallen through the atmosphere and absorbed the gases present, to include CO_2 and SO_2. Industrial areas burning large volumes of fuel tend to increase the amount of gases within the environment and therefore increase the amount of acid rain. Owing to its high acidic nature, acid rain damages the environment in several ways including damage to trees, freshwater life and the erosion of stone buildings.

ACIDIC
A term used to mean having the properties of an acid. When dry this is of no significance but in the presence of water this property may cause chemical reactions to take place. Few building materials are acidic in nature but some timbers such as oak and red cedar are. When water is said to be acidic it implies that it has a corrosion tendency like 'acid'. Rain water which has fallen through the atmosphere absorbs small amounts of gases such as carbon dioxide and when it reaches the Earth it is no longer pure water but a mixture of water and carbon dioxide, in this state it could be called a weak carbonic acid. This water could then percolate through decaying vegetation and peat, introducing more carbon dioxide thus becoming more acidic and more aggressive to metals. Acidic water is often called SOFT WATER, see also ALKALINE and pH VALUE.

ACRYLIC (PERSPEX)
A THERMOPLASTIC which is extensively used in the manufacture of sanitary appliances. It is very tough and has a good resistance to abrasion.

ACRYLONITRILE BUTADIENE STYRENE (ABS)
A common THERMOPLASTIC material used extensively for small diameter waste and discharge pipes. It can withstand high water temperatures for long periods of time and also retains its strength in cold conditions, unlike PVC (POLYVINYL CHLORIDE). It is distinguished from PVC by the fact that it has a dull matt surface unlike PVC which is shiny, also ABS will burn if ignited, PVC will not.

ACTIVE FLUX   See FLUX.

ACTUAL CAPACITY
The volume or capacity of a cistern measured up to the water line. See also NOMINAL CAPACITY.

ACTUATOR
A device that converts an electrical signal into a physical action. For example, the operation of a motorised valve could be triggered in this way. The valve may receive a signal between 0 and 5 V, which in effect corresponds to the fully closed and open positions. Thus a signal of 2.5 V would allow the valve to open only half way. See also COMPENSATOR.

ADAMS SEWAGE EJECTOR   See PNEUMATIC EJECTOR.

ADJUSTABLE SPANNER   See SPANNER.

ADVENTITIOUS AIR/VENTILATION

Air that enters a building through gaps in doors, floors and window openings. In calculating the size of a VENTILATION GRILLE, adventitious air is taken into account when allowing for a supply of combustion air.

AERATED BURNER

A gas burner which has had air mixed with the gas prior to ignition; the most common example would be a Bunsen burner. The type of burner used for NATURAL GAS should be of this type otherwise the gas will not completely burn and will be unstable and rather smoky. For example, see GAS BURNER.

AERATED FLAME (BUNSEN FLAME)

A flame which burns at the point of combustion when using an AERATED BURNER. The flame tends to be noisy and has a bluish colour. Because air is added to the gas before ignition at the burner jet, it is possible that LIGHT BACK may occur. This can result if the gas velocity is too low.

AEROMATIC CYLINDER

A design of single feed cylinder, see SINGLE FEED HOT WATER SUPPLY.

AGA

A proprietary name for a continuous burning cooker, fuelled usually by either solid fuel or oil. These cookers are popular in rural districts where gas is unavailable; the cooker also provides a hot water supply and limited heating.

AGRÉMENT CERTIFICATE   See BRITISH BOARD OF AGRÉMENT.

AIR

A mixture of gases that surrounds the Earth forming the atmosphere in which we live.

Composition of air (approximate)

Gases	Percentage
Nitrogen	78.084
Oxygen	20.946
Argon	0.93
Carbon dioxide	
Helium	
Hydrogen	
Krypton	
Methane	0.04
Neon	
Nitrous oxide	
Xenon	
Water vapour	
Total	100%

A

AIR ADMITTANCE VALVE

A large valve fitted to the top of ventilating pipes which terminate internally within the building. It is designed to allow fresh air into the pipe but prevents odours from escaping. Air admittance valves should not be used on discharge stacks connecting to a drain which has an INTERCEPTING TRAP or to drains which are subject to SURCHARGING as it could result in the trap seal loss of appliances. There should only be a limited number of dwellings within the local vicinity fitted with these valves thus ensuring a good air circulation within the drain.

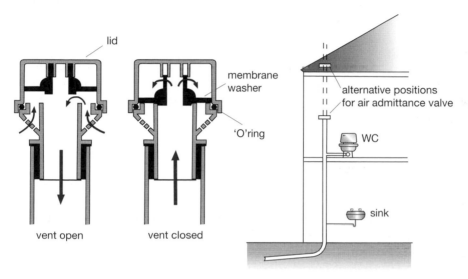

Air admittance valve

AIR BAG STOPPER   See INFLATABLE STOPPER.

AIR BLAST BURNER   See FORCED DRAUGHT BURNER.

AIR BREAK   See AIR GAP.

AIR CHANGE

The rate at which air in a room is changed by ventilation. A good healthy living room condition requires about 17 m³ of air per person per hour. To achieve this, a room requires about two air changes every hour; any more could reduce the heat of the room and cause cold draughts; any less might cause one to feel rather stiff and tired.

AIR COCK   See AIR RELEASE VALVE.

AIR COCK KEY (RADIATOR KEY)

A small key used to operate a manually operated AIR RELEASE VALVE.

AIR COMPRESSOR
A machine which is designed to compress air, thus giving it a pressure which can be used as a force to assist in the operation of lifting liquids such as in PNEUMATIC CYLINDERS and EJECTORS. The compressed air can also be used to operate pneumatic power tools or simply used to fill up car tyres.

AIR CONDITIONING
A process in which the air within an environment is monitored to provide the correct temperature, moisture content and air movement to maintain the well being of the occupants or equipment used, such as computer equipment. The air may be either cooled or heated as necessary within an air-handling unit and filtered to remove air-borne dust or pollen.

AIR CURTAIN
A system in which a stream of warmed or cooled air is blown directly downwards or across a door opening to prevent any sudden in-rush of un-conditioned air from outside. Air curtains are typically found at the entrances to shops.

AIR CUSHION   See AIR VESSEL.

AIR ELIMINATOR
An automatic air release valve fitted in steam heating systems. Should water be present, a float inside the fitting seals off the air outlet; should steam be present, a bellows containing a VOLATILE FLUID expands and likewise seals off the air outlet.

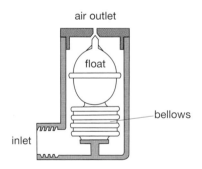

Air eliminator

AIR GAP (AIR BREAK)
The vertical distance between the water inlet or feed pipe to an appliance and the highest possible level the water could rise in that appliance. For example, see BACK-SIPHONAGE.

AIR LOCK
The entrapment of air in a pipe stopping or reducing the flow of water. Air locks are only found in water systems fed via a cold water storage cistern with little pressure and are the result of bad pipework installations. All pipes fed from storage cisterns must be run truly horizontal or with a slight fall allowing the air to escape from the system when filling.

A

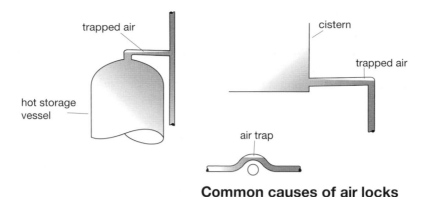

Common causes of air locks

AIR PURGER (AIR SEPARATOR)
As water pushes through a pipe it forms small air bubbles which can cause CAVITA-TION and air pockets. An air purger is a device which when fitted into the pipeline of a CLOSED CIRCUIT causes turbulence of the water and allows the air to rise to the top of the fitting where it can escape from the system via a vent pipe or an automatic air release valve. See also DE-AERATOR.

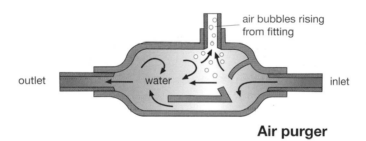

Air purger

AIR RELEASE VALVE (AIR COCK)
A manually operated valve used to release trapped air from pipework or vessels. These valves are located in the top of radiators or any high points in low pressure pipework where air might be trapped. The valve is operated by the use of a special air cock key.

AIR SEPARATOR   See AIR PURGER.

AIR TEST (PNEUMATIC TEST)
For gas installations see TIGHTNESS TESTING.

A pressure test applied to above and below ground drainage installations to check for leaks. An air test will find the smallest of holes therefore it is advisable to test in sections as you install your work to save a lot of time and trouble searching for leaks on completion.

The procedure for testing above ground drainage is as follows: fill all traps and stop up with DRAIN PLUGS any open ends remaining, then seal off the plugs with water; the water will block up the smallest of holes through which air could escape. The MANOMETER should now be connected to the pipework and air is pumped into the system. When the manometer reads 38 mm the hand pump valve is shut off (see sketch), and the pressure should hold for at least three minutes.

For below ground drainage the procedure is very similar except that there are no traps to fill and in most cases it is not possible to seal off the drain plugs with water, due to the pipe being horizontal. Air is now pumped into the pipework until a pressure of 100 mm reads on the gauge. Over a period of five minutes there must not be a pressure drop of more than 25 mm. Should a leak be evident the joints can be smeared with soapy water, which would bubble up as the air is escaping. See also SMOKE TEST and WATER TEST.

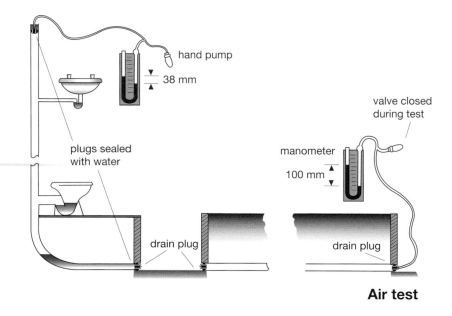

Air test

AIR VALVE   See AUTOMATIC AIR RELEASE VALVE.

AIR VENT
A non-adjustable grille or duct, that will allow the passage of air at all times. See VENTILATION GRILLE.

AIR VESSEL (AIR CUSHION/SURGE VESSEL)
An enlarged pipe or vessel fitted in a pipeline or on an appliance to overcome problems caused by WATER HAMMER. Shock waves created by pressure surges in water pipes are taken up compressing the air, in the vessel, thus it acts as a cushion. Because water absorbs air, the vessel has to be periodically recharged with air. This is done by turning the isolation valve off and opening the pet cocks draining out the water. It must be noted that if the air vessel is fitted it must be installed in the vertical position, entrapping the air. Also shown in the sketch is a method sometimes employed by plumbers when connecting to a ballvalve in the roofspace; the upstanding pipe traps air to form a cushion. A disadvantage of using an air vessel is that the water which absorbs the air over a period of time can give rise to corrosion problems. Another method which does not allow gases to be absorbed by the water but basically works on the same principle is a HYDROPNEUMATIC ACCUMULATOR. An air vessel will also be found on the delivery side of a LIFT or FORCE PUMP to absorb the force of the lifting action of pumping.

A

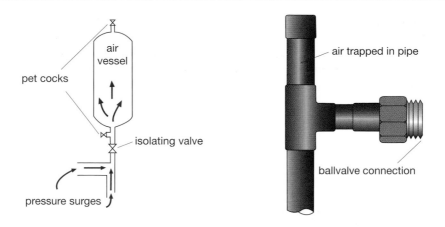

pet cocks

air vessel

isolating valve

pressure surges

air trapped in pipe

ballvalve connection

Types of air vessels

ALDEHYDES

A gas produced as the result of incomplete combustion of fuel. Aldehydes are basically oxidised alcohol and, when present within the atmosphere due to SPILLAGE of combustion products, cause irritation to the eyes.

ALIGNMENT TEST

A test to inspect drains for internal obstructions. One of the most common tests is the reflection test in which a light is reflected along the drain; the internal bore can be inspected by looking in another reflector placed further along the drain. Another test is the obstruction test in which a ball, 13 mm smaller than the diameter of the pipe, is inserted into the pipeline, in the absence of any obstruction it should roll freely down the INVERT of the drain.

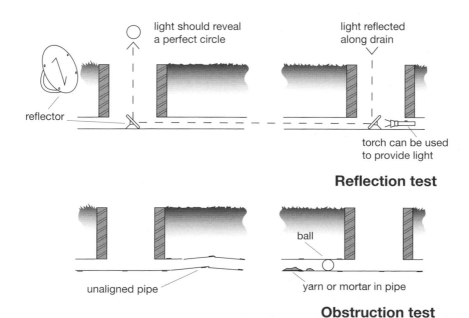

light should reveal a perfect circle

light reflected along drain

reflector

torch can be used to provide light

Reflection test

unaligned pipe

ball

yarn or mortar in pipe

Obstruction test

ALKALINE

A term used to mean having the properties of an alkali. When dry this is of no significance but in the presence of water this property may cause chemical reactions to take place. Alkaline building materials include cement and lime. When water is said to be alkaline, it means that it contains calcium bicarbonate, calcium sulphate and magnesium sulphate and is often referred to as HARD WATER. When rain water, which contains dissolved carbon dioxide, falls upon and soaks into soils containing limestone or chalk, the lime or chalk is dissolved and taken into suspension. See TEMPORARY HARD WATER and PERMANENT HARD WATER. Alkaline waters are not as aggressively corrosive as ACIDIC waters.

ALLEN KEY

A small hexagonal steel rod used as a tool to turn the recessed hexagonal head found in an 'Allen screw'. Generally these tools are supplied as a set to cover a range of sizes.

ALLOY

An alloy is a mixture of two or more metals or a metal mixed with a non-metallic element.

Some common alloys

Alloy	Components
Brass	Zinc and copper
Bronze	Copper and tin
Cast iron	Iron and 2–4% carbon
Gunmetal	Copper, tin and zinc
Mild steel	Iron and approx. 0.2% carbon
Pewter	Tin and lead
Solder	Lead and tin
Solder (lead free)	Tin and copper
Solder (lead free)	Tin and silver
Solder (lead free)	Tin and antimony
Stainless steel	Iron, chromium and nickel

ALTERNATING CURRENT (AC)

An electrical charge that changes direction of the electron flow from positive to negative, to positive again, etc. continually, unlike in a direct current (DC) where they always pass from positive to negative. In the UK this change of direction occurs at a rate of approximately 50 times per second (50 hertz). An alternating current is produced by a coil of wire rotating between the poles of a permanent magnet such as that found in an alternator at a power station. Direct current, on the other hand, is produced by items such as batteries, THERMOCOUPLES or dynamos.

ALTITUDE GAUGE

A gauge fitted on boilers of the larger type of central heating system to show the HEAD of water in metres. For an example of its workings see BOURDON PRESSURE GAUGE.

ALUMINIUM

A NON-FERROUS METAL produced for many uses, a plumber is most likely to come across it in sheet form. See also ALUMINIUM SHEET AND ROOF COVERINGS.

Chemical symbol	Al
Colour	bluish-white
Melting point	660°C
Boiling point	2467°C
Coefficient of linear expansion	0.0000234/°C
Density	2705 kg/m^3

ALUMINIUM SHEET AND ROOF COVERINGS

Sheet aluminium can be obtained in rolls with various widths ranging from 150 mm–900 mm. There are various thicknesses of the material available between 0.6 mm and 1 mm. Only two grades of aluminium are available for roofwork: super purity which contains 99.9% aluminium and commercial purity which contains 99% aluminium, the rest being made up of other elements. Aluminium has similar working properties to those of copper and in general aluminium roof details would be carried out along the same lines as those for COPPER SHEET AND ROOF COVERINGS. Aluminium does not CREEP but will become WORK HARDENED if COLD WORKED too much.

ALUMINIUM WELDING

To produce a good quality and well-contoured aluminium, weld cleanliness is essential. Aluminium melts at 660°C but the aluminium OXIDE which is present on its surface melts at 3000°C, therefore it is important to remove this film thoroughly with a wire brush and if necessary a degreasing fluid. When welding, a special BORAX and SILICON flux is required to dissolve and prevent any more oxide forming on the surface, making welding impossible. When welding aluminium the LEFT-WARD WELDING TECHNIQUE should be adopted and the blowpipe should produce a very slight CARBONISING FLAME. The work should be fully supported to prevent the molten metal falling away from the joint. To weld material 3 mm thick and over, the joint should be 'veed' to an angle of about 85° in order to give good penetration. The filler rods used are aluminium or an aluminium ALLOY with or without silicon. To weld aluminium successfully the TUNGSTEN INERT GAS WELDING approach is often adopted, which uses a gas shield to protect the surface from OXIDATION. See also VERTICAL WELDING.

AMBIENT AIR

The surrounding air.

AMMETER   See MULTI-METER.

AMPERE (AMP)

The unit used to measure the amount of electrical current flowing through a conductor.

amperes = WATTS divided by VOLTS

ANACONDA   See SEMI-RIGID STAINLESS STEEL.

ANCHOR BRACKET   See PIPE SUPPORT.

ANEMOMETER
A device used to measure the velocity of airflow. A simple anemometer consists of a set of rotating vanes that, via a set of gears, conveys the air speed to a dial measuring in m/s.

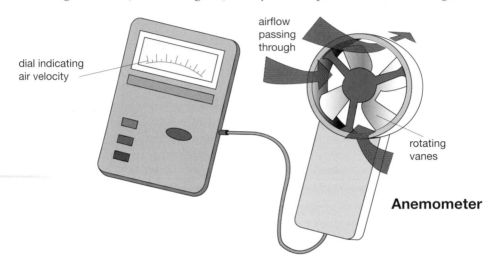

airflow
passing
through

dial indicating
air velocity

rotating
vanes

Anemometer

ANEROID GAUGE
A gauge designed to measure gas pressures within a pipeline up to a pressure of approximately 300 mbar (aneroid means 'not liquid'). The gauge consists of a small flexible bellows that expands and contracts in response to the pressure within the supply. The movement of the bellows is transferred through a lever to a pointer that registers on a scale.

ANGLE BASIN   (CORNER BASIN)   See WASH BASIN.

ANGLE BRANCH   (SPLAY BRANCH)
A branch in which the pipe or channel joins the main run of pipe at an angle of less than 90°. See BRANCH for example.

ANGLE WELDED JOINT   (FOR LEAD)
The welding of two sheets of lead lying at different angles. The weld is achieved by first fusing the two surfaces together (burning in) not using any filler rod, then applying two reinforcing welds. Where one sheet is in the vertical position (sometimes known as fillet welding) care must be taken to avoid UNDERCUTTING the metal and if at all possible it is easier to incline the work supporting it in a jig as shown. See also LEAD WELDING.

(See illustration opposite)

ANNEALING
The treatment of a metal or alloy to reduce its brittleness and improve ductility. Annealing is often referred to as softening of a metal. If a metal becomes WORK

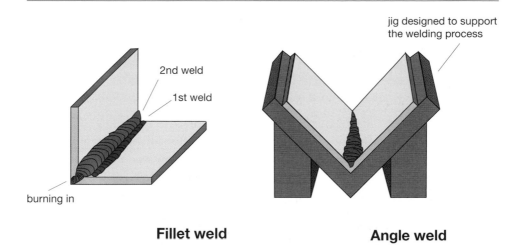

Fillet weld   **Angle weld**

HARDENED it may require softening before work is continued, otherwise it might fracture. Annealing is achieved by the application of heat. All metals are annealed at different temperatures and a method of gauging when the correct temperature is achieved is described below.

Annealing of metals

Aluminium	Heat until when a matchstick is drawn across the surface of the metal it leaves a charcoal line.
Copper	Heat until a dull red colour is seen then allow the metal to cool or quench it in cold water.
Iron	Heat to a cherry red colour, then allow it to cool slowly.
Lead	Heat the metal for a short period then 'spit' on it, if the spittle is seen to fly off the lead it is fully annealed, but if it just bubbles on the surface the right temperature has not yet been achieved.
Zinc	Heat until the metal is warm to the touch. Do not overheat.

ANNULAR SPACE
The void between two pipes or surfaces, such as that found between the walls of a twin walled flue pipe or between a flue lining and the chimney. See also ANNULUS.

ANNULUS
(1) A type of HEAT EXCHANGER fitted inside a hot storage vessel. For example, see INDIRECT CYLINDER.
(2) The space between two concentric circles (circles with the same centre but with different diameters).

ANODE
A metal which would be destroyed by a cathode in ELECTROLYTIC CORROSION. Copper would be an anode to silver whereas it would be a cathode to aluminium, see ELECTROMOTIVE SERIES.

ANODISING

A process in which a protective metal OXIDE coating has been applied to the surface of metals such as aluminium or magnesium in order to give CATHODIC PROTECTION. See also SHERADISING.

ANTHRACITE (HARD COAL)

A SOLID FUEL which consists of a black hard coal, representing the final product of the natural coal-forming process within the ground. Anthracite burns at a relatively slow speed with a pale-blue flame and intense heat but produces very little smoke. It is, therefore, particularly suitable for domestic use. Anthracite is sometimes mixed with the more bituminous coals to reduce smoke production.

ANTI-CAPILLARY GROOVE or GAP (CAPILLARY GROOVE)

A groove or gap which has been cut into the vertical timber face of a lead drip to prevent CAPILLARY ATTRACTION from taking place resulting in water possibly entering the building. For example, see LEAD SHEET AND ROOF COVERINGS.

ANTI-CORROSIVE BANDAGE

A wrap round protective tape used on underground pipes to protect them from CORROSION. The bandage is made from cotton or hessian impregnated with petroleum jelly. 'Denzo tape' is a well-known trade name for this type of wrapping.

ANTI-DRAUGHT FLAP (DRAUGHT FLAP)

A flap which is sometimes found fixed to the end of overflow and warning pipes to prevent cold draughts from blowing up the pipe into the building and storage cistern. The flap is usually made of copper and soldered to copper overflows. The biggest problem with these devices is due to continued coats of paint, with lack of use they become seized up and block up the pipe. With the introduction of plastic pipe these flaps are now rarely fitted. To prevent draughts from blowing up the overflow pipe it is a requirement to turn the overflow pipe down into the water at its connection to the storage cistern.

ANTI-FLOOD INTERCEPTOR   See ANTI-FLOOD VALVE.

ANTI-FLOOD VALVE

A drainage fitting designed to prevent SURCHARGING of the drain. Anti-flood valves are often only recommended for use with SURFACE WATER drains. With the ball type shown, when backflow or surcharging occurs the ball rises up to the underside of the access cover until it finally sits into the rubber seating, preventing any further flow in either direction until the flood water subsides. With the trunk valve any backflow causes the float to rise, thus turning the flap valve to close.

(See illustration opposite)

ANTI-FLOODING GULLY (TIDAL GULLY)

A gully fitted with a ball which floats and blocks its inlet should SURCHARGING of the drain occur. See ANTI-FLOOD VALVE.

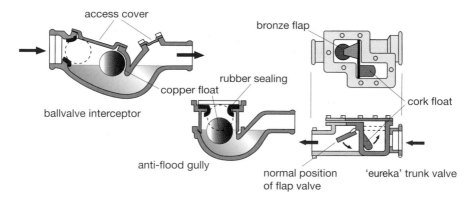

Anti-flood valves

ANTI-GRAVITY LOOP

A pipe which runs up vertically to a minimum height of one metre then down again, sometimes fitted in a non-circulating pipe to overcome the problems of unwanted GRAVITY or ONE PIPE CIRCULATION. At the top of the loop an automatic air release valve should be fitted to let out any trapped air.

ANTI-GRAVITY VALVE

A valve fitted vertically in the pipeline, designed to overcome unwanted GRAVITY CIRCULATION in central heating pipes. During the summer months when the boiler is used to heat the domestic hot water the radiators will sometimes get hot due to gravity circulation; by fitting an anti-gravity valve in the flow pipe to the radiators this problem can be overcome. The valve will only open when pressure is created by a pump. The pressure exerted by CONVECTION CURRENTS is insufficient to cause the valve to lift, see sketch.

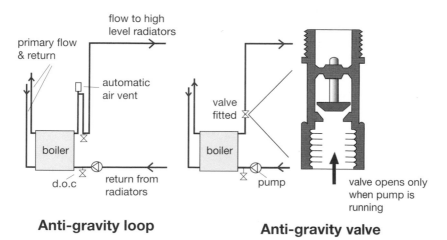

Anti-gravity loop   **Anti-gravity valve**

ANTI-LEGIONELLA VALVE

A special valve sometimes found installed on the inlet pipe to an expansion vessel. It is designed to ensure that fresh water is drawn into the vessel as it enters the system, rather than the existing system water, thereby minimising the possibility of bacteriological growth within the vessel itself.

ANTIMONY

A silvery white-coloured metal which expands slightly when it solidifies. It is used in some lead/tin solders to give the joint a certain degree of hardness. It also makes the joint more resistant to corrosion.

ANTI-SIPHON PIPE   See TRAP VENTILATING PIPE.

ANTI-SIPHON TRAP   See RESEALING TRAP.

ANTI-SPLASH FLOOR CHANNEL

A block channel designed to overcome the problems of spillage over its edges. The cross-section of its waterway is more than half a circle, for example, see URINAL.

ANTI-SPLASH RAINWATER SHOE   See RAINWATER SHOE.

ANTI-SUCTION VALVE

A valve sometimes found fitted to the inlet to a gas compressor. It is designed to cut off the supply of gas should the pressure within the pipeline fall to a set amount above ATMOSPHERIC PRESSURE, one purpose being to prevent the gas meter imploding should the suction be too great. Today these valves are generally being superseded by pressure switches, which close some form of SAFETY SHUT OFF VALVE.

ANTI-THEFT WASTE PLUG (CAPTIVE PLUG)

A waste fitting designed so that the plug cannot be completely removed.

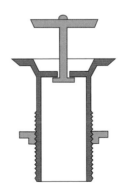

Anti-theft waste plug

ANTI-VACUUM VALVE (VACUUM BREAKER)

A valve designed to open a pipe to ATMOSPHERIC PRESSURE should the pressure within the pipe line drop, to below that created by the atmosphere, see sketch. There are basically two types of anti-vacuum valves; the 'atmospheric' type which opens by the force caused by atmospheric pressure and the spring loaded 'pressure' type. Anti-vacuum valves can be of assistance to overcome problems such as BACK-SIPHONAGE, and would be fitted downstream to a CHECKVALVE. Should there be a negative pressure within the pipe the valve would open admitting air, thus breaking the SIPHONIC ACTION. Anti-vacuum valves, when fitted, must be installed at a suit-able highpoint at least 150 mm above the highest possible contaminated water level. In addition to being fitted to pipes and on vessels anti-vacuum valves are also used on appliances such as some types of RE-SEALING TRAPS.

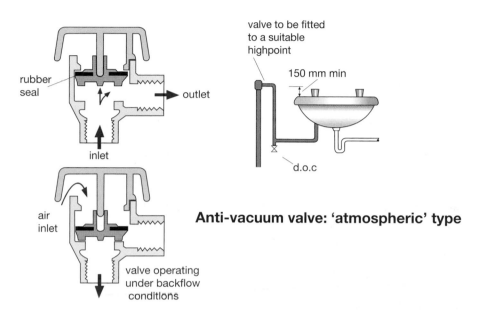

Anti-vacuum valve: 'atmospheric' type

ANTI-VIBRATION MOUNTINGS
Devices, usually made of rubber, that are designed to prevent the transfer of noise and movement from plant such as motors or pumps.

APPLIANCE SHUT OFF DEVICE
Term sometimes used referring to an ISOLATION VALVE located next to an appliance to facilitate turning off the gas or water supply.

APPRENTICESHIP
An agreement (known as an indenture) between an employer and an employee in which the employee agrees to work under a skilled craftsman on reduced pay for a number of years on the condition that the employer teaches him/her a craft.

APRON FLASHING
A COVER FLASHING which lays out over a sloped roof structure as it meets an ABUT-MENT. It is designed to prevent the entry of water into the building. For example, see CHIMNEY FLASHING.

ARC WELDING
A method of welding metals by using a heat source which is generated by an electric current made to 'arc' across from a positive to a negative terminal. Arc welding is carried out by first connecting the negative terminal to the work to be welded, then a welding rod (welding electrode) is fixed into the insulated electrode holder which is the positive terminal. When the welding rod is brought down to a close proximity to

the work to be welded, an electric current arcs across the two and fuses, welding the joint. The welding rod has a special coating which shields the molten metal from OXIDATION during welding. This coating burns to form a slag which is less dense than the molten metal and thus floats to the surface. As the weld cools the slag solidifies and is easily chipped off. Once one has mastered the art of maintaining the correct gap the process of welding is quite easy to carry out. As a safety precaution always use the correct eye shield with this type of welding as the intensely bright flame can burn your eyes. See also TUNGSTEN INERT GAS WELDING and METAL ARC GAS SHIELDED WELDING.

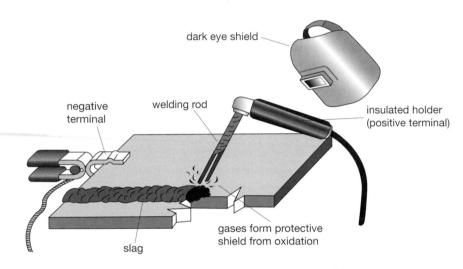

Arc welding

ARGON ARC WELDING   See TUNGSTEN INERT GAS WELDING.

ARTESIAN WELL
A well or borehole driven through the impervious strata of the Earth's surface into strata that receives water from a higher level than the well. With a true artesian well there is sufficient pressure of water to force the water flow up the well without the need for any pumping arrangement. See WATER TABLE and WATER CYCLE.

ASBESTOS
Fibrous material used in the form of cloth and rope or mixed with cement to form asbestos/cement pipes, cisterns and sheets. It is highly resistant to heat and was often used in the manufacture of flue pipes and fittings. The use of asbestos can be detrimental to health, therefore precautions should be taken when working with it.

For a comprehensive guide to working with asbestos, referral should be made to the *Asbestos Regulations* (HMSO), but as a general guide one should at least observe the following basic precautions:

(1) Avoid making dust and breathing it in. This can be achieved by dampening the material.
(2) Do all the cutting of material outside.
(3) Upon completion, change clothes and wash or shower as necessary.

ASBESTOS CEMENT
A mixture of cement and about 10% asbestos fibre. See ASBESTOS.

ASBESTOS-CEMENT CORD (PC4)
Jointing material which was sometimes used to join large diameter socketed drainage pipes, it was packed into the joints in a wet condition to bring about the chemical reaction needed to unite the asbestos fibres with the cement.

ASBESTOS CEMENT PIPES
A material once used for drainage and large diameter water mains pipework. It comes in long lengths and thus reduces the amount of joints required. Some soils tend to be aggressive to this material and destroy it, therefore protection of the pipe may be required. Asbestos cement pipes were also used for flue pipes to some gas appliances in which case socket and spigot joints are used, the joint being made with fire cement and the socket fitted uppermost, see FLUE PIPES.

ASBESTOS MAT
A specially designed piece of matting used to pass around pipes when making joints which require a flame. The mat protects the decorations, other pipework or more flammable material which might come into contact with the flame.

ASCENDING SPRAY (DOUCHE SPRAY)
A rose which sprays upwards, unlike shower roses which spray water downwards. Ascending sprays are found in sanitary appliances such as BIDETS and EYE WASH FOUNTAINS.

ASIATIC CLOSET (ASIAN CLOSET)   See SQUATTING WC PAN.

ASTRAGAL JOINT
A socketed joint used on lead rainwater and discharge pipes which incorporates an ornamental moulding called an astragal.

ATMOSPHERE
The envelope of gases and vapour, surrounding the Earth. See ATMOSPHERIC PRESSURE.

ATMOSPHERE SENSING DEVICE (ASD)
A device that senses the condition of the atmosphere in order to determine whether it is safe to allow the continued operation of a gas burner. Should continued spillage be occurring with an open flued gas burning appliance, the environment would eventually lack oxygen (i.e. become vitiated) and CARBON MONOXIDE would be produced leading to the possible death of the occupants of the room.

Two types of ASD are found:

(1) Those that sense the temperature of the area around the draught diverter. Where spillage occurs a THERMISTER positioned there becomes hot and breaks the electrical circuit, usually via a THERMOCOUPLE INTERRUPTER.
(2) Those that detect vitiated air. These cause the pilot flame to burn incorrectly and away from the thermocouple tip. Hence it cools and the THERMOELECTRIC VALVE closes.

ATMOSPHERIC BURNER (NATURAL DRAUGHT BURNER)
A burner design that obtains all the primary and secondary air that is used for combustion directly from the atmosphere, drawing in the air due by natural means. These are unlike those burners in which an electric fan is used to suck and blow in the air; these are referred to as FORCED DRAUGHT BURNERS.

ATMOSPHERIC CORROSION
Corrosion caused by the gases and moisture in the atmosphere. FERROUS METALS corrode by rusting. This is a reaction brought about by the oxygen in the air combining with the metal to form an oxide coating commonly called rust. Rust falls away exposing fresh metal underneath and the process continues until it rusts away completely. NON-FERROUS METALS are attacked by such gases as sulphur dioxide and carbon dioxide, which are present in the air, but the film which forms on the surface of non-ferrous metals does not flake off like rust, instead it tends to protect the metal from further corrosion. In areas around factories and towns corrosion is much more of a problem than in country areas, because of the increased level of gases, such as sulphur dioxide, in the atmosphere. Along the coastline where there is a lot of salt in the air aluminium should not be used, as the rain water would tend to be a strong alkali which destroys this metal.

ATMOSPHERIC PRESSURE
The pressure created by the weight of the atmosphere pushing down on to the Earth. The pressure at the top of a mountain is different from that in a valley below sea level. The pressure created at sea level would be 101.3 kilonewtons per metre squared (101.3 kN/m^2).

ATOM
The smallest chemical particle into which a substance can be divided. An atom, however, is made up of smaller parts, namely a central nucleus and electrons that orbit around it. See also MOLECULE and CONDUCTOR.

ATOMISATION
The break-up of a liquid into a fine spray. It is generally brought about by passing the liquid through a small hole under pressure. See ATOMISING BURNER.

ATOMISING BURNER
A burner jet used on oil fired boilers. Basically the burner works on the principle of breaking up the oil into a very fine misty spray (atomisation) which can be ignited by an electrode. The more common type of atomising burner is the pressure jet type, for example, see PRESSURE JET BURNER. See also VAPORISING BURNER.

AUGER (PIPE GOUGE/SCALLOPE)
A tool which has a conical shaped cutting edge and is used to make holes in lead sheet and pipe; for example, see GIMLET. See also DRAIN AUGER.

AUTOGENOUS WELDING
A welding process in which the two surfaces being welded are of the same material; the filler rod, if used, is also of the same material. A prime example would be LEAD WELDING.

A

AUTOMATIC AIR RELEASE VALVE (AUTOMATIC VENT)

A valve designed to discharge air from and admit air into a water pipe, sometimes fitted to a high point on a low pressure plumbing system to overcome the problem of an AIR LOCK. The type shown works on the principle that should water be evident in the pipe the float would rise due to its buoyancy and seal off the opening. See also AIR ELIMINATOR.

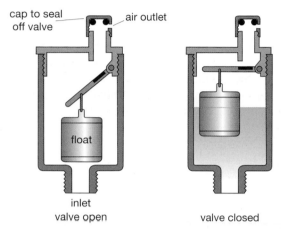

cap to seal off valve   air outlet

float

inlet
valve open   valve closed

Automatic air release valve

AUTOMATIC BYPASS

A device that opens or closes automatically depending upon the pressure build-up within a fully pumped central heating system. As the motorised valves or thermostatic radiator valves (TRVs) close, the pressure within the pipework would increase, causing the spring-loaded valve to open, thereby maintaining the correct water flow rate through the boiler and minimising noise build-up.

AUTOMATIC CHANGE-OVER VALVE   See REGULATOR.

AUTOMATIC FLOW CUT OFF DEVICE

A valve fitted in the pipe line to automatically shut down the flow of water. Such a device could be either an electrically operated MOTORISED VALVE, being controlled from a time clock, or the valve could work only when the pressure builds up in the pipeline, such as a HYDRAULICALLY OPERATED VALVE which might be used to conserve water feeding an AUTOMATIC FLUSHING CISTERN.

AUTOMATIC FLUSHING CISTERN

A flushing cistern designed to discharge its contents of water at regular intervals into a urinal. The rate at which the water will flush depends upon the rate at which the water is fed into the cistern and for a single installation this should not exceed ten litres per hour. To prevent wastage of water from these cisterns, during times when the cistern is not used such as at weekends, an AUTOMATIC FLOW CUT OFF DEVICE should be fitted. See also DISC FEED.

The automatic flushing cistern shown operates as follows:

(1) The water upon filling rises equally inside and outside the siphon until the air hole is reached.
(2) As the water rises the air is trapped inside the dome thus becoming compressed and eventually, when enough pressure is created, the water is forced out of the 'U' tube and the air pressure inside the dome becomes reduced.
(3) The reduced air pressure immediately allows SIPHONIC ACTION to start, thus flushing the appliance.
(4) When the flush is finished, the water in the upper well is siphoned out through the siphon tube and refills the lower well and 'U' tube.

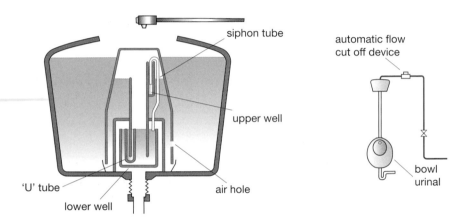

Automatic flushing cistern

AUTOMATIC FLUSHING TANK

A vessel used to hold a quantity of water which is periodically discharged into a system of drains giving them a good flush through. They are used on drainage systems which do not have a sufficient fall to give them the required self-cleansing gradient. The tank is supplied via a drip feed from the water authorities main, or, if accessible, rainwater, which is free from foreign matter and will not block up the siphon. The automatic flushing tank works as follows:

(1) The water upon filling rises up inside the tank and compresses the air inside the 'bell'.
(2) When the water rises to a certain height it creates enough pressure to force the water out of the water trap. Air is then forced out allowing SIPHONIC ACTION to start immediately.
(3) As the water level falls to the bottom of the bell, air is admitted and breaks the siphonic action.

(See illustration opposite)

AUTOMATIC PNEUMATIC COLD WATER SUPPLY

A system of cold water supply to a high rise building above the height to which the water main will travel. In this system water is pumped to the highest point. During

A

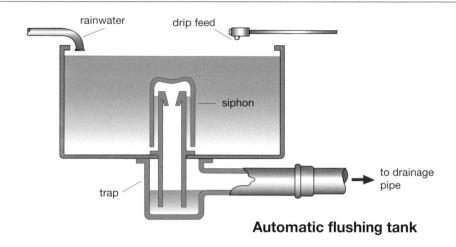

Automatic flushing tank

the on/off cycles of pumping an AUTOMATIC PNEUMATIC CYLINDER is used to give a constant pressure to all appliances and draw off points (see sketch). See also BOOSTED SYSTEM OF COLD WATER SUPPLY.

AUTOMATIC PNEUMATIC CYLINDER
A specially designed cylinder or tank used to give a constant pressure to the cold water supply in high rise buildings during the on/off cycles of pumping. Water is pumped into the bottom of the cylinder and compresses the air inside; when the water has filled to a predetermined level the pump switches off. The compressed air now acts as a force to push the water back out of the cylinder as it is required at the draw off points. Eventually the water level inside the cylinder falls to such a level that the pressure switch cuts in the pump to switch on again, re-pressurising the cylinder. After a period of use, some of the air inside the cylinder becomes absorbed into the water and therefore has to be replaced. This is achieved by an air compressor switching on when the water is at its highest level, as indicated by the FLOAT SWITCH. The compressor continues to run until the required air volume has been achieved. Modern systems employ the use of a large rubber bag into which the water flows. This separates the water from the air and thus prevents the air from being absorbed which also results in the compressor no longer being required. To conserve the air pressure inside the cylinder, a DELAYED ACTION BALLVALVE is fitted in the storage cisterns; this also reduces the frequency of pumping. See also AUTOMATIC PNEUMATIC COLD WATER SUPPLY.

(See illustration over.)

AUTOMATIC VENT   See AUTOMATIC AIR RELEASE VALVE.

AUXILIARY CIRCULATORS
A method of installing an alternative means of heating the domestic hot water supply during the summer months. During the winter the water could be heated by a BACK BOILER and during the summer it could be heated by a GAS CIRCULATOR.

AUXILIARY TANK   See SUPPLEMENTARY STORAGE SYSTEM.

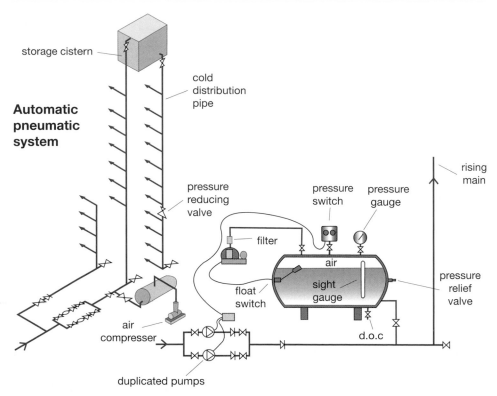

**Automatic
pneumatic
system**

storage cistern

cold
distribution
pipe

rising
main

pressure
reducing
valve

pressure
switch

pressure
gauge

filter

air

pressure
relief
valve

float
switch

sight
gauge

air
compresser

d.o.c

duplicated pumps

Automatic pneumatic cylinder

AXIAL FLOW FAN

A fan in which the impeller fins are connected directly on to the drive shaft of the motor, which in effect means that the air flow passes over the motor which is parallel to the axis of the fan. An example will be seen under the entry of CENTRIFUGAL FAN.

BACK BOILER (BLOCK BOILER/RANGE BOILER)

A boiler fitted behind or in a domestic fire; in the case of solid fuel the heat from the fire heats up the room as well as the water. See BOILER.

BACK or SIDE INLET GULLY

A type of GULLY that has an additional connection entry point for a waste pipe.

BACK-DROP CONNECTION (TUMBLING BAY)

A section of vertically run drain pipe joining the INVERT LEVEL in an INSPECTION CHAMBER and a drain pipe at a higher level. A backdrop would be used if the vertical distance between invert levels of the drain exceeded 680 mm. The backdrop could be installed inside the inspection chamber if it is run in cast iron or plastic, but should clayware pipes be used, it must be fitted externally to the inspection chamber and encased in concrete. See also RAMPS.

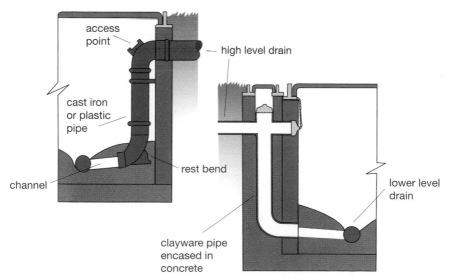

Back-drop connection

BACK-PLATE ELBOW (BIBTAP WALL FLANGE)

A pipe fitting designed to be fixed to the wall. It has one connection for a pipe and one connection has a female thread designed to receive a BIB TAP.

(See illustration over.)

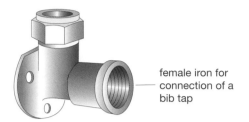

Back-plate elbow

female iron for connection of a bib tap

BACK PRESSURE
Term sometimes used to mean compression, see TRAP SEAL LOSS.

BACK-PRESSURE VALVE
A NON-RETURN or FLAP VALVE.

BACK-SIPHONAGE
The backflow of water by SIPHONIC ACTION from an appliance or storage cistern into the pipe feeding it, thus contaminating the water supply. To overcome the problems caused by back-siphonage, the water authority lay down strict guide lines which must be observed and any pipe on mains supply connected to appliances and cisterns must have some means of backflow prevention. This could be achieved by ensuring that where an outlet discharges allowance is made for an air gap of at least that shown in the chart (*over page*).

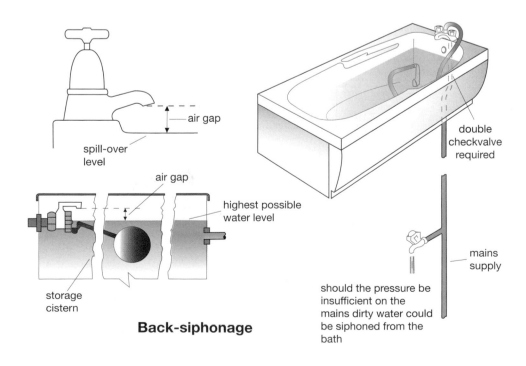

air gap

spill-over level

air gap

highest possible water level

storage cistern

double checkvalve required

mains supply

should the pressure be insufficient on the mains dirty water could be siphoned from the bath

Back-siphonage

Internal pipe diameter	Vertical distance (air gap) between outlet and highest possible water level
up to 14 mm	20 mm
15 mm–21 mm	25 mm
22 mm–41 mm	70 mm
over 41 mm	twice the internal bore of the inlet pipe

If a specified air gap cannot be achieved some other means of backflow prevention must be catered for. In the case of hose pipes and shower hoses connected to the mains supply a DOUBLE CHECK VALVE ASSEMBLY or similar arrangement such as a CHECKVALVE followed by an ANTI-VACUUM VALVE must be used.

BACKFIRE

The pre-ignition of fuel gases in the mixing chamber of a blowpipe when using oxy-acetylene equipment. See FLASHBACKS and LIGHT BACK.

BACKFLOW

Water flowing in a direction contrary to the intended direction of flow.

BACKFLOW PREVENTION DEVICE

A fitting or combination of fittings specifically designed to overcome BACKFLOW. For examples see DOUBLE CHECKVALVE ASSEMBLY and ANTI-FLOOD VALVE.

BACKGUIDE

A guide used to support the back of the pipe in the process of bending. See BENDING MACHINE.

BACKGUTTER

The IMPERVIOUS roofing material fitted behind a chimney or other penetration which passes through the roof. It is designed to prevent the entry of rainwater into the building. See CHIMNEY FLASHING for an example of its application.

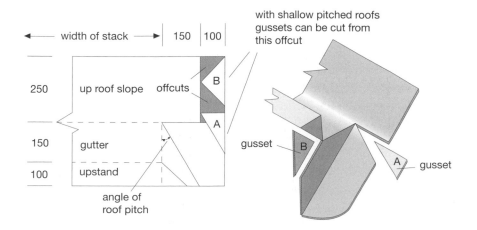

Setting out detail for lead welded backgutter

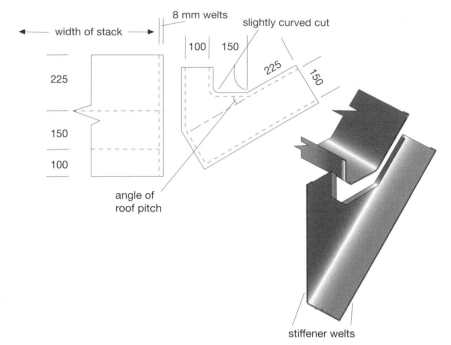

Setting out detail for copper backgutter

BACKHAND WELDING   See RIGHTWARD WELDING TECHNIQUE.

BACKNUT (LOCKNUT OR JAMNUT)
The nut which is screwed onto pillar taps, ballvalves or other pipe fittings, its purpose is to secure them to a fixture, i.e. taps to sinks, etc. See PILLAR TAP and BALLVALVE for an example. Some backnuts are dished on one face to retain a GROMMET and are used to assist in making water tight joints; for example, see LONG SCREW.

BADGER
When clayware pipes are joined by means of cement joints a badger is used to clear out any cement droppings which may have fallen through into the internal surface of the pipe, possibly causing a blockage. The badger is a homemade tool consisting of a piece of semi-circular timber (the radius of the pipe) nailed onto the end of a pole. It is passed into the pipe, lifting its semi-circular head over the cement droppings, and held down onto the INVERT and withdrawn.

BAFFLE
A device fitted to an assortment of fixtures and fittings. A baffle is basically a device to divert the flow of liquid or gas from its normal course of flow. For example, see DOWN DRAUGHT DIVERTER and ELECTRIC WATER HEATING.

BAG TRAP
An old-fashioned type of trap used on vertical waste pipes where space was restricted. For example, see TRAP.

BALANCED COMPARTMENT

A situation where an open flued appliance has been installed into a large cupboard and is, in effect, 'room sealed'. The air supply to the compartment is taken from the external environment at a point adjacent to the flue gas extract terminal. The door to the compartment must be self-closing and be fitted with some form of draught excluder to ensure the compartment remains sealed. A switch should be incorporated to make the appliance inoperable if the door is left open.

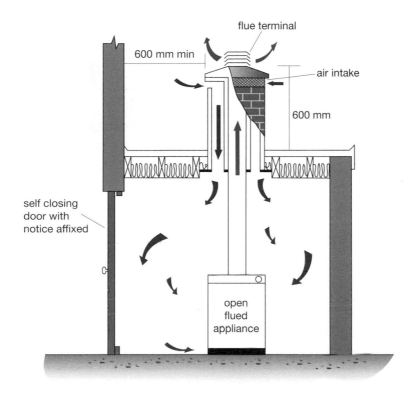

Balanced compartment

BALANCED FLUE

A room sealed gas, or oil fired appliance, in which the air supply and flue gas extract are at the same pressure. The balanced flue appliance takes its air from outside and also discharges its combustion products outside. Most balanced flued appliances today use a fan to assist the extraction of flue products from the building, allowing for greater freedom in siting as well as improving appliance efficiencies. Older natural draught balanced flued appliances can also be found and for installation purposes they are generally restricted to installation upon an outside wall. When siting the terminal thought must be given to its location in order to prevent problems and avoid flue gases re-entering the building through openings. See FLUE TERMINAL for suggested location positions. If the terminal is located less than 2 m above an accessible area, a flue guard will be required to prevent someone touching it and getting burnt. See also ROOM SEALED APPLIANCE.

(See illustration over.)

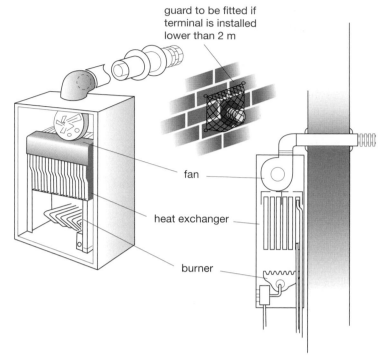

guard to be fitted if
terminal is installed
lower than 2 m

fan

heat exchanger

burner

Fan-assisted balanced flue

BALANCED PRESSURE
The equal or practically equal pressures at the hot and cold inlet points supplying a MIXING VALVE or fitting.

BALANCING
The adjustment of the HEAT EMITTERS in a central heating system so that approximately equal amounts of heat are distributed around the system. In the case of systems heated by radiators the LOCK-SHIELD valve is adjusted until the correct temperature is achieved. It is not uncommon for the first radiator in the system to have its lockshield valve only opened a little in order to achieve equal balancing to all the radiators. One method to check the temperature when balancing is to attach a temporary dial thermometer to the return pipe from each heat emitter.

BALCONY OUTLET
A rainwater fitting connected to a vertical rainwater pipe. It is designed to collect the rainwater from a flat roof or balcony.

BALE TACK (CLEAT)
A fixing clip used to hold the free edge of COVER or STEP FLASHINGS. The bale tack consists of a piece of copper or lead 40 mm wide which is fixed or fitted over the upstand of the sheet weathering material prior to fixing the flashing. See SHEET FIXINGS.

BALL TEST   See ALIGNMENT TEST or OBSTRUCTION TEST TO DRAINS.

BALLAFIX
The trade name for one design of BALLVALVE, being a quarter turn valve fitted in a pipeline to isolate an appliance.

BALLFEED
A device which is sometimes found connected into domestic FEED AND EXPANSION CISTERNS to control the flow and level of the water. It is designed to replace the normal float operated BALLVALVE and give a trouble free service; because the normal ballvalve has very little use it often becomes seized up. The ballfeed consists of a flexible inlet tube with a float at its outlet nozzle. The ballfeed is supplied with water from a storage cistern and not directly from the incoming main; basically it operates as follows:

(1) Upon installation, the F and E cistern must be carefully positioned so that point 'A' on the ballfeed is in line with the normal water level in the storage cistern.
(2) When the water is supplied, the storage cistern fills and water flows to the ballfeed fitted in the F and E cistern. As the float raises the inlet arm to the level of that of the water in the storage cistern, the water ceases to flow and no more water can pass through the ballfeed, thus the water settles at a lower level.
(3) When the water in the heating system expands, the water rises and lifts the float; conversely, should water be lost from the system the float lowers to allow more water through the ballfeed.

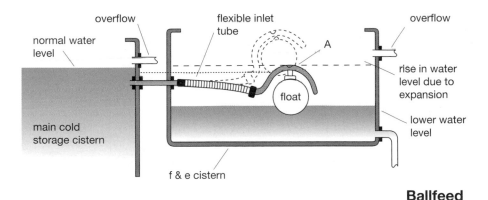

Ballfeed

BALLFLOAT   See FLOAT.

BALLJOINT
A Scottish term sometimes used meaning a wiped underhand ROLLED JOINT.

BALLOON GRATING (BIRD OR DOMICAL CAGE)
A wire or plastic guard fixed over an open VENTILATING PIPE to prevent birds from nesting there. Balloons are also sometimes fitted in a gutter outlet to reduce the likelihood of blockage due to leaves, etc.

BALLPEIN HAMMER   See HAMMER.

BALLVALVE (FLOAT OPERATED VALVE)

A valve used to shut off the supply of water in a cistern to a predetermined level. The ballvalve works on the principle of a lever. The valve is open when the float is in the lower position. As the cistern fills the float becomes buoyant and rises with the water level; as the float and arm rise they pivot at the split pin, thus pushing the washer towards the seating closing off the supply of water.

The ballvalves shown are of the DIAPHRAGM and PORTSMOUTH types but also see BROADSTONE, CROYDON and EQUILIBRIUM BALLVALVES. The sketches show that the water passes through a hole in the seating when filling. The water supply to a ballvalve could be at any pressure, e.g. on mains supply (high pressure) or fed from a storage cistern (low pressure). It would be pointless to have a small hole in the seating if the supply of water was poor, as it would take too long to fill; conversely if a seating with a large hole was used on a high pressure supply the water pressure would be too strong against the upward thrust of the ballfloat, consequently the supply would not shut off. In many cases the seating is interchangeable and labelled HP, MP, or LP.

Under current water regulations one must ensure that an SIR GAP is achieved to overcome any problems of BACKSIPHONAGE; in most cases this is achieved by fitting a ballvalve with a top entry as in the diaphragm type shown. See also DELAYED ACTION BALLVALVE.

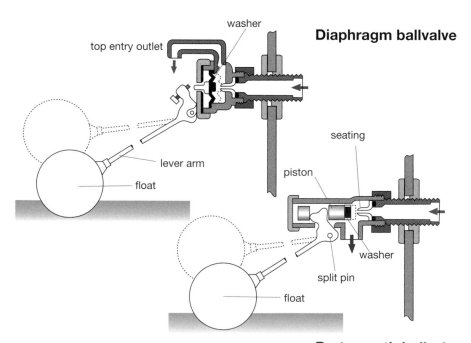

Diaphragm ballvalve

Portsmouth ballvalve

Ballvalve closing pressures		
Seating type	Diameter of hole mm	Max pressure kN/m^2
High pressure (HP)	3	1380
Medium pressure (MP)	4–5	690
Low pressure (LP)	6	276

BALLVALVE

The name given to a special quarter turn valve which differs from a PLUG COCK. Instead of a tapered plug through which water passes, a specially designed ball joint is used.

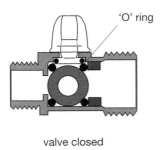

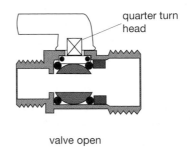

valve closed   valve open

Ballvalve

BALLVALVE MURMUR   See WATER HAMMER.

BALLVALVE RESEATER TOOL   See RESEATING TOOL.

BAR (ONE ATMOSPHERE)

A way of expressing pressure in steam, water and gas installations. One bar equals 100 kN/m². It is worth noting that atmospheric pressure is 101.3 kN/m² therefore one bar is often referred to as one atmosphere.

BARGE

Alternative name for a FRONT APRON.

BARREL

Term meaning LOW CARBON STEEL PIPE.

BARREL NIPPLE (SPACE NIPPLE)

A short piece of LOW CARBON STEEL PIPE (barrel) with a MALE THREAD at each end. Sometimes these nipples are made by the plumber who cuts a long thread onto a piece of pipe and cuts it to the length required. This type of nipple is then often called a running nipple.

BARRIER PIPE

A plastic pipe system that is manufactured to ensure air will not be drawn through the molecular structure of the plastic material. This is achieved by the inclusion of an aluminium inner structure or cross-linking the polythene.

BARRON BEND (SLIPPER BEND/THREE QUARTER BEND)

A channel bend seen inside some INSPECTION CHAMBERS. A barron bend is often called a three-quarter bend because only one-quarter section is missing from the pipe. For example, see CHANNEL BEND.

BASE (ION) EXCHANGE WATER SOFTENER

A water-softening process where HARD WATER is allowed to pass through a pressure vessel containing ZEOLITES, or a resin which absorbs the calcium (and magnesium) salts in the water. After a period of time the zeolites become clogged with calcium and need to be regenerated with common salt (sodium chloride). This is done by diverting the inlet water, closing valves A and opening valves B. This backwashes the water and salt solution (brine) through the zeolite bed; this water then goes to waste. After a suitable period of time the backwash ceases and both valves B close and the A valves open to put the softener back into operation. This process of back-washing the water is carried out automatically and usually timed via a timeclock to be done about 3 am giving no inconvenience to the householder. Prior to the softener inlet connection a branch pipe should be taken from the mains to provide a drinking water supply. See also ZEOLITES, WATER CONDITIONER and CLARKS PROCESS.

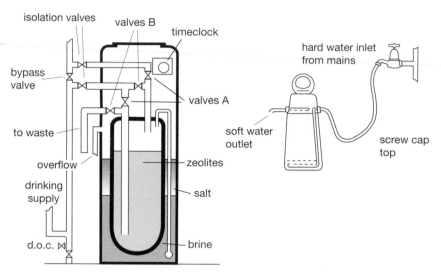

Base exchange water softener

BASIN   See WASH BASIN.

BASIN SPANNER (BASIN WRENCH/LAVATORY UNION KEY)

A tool designed to reach the backnuts on PILLAR TAPS behind baths, washbasins and sinks, etc. There are two basic designs which are available in different sizes. One type works on the principle of a SPANNER and is often known as a 'Shetack', this being a trade name, the other type works on the principle of a WRENCH.

(See illustration over.)

BAT   See WEDGES.

BATH

A SANITARY APPLIANCE in which the human body can be immersed and cleansed. Baths are available in enamelled cast iron, pressed steel or in various types of plastics

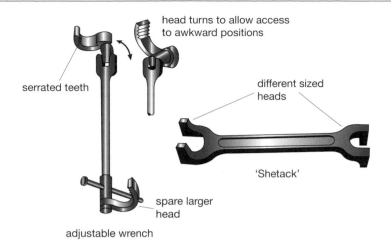

head turns to allow access to awkward positions

serrated teeth

different sized heads

'Shetack'

spare larger head

adjustable wrench

Basin spanners

and acrylics. When fixing a bath it is most important to ensure that you follow the makers' recommended methods of securing the bath in position, otherwise with all the weight they have to support they will move. Baths in general should be fixed as low as possible to assist getting in and out, but in many cases the height is deter- mined by the bath side panel. Baths no longer only come in the standard oblong shape, but are available in an assortment of shapes and sizes. The standard length of a rectangular bath is 1.7 m. It must be noted that due to the thickness of material used in the production of some baths a TOP HAT may be required when securing PIL- LAR TAPS. There are two basic designs of oblong bath; the 'tub' pattern and the 'magna' pattern. The tub pattern of bath is supplied with fixed feet which cannot be adjusted, the tops have rolled edges and the end of the bath is rounded. The magna pattern, on the other hand, has adjustable feet, shelf-like top edges and is rectangu- lar in shape. The tub pattern is no longer made. For the waste connections to baths, see WASTE FITTING. See also WHIRLPOOLS AND SPAS.

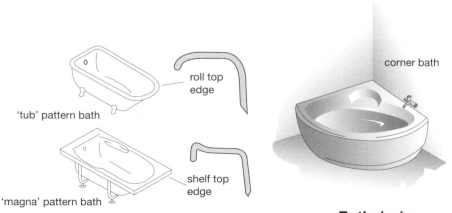

'tub' pattern bath

roll top edge

'magna' pattern bath

shelf top edge

corner bath

Bath design

BATH PANEL
A panel made of plastic or timber to hide the underside and pipework behind a bath. When fitting a bath panel, it should be designed so that it can be removed to gain access for inspection or repair work.

BATTEN ROLL
An expansion joint used on copper and zinc roofs. See COPPER OR ZINC SHEET AND ROOF COVERINGS. See also WOOD-CORED ROLL.

BATTER
A steep slope, near to VERTICAL.

BATTING IRON
A cold chisel with a blunted end. It is used to knock the lead WEDGES (bats) into the brickwork. In Scotland the term is also used to mean a CAULKING TOOL.

BAY
One of several uniform divisions of a flat roof or wall CLADDING, such as the distance between DRIPS and ROLLS. For example, see LEAD SHEET AND ROOF COVERINGS.

BAYONET CONNECTOR
A device that allows for the quick release to a hose system as found in air lines and to the connection of a gas cooker. When the hose has been removed, the bayonet connector valve remaining on the supply pipe holds the system sound.

BEACON WASTE
A SKELETON WASTE FITTING used for lead waste pipes.

BEADED EDGE   See STIFFENING BEAD.

BEADS
Alternative name for BOBBINS and FOLLOWERS.

BED PAN SINK (BED PAN SLUICE)
A sink in which bed pans and urine bottles are emptied and washed. See also BED PAN WASHER, SLOP SINK and HOSPITAL APPLIANCES.

BED PAN WASHER
A sanitary appliance used in hospitals in which bed pans and urine bottles are emptied and washed. The bed pan washer differs from a BED PAN SINK in that the bed pan is inserted into the appliance and a door is shut prior to cleansing. For example, see HOSPITAL APPLIANCES.

BEDDING
The material used to give support and protection to below ground drainage pipes. There are different types of bedding used for different circumstances, such as the size of the pipe and the depth to which it is laid. The bedding can either be for rigid or flexible pipes (rigid pipe materials include asbestos, clayware, concrete and iron;

flexible pipe includes various types of plastic). If the ground is stable some authorities will permit drainage pipes to be laid directly onto the trench bottom although
the main difficulty is then ensuring a steady gradient. The purpose of the granular
material is to distribute any excessive loads more evenly around the surface of the
pipe, preventing distortion or damage. Traditional methods of bedding should be
avoided; if the soil moves, due to shrinkage, the concrete will crack often resulting in
the cracking of the drain pipes.

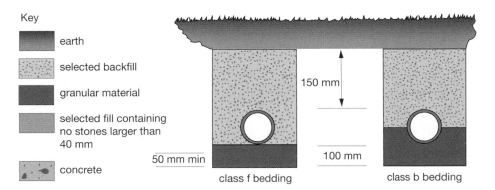

Key

earth

selected backfill

granular material

selected fill containing
no stones larger than
40 mm

concrete

50 mm min

class f bedding

150 mm

100 mm

class b bedding

Bedding for rigid pipes

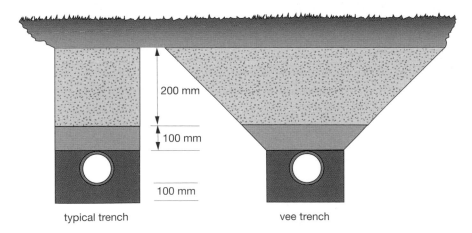

200 mm

100 mm

100 mm

typical trench

vee trench

Bedding for flexible pipes

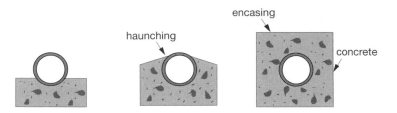

haunching

encasing

concrete

Traditional methods of bedding

BELFAST SINK   See SINK.

BELFAST WASTE
A SKELETON WASTE FITTING used on lead waste pipes.

BELL FLUSHING CISTERN   See FLUSHING CISTERN.

BELL JOINT   (TAFT OR CUP JOINT)
A method of making a bronze welded joint to copper tube. For example, see
BRONZE WELDING. See also TAFT JOINT.

BELLY TRAP
A running trap.

BELOW GROUND DRAINAGE
A system of pipes which conveys surplus water or liquid sewage away from the
building in the most speedy and efficient way possible to the sewer or other discharge
point without any risk of nuisance or danger to health and safety. There are three
types of below ground drainage systems: the COMBINED, the SEPARATE and the
PARTIALLY SEPARATE systems, each of these headings should be sought for further
study.
   When designing any drainage system one must observe the following principles:

(1) Provide adequate access points.
(2) Keep pipework as straight as possible between access points and for all bends
     over 45° an access point should be provided.
(3) Ensure all pipework is adequately supported (see BEDDING).
(4) Ensure the pipe is laid to a self-cleansing gradient (see MAGUIRE'S RULE).
(5) Ensure the drainage system is well ventilated (see VENTILATION OF DRAINS).
(6) The whole system must be water tight including inspection chambers.
(7) Drains should not be run under buildings, unless this is unavoidable or by so
     doing it would considerably shorten the route of pipework.

   All drains which convey FOUL WATER must be trapped. Should the system only
carry SURFACE WATER, no trap would generally be required. In the construction of
below ground drainage the most common materials used are plastic, or clayware,
but other materials are often found or used such as asbestos cement, cast iron, concrete and pitch fibre. Each of these materials have their own method of jointing, and
the fittings used vary tremendously. When jointing one material to another, in most
cases, a proprietary jointing fitting should be used. See also TESTING and INSPECTION
OF DRAINS; VENTILATION OF DRAINS AND DISCHARGE PIPES.

(See illustration over.)

BENCH MARK   See SITE DATUM.

BENCHING
The sloping sides constructed at the base of an INSPECTION CHAMBER to prevent the
accumulation of solid deposits.

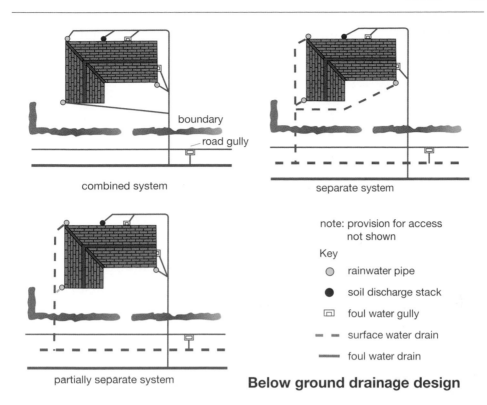

boundary

road gully

combined system

separate system

note: provision for access
not shown

Key

○ rainwater pipe

● soil discharge stack

▣ foul water gully

– – surface water drain

—— foul water drain

partially separate system

Below ground drainage design

BENCHMARK

A code of practice designed to ensure that operatives install appliances in accordance with manufacturers' instructions. To meet this aim, manufacturers provide a logbook and certificate to be kept with an appliance throughout its life.

BEND

A fitting which changes the direction of a pipe or channel. Bends are made in standard angles of 90°, 135° (45°) and 157^1/$_2$° (22^1/$_2$°) and are often referred to as quarter,

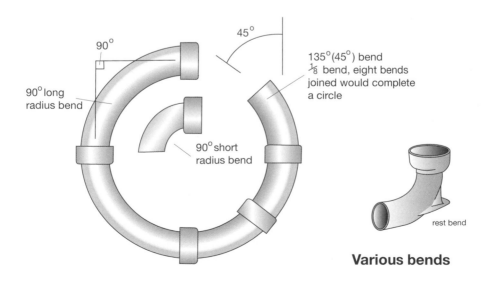

90°

45°

135°(45°) bend
^{1}/$_8$ bend, eight bends
joined would complete
a circle

90° long
radius bend

90° short
radius bend

rest bend

Various bends

eighth and sixteenth bends respectively; this refers to the number of bends required to make a full circle.

BENDING MACHINE

A machine used to bend pipes as necessary. There are two types of bending machine: those which use HYDRAULIC POWER and are used to bend heavy duty pipes such as low carbon mild steel (for these bending machines, see HYDRAULIC PRESS BENDER); and those which are designed to work on the principle of the lever and are used to bend light gauge tube such as copper. These benders are available ranging from hand-held machines, which only bend one or two small sizes, to much larger machines which bend various sizes of pipe. With the larger machines different-sized formers and backguides are used for various sizes of pipe and the roller pressure is set by turning a screw on the handle as required. Incorrect adjustment of the roller has one of two effects; if it is adjusted too low, it will be too tight and will cause excessive THROATING to the finished bend; if it is adjusted too high, it will be too loose and ripples on the pipe will be produced on the finished bend. With the hand-held machine used for 15 mm and 22 mm pipe no roller adjustment is possible; this can give problems due to wear and tear. To use these machines accurately, see PIPE BENDING TECHNIQUE (light gauge tube).

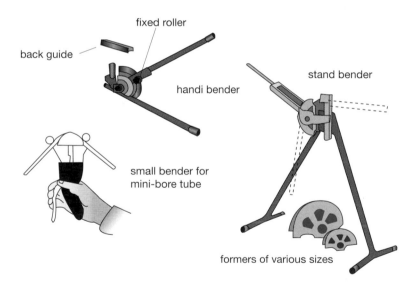

Portable 'lever' type bending machines for light gauge tube

BENDING SPRING

Tool used to assist the bending of copper, lead and plastic pipes, it is so designed to prevent the pipe from collapsing during the bending process. There are two types, those which go internally into the pipe and those used externally. To assist the removal of the bending spring the bend should be pulled to a degree or two more than is required, then bent back. On removing the internal spring, it should be turned in a clockwise direction. This tightens up the spring giving it a smaller diameter.

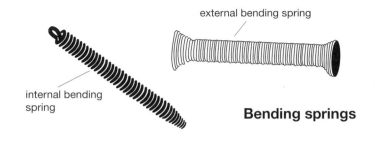

external bending spring

internal bending spring

Bending springs

BENDING STICK (BENDING OR SPOON DRESSER)

A tool made of BOXWOOD or high density plastic and used to assist in the process of dressing lead around to the back of a bend on lead pipe when bending.

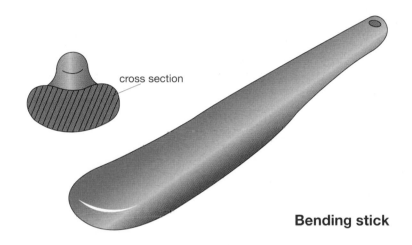

cross section

Bending stick

BENT BOLT (BENT PIN)

Tool used to open out a branch on lead or copper pipes.

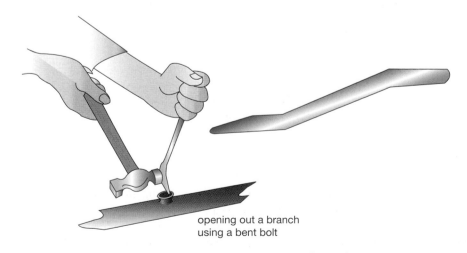

opening out a branch using a bent bolt

Bent bolt

BERGE
A Scottish term meaning an APRON FLASHING.

BIB TAP (BIB COCK OR VALVE)
A valve with a horizontal male thread as its inlet. This type of valve screws directly into a fitting which in effect supports the tap. These taps are commonly found over BUTLER SINKS. Bib taps are often used externally for outside use. The tap would normally be of a design which incorporates a threaded nozzle for the attachment of a hose union. If hose union connections are used on mains supply precautions to prevent BACK-SIPHONAGE must be taken.

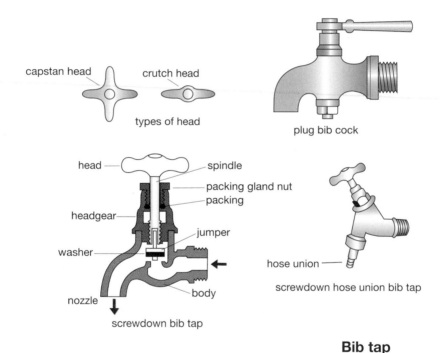

Bib tap

BIB TAP WALL FLANGE   See BACK-PLATE ELBOW.

BIDET
A sanitary appliance primarily designed to wash the excretory organs. Bidets which are fitted to mains supply must be of a design which is not equipped with a submersible ASCENDING SPRAY, but must be of the over-rim type, i.e. PILLAR TAPS. Those with submersible sprays must only be supplied with water from a cold water storage cistern and the cold feed should have no other draw-off points and be run independently. Also if a bidet is fitted with a submersible spray, the hot supply must be run independently from the hot storage vessel. The waste connections for bidets are the same as for wash basins.

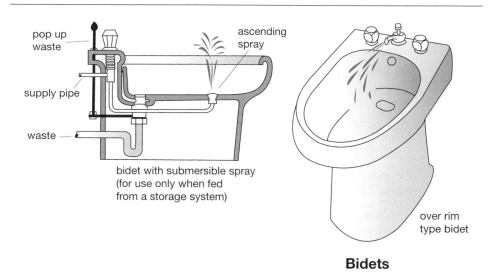

Bidets

BIMETALLIC STRIP

A strip of two metals with different expansion rates, such as brass and invar metal, which have been fixed together. When heat is applied to the strip it bends due to the fact that one metal wants to expand and one does not, thus it becomes distorted. The bimetallic strip is put to advantage in the operation of a BIMETALLIC STRIP FLAME FAILURE DEVICE.

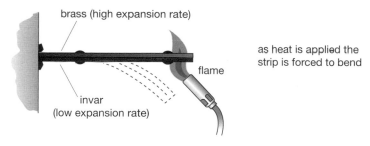

Bimetallic strip

BIMETALLIC STRIP FLAME FAILURE DEVICE

A FLAME FAILURE DEVICE used on gas water heaters and circulators. When the pilot light of an appliance is lit the flame plays onto the BIMETALLIC STRIP which bends and opens the gas valve, allowing gas to flow to the main burner which in turn is lit by the pilot light. A disadvantage with this device is that should the pilot light be extinguished, the bimetallic strip cools and only closes the main gas valve; the gas will still be discharged into the combustion chamber of the appliance via the pilot jet giving rise to a minor explosion when it is re-lit. See also INSTANTANEOUS WATER HEATERS.

(See illustration over.)

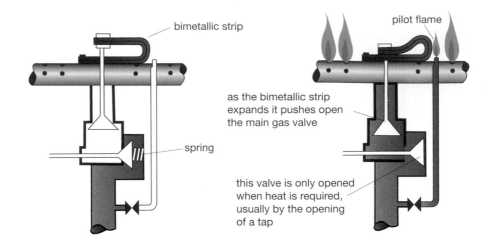

bimetallic strip

pilot flame

as the bimetallic strip
expands it pushes open
the main gas valve

spring

this valve is only opened
when heat is required,
usually by the opening
of a tap

Bimetallic strip flame failure device

BLACK BULB SENSOR
A thermostat used with a system of RADIANT HEATING. A bimetallic strip or electronic thermostat is enclosed within a black bulb-shaped plastic hemisphere. The black bulb absorbs the RADIANT HEAT (matt black being the best absorber of heat), which in turn warms the trapped air inside. This air now acts on the enclosed thermostat to provide the control over the heating panels.

BLACK WATER
A name sometimes used to indicate foul water, containing faecal matter or urine. See also GREY WATER.

BLADDER
An INFLATABLE STOPPER used for testing drains.

BLOCK BOILER   See BACK BOILER.

BLOWLAMP
An essential tool used by the plumber to produce heat for various uses such as making solder joints and melting lead, etc. The method used by the plumber to produce heat has changed many times over the years and in the past blowlamps were fuelled by liquids such as gasoline, methylated spirits and paraffin. Today, however, the fuel used is generally propane or butane gas.

BLOWPIPE (BLOWTORCH)   See WELDING BLOWPIPE and CUTTING BLOWPIPE.

BLUEING
A corrosion resistant surface obtained on the surface of steel. To achieve this, the metal is heated to about 900°C and then steam is introduced, which produces a thin blue oxide film on the surface.

BOBBINS

Oval shaped beads, made of BOXWOOD, with a hole through their centre to enable a cord to be passed through. A bobbin is exactly the same size as the nominal bore of the pipe and is either pushed by FOLLOWERS or pulled by a SNATCH WEIGHT through lead pipe to restore it to its true bore after the pipe has been flattened due to bending.

B

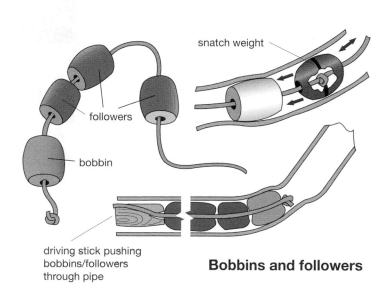

snatch weight

followers

bobbin

driving stick pushing
bobbins/followers
through pipe

Bobbins and followers

BOILER

An enclosed vessel in which water is heated by the application of heat, which could be either solid fuel, gas, oil or electricity. There are many types, sizes and shapes of boiler therefore the examples shown below are but a few. See also BACK BOILER, GAS BOILER, OIL FIRED BOILER, ELECTRIC BOILER, SECTIONAL BOILER, COMBINATION BOILER and CONDENSING BOILER.

(*See illustration over.*)

BOILER FEED TANK   See STEAM HEATING.

BOILER INTERLOCK

A situation that ensures that a boiler will not fire up unless there is a demand for heated water. Boiler interlock is in effect the making or breaking of a cylinder or room thermostat calling for heat. It is one of these thermostats that allows an electrical current to flow to the boiler, allowing it to operate. Boiler interlock is designed to prevent the boiler continuously going on and off all day long (short cycling) and therefore wasting heat.

BOILER MOUNTINGS

The accessories which are fitted to a boiler to ensure that it works safely and efficiently. These accessories could include devices such as THERMOSTATS, SAFETY VALVES and DRAIN-OFF COCKS, etc.

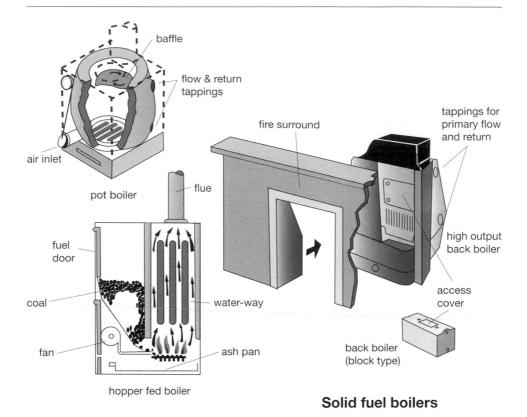

Solid fuel boilers

BOILER NOISES

Noises created in a boiler are often due to the formation of scale or the movement of loose rust and debris deposits. In the case of rust and debris a good flush through of the system should cure the problem. Should scale build up, especially in the flow pipe, trapping air within the boiler, a kind of boiling noise (often called kettling) will result, caused by steam forming and condensing. The scale can generally be removed by treating the system with a descaling solution. See also NOISES IN HOT WATER PIPEWORK.

BOILER RATING

Term denoting the amount of power a boiler will pass through 1 m² of heating surface. It is expressed in kW/m². For the plumber this is of little help when designing a simple hot water supply system. The plumber is mainly concerned with the rated boiler output which can be defined as the boiler rating multiplied by the amount of square metres of heating surface in the boiler. Manufacturers of boilers give the rated boiler output in their catalogues which will be found expressed in both kW and BTU/h.

To convert BTU/h into kW divide by 3412.
To convert kW into BTU/h multiply by 3412.

Example 50 000 BTU/h = ?kW
$$= 50\ 000 \div 3412 = 14.65\ \text{kW}$$

BOILING POINT

The temperature at which a liquid boils and changes to a gas or vapour. Water boils at 100° Celsius. It expands approximately 1600 times while changing to steam. Liquid changing to a gas in this way is known as a physical change. The boiling point is effected by the pressure which is acting upon the liquid, for example as stated, water boils at 100°C, this is at atmospheric pressure i.e. 1 bar, but should the pressure be increased or decreased the water would boil at a different temperature, see table.

Boiling points of water at different pressures

Absolute pressure bars	Boiling temperature °C
0.5	82
1	100
2	121
3	134
4	145
5	153
10	181

BOLSTER (FLOOR BOARD CHISEL)   See CHISEL.

BOMB

The PRESSURE REDUCING FITTING which is fitted in close coupled siphonic WC suites between the cistern and the pan.

BONDING   See CONTINUITY BONDING.

BONING ROD (TRAVELLER)   See SIGHT RAIL AND BONING ROD.

BOOSTED SYSTEM OF COLD WATER SUPPLY

A method of raising cold water in high rise buildings above the height to which the water mains would supply. There are three basic types of boosted supplies these being:

(1) Directly boosted system.
(2) Indirectly boosted system.
(3) Automatic pneumatic system.

With the directly boosted system the water is pumped directly from the water main up to the highest points in the system where it supplies water to any necessary storage cisterns and also to a drinking water cistern or a HEADER PIPE. The capacity of the water stored in the drinking water cistern or header pipe must not exceed 4–5 litres per dwelling to prevent it from becoming stagnant. With this type of system, if the building is large, a serious drop in the mains pressure to other buildings can be caused while the pumps are running. Therefore some water authorities insist on BREAK CISTERNS being fitted which serve as a pumping reservoir at low level. Should such a cistern be required the system is known as an indirect boosted supply. Pumps cannot be made to run continuously against a force, such as would be caused by the system when full with water, therefore the pumps only run when required. The indication for this is given by float switches or some other type of probe switch which would cut off the power to the pump when the water rises to a predetermined level; conversely the pump is switched on as the water level falls. Pumps should always be fitted in duplicate to ensure a supply of water is available during mainten-

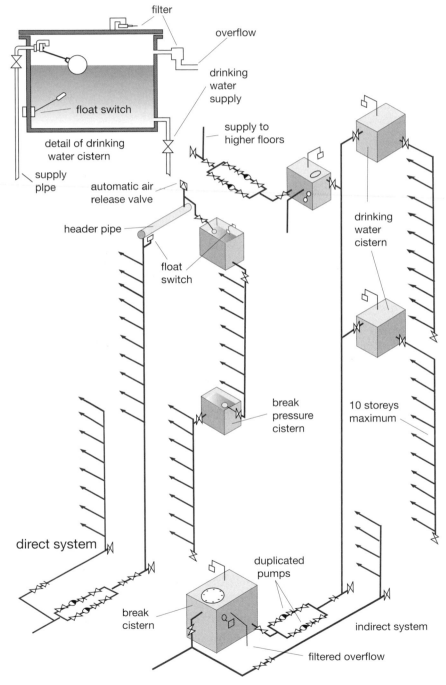

filter

overflow

drinking
water
supply

float switch

detail of drinking
water cistern

supply to
higher floors

supply
pipe

automatic air
release valve

header pipe

float
switch

drinking
water
cistern

break
pressure
cistern

10 storeys
maximum

direct system

duplicated
pumps

break
cistern

indirect system

filtered overflow

Boosted cold water supplies

ance. For buildings over 20 storeys, a second set of pumping equipment may be required on the twentieth floor to push the water even higher. When supplying water from storage cisterns and header pipes to draw-off points, the head of water should be limited to 30 metres as this will avoid excessive pressures. For the third type of boosted cold water supply see AUTOMATIC PNEUMATIC COLD WATER SUPPLY.

B

BOOSTER PUMP
A PUMP used to raise the PRESSURE and VELOCITY of water flowing through a pipe. For example one might use a booster pump on the pipe line going from the SHOWER MIXING VALVE to the shower head and rose, thus giving a more powerful shower spray.

BORAX (HYDRATED SODIUM BORATE)
A type of flux used for welding and soldering, it is a SOLVENT of OXIDE slags.

BOSS
A means of facilitating the connection of a pipe or fitting to another pipe, fitting or vessel. The boss could form an integral part of the installation or be attached by a mechanical fixing. See also IMMERSION HEATER BOSS.

bossed branch

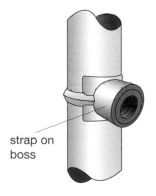
strap on boss

Bossed connections

BOSS WHITE   See JOINTING COMPOUND.

BOSSING
A term used meaning the general shaping of a MALLEABLE metal, in particular lead. Lead can be worked into many complicated shapes although a little time, patience and practice is needed to master the skill, which no book could possibly demonstrate, but the following gives a few guide lines on the technique. Firstly, there are two golden rules: one, remember the object is to move lead from one place to another, either to gain lead or to lose it; not to try to squeeze it into one area or stretch it out to fill another. So how is it moved? This gives rule two: set up nice curves, like waves which will flow.

(See illustration over.)

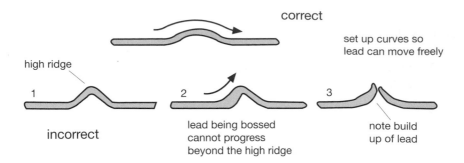

Bossing lead

Losing lead (e.g. internal corner)
First the work is set out and the surplus lead cut off, as shown. The creasing lines should be set in with a SETTING IN STICK and the sides pulled up to the required angle. After ensuring that the work has no high ridges (if so they should be rounded) with the BOSSING MALLET a few blows are directed in a downward direction to set in the base of the corner (see sketch). Now about 50 mm from the edge of the corner two temporary stiffening creases are put into the sheet, this helps the corner keep its shape while bossing. The corner is then worked up with a mallet being held on the inside supporting the work and aiming blows from the BOSSING STICK. The blows should be directed at the metal, drifting it outwards progressively towards the top edge of the corner (watch those high ridges); it is finished by trimming off the surplus lead.

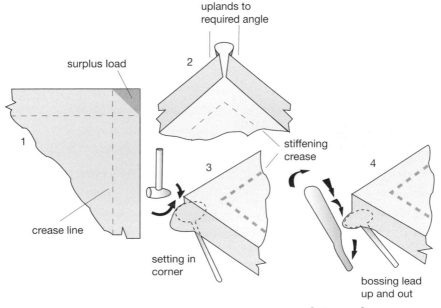

Internal corner

Gaining lead (e.g. external corner)

The lead should first be folded along its length to the angle required. This will cause a hump to occur which if it forms a ridge must be curved, then the metal is bossed around, in and down aiming all blows towards the new corner being formed, taking care not to stretch the lead.

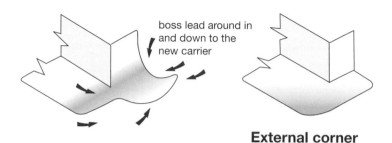

External corner

BOSSING MALLET (PEAR SHAPED MALLET/PLUMBER'S MALLET).

A useful tool used to assist the working of sheet roofing materials. The head is made of hard wood and the handle of Malacca cane to give it a spring effect and absorb the blows. For example of a bossing mallet see BOSSING.

BOSSING PIN

A short handed DUMMY.

BOSSING STICK

A tool used mainly to assist the process of BOSSING of corners in sheet lead. It is made of boxwood or high density plastic.

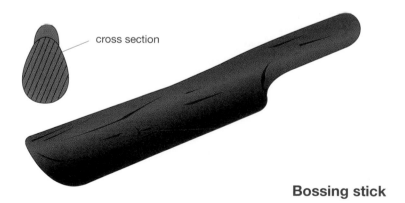

Bossing stick

BOTTLE TRAP   See TRAP.

BOTTLED GAS

Term associated with BUTANE and PROPANE, which are stored as liquids under pressure in refillable cylindrical vessels.

BOUNDARY WALL GUTTER   See PARAPET GUTTER.

BOURDON PRESSURE GAUGE

A gauge fitted on boilers of central heating systems to show the INTENSITY OF PRESSURE in kN/m² acting on the gauge. The gauge works by allowing water to enter a flattened circular loop tube (the Bourdon tube) which tends to open out as pressure is applied. As it opens it pulls on a toothed cog which rotates the pinion of the pointer registering the pressure. A gauge which registers the head in metres instead of pressure is known as an altitude gauge. See also ANEROID GAUGE.

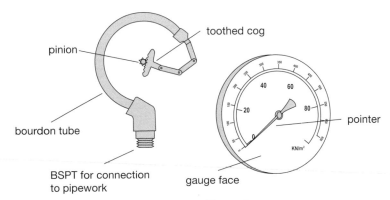

Bourdon pressure gauge

BOWER-BARFFING PROCESS

Heat treatment carried out on low carbon steel or cast iron to make it more resistant to corrosion. The metal is heated to a high temperature then steam is blown across its surface which forms a thin film of adherent magnetic iron oxide, giving protection.

BOWL URINAL

A bowl-shaped sanitary appliance which is fixed at a convenient height above the floor (about 610 mm to the front lip). Designed to receive and drain away urine. The bowl is flushed periodically from an AUTOMATIC FLUSHING CISTERN. For waste connections to bowl urinals, see WASTE FITTING. For example, see also URINAL.

BOX GUTTER   See GUTTER.

BOX HEAD

A rainwater HOPPER HEAD.

BOX'S FORMULA

A formula developed by Thomas Box used for determining the sizes of pipes and water flowing through them. The formula is expressed as follows:

$$q = \sqrt{\frac{d^5 \times h}{25 \times l 10^5}}$$

q = discharge of water through the pipe, in litres/second
d = diameter of the pipe in millimetres
h = head of water in metres
l = length of pipe in metres

Example

Find the amount of water in litres which would be discharged through a 50 mm bore pipe 16 m in length and supplied from a head of water of 6 m.

$$q = \sqrt{\frac{d^5 \times h}{25 \times l10^5}}$$

$$\therefore q = \sqrt{\frac{50^5 \times 6}{25 \times 16 \times 10^5}} = 6.8$$

q = approximately 6.8 litres per second

See also PIPE SIZING and LOADING UNITS.

BOXWOOD

Hardwood timber from the box tree, used for many plumber's sheet-forming tools such as FLAT DRESSERS and BOSSING STICKS.

BOYLE'S LAW

A relationship between volume and pressure. Providing that the temperature remains constant, the volume of a gas is inversely proportional to its absolute pressure. In other words, as one quantity increases the other decreases. Thus if the absolute pressure on a volume of gas were increased two-fold, the space that the gas would then occupy would be reduced to a half. Further increase the pressure four-fold and the volume would be reduced to a quarter. Thus with increased pressure the volume decreases; conversely decrease the pressure and the volume increases. Putting this law into practice, we could determine that if the gas pressure was reduced from 21 mbar at the meter to 6 mbar at the appliance an increase in volume would occur. Therefore the consumers get more gas than they have paid for. See also CHARLES' LAW.

BRANCH

The 'tee' connection of a pipe or channel joining the main flow. There are various designs of branch and tee connections, many of which are illustrated.

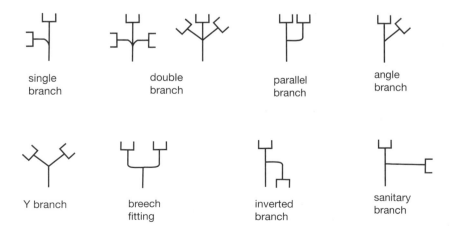

| single branch | double branch | parallel branch | angle branch |

| Y branch | breech fitting | inverted branch | sanitary branch |

Various branches

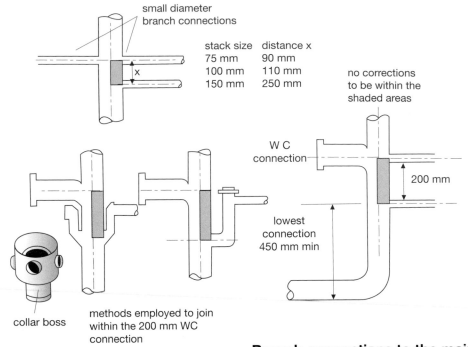

small diameter
branch connections

stack size	distance x
75 mm	90 mm
100 mm	110 mm
150 mm	250 mm

no corrections
to be within the
shaded areas

W C
connection

200 mm

lowest
connection
450 mm min

collar boss

methods employed to join
within the 200 mm WC
connection

**Branch connections to the main
stack to avoid crossflow**

BRANCH DISCHARGE PIPE

A DISCHARGE PIPE which connects the various sanitary appliances to the main discharge stack. When making connections to the main stack the pipe should not be joined in such a way as to cause a CROSSFLOW into other pipes, also it should not join the main stack lower than 450 mm above the INVERT at the base of the discharge stack thus avoiding COMPRESSION and trap seal loss. Ground floor appliances may discharge directly into the drain or into a STUB STACK. The sizes of branch discharge pipes should be at least the same diameter as the appliance trap. Oversizing the pipe to avoid SELF-SIPHONAGE could prove uneconomical and lead to an increased rate of deposit accumulation. Bends should be avoided if possible and, where they cannot be avoided, long radius bends should be used. The gradient of branch discharge pipes should be between 1°–1¹/4° (18–22 mm drop per metre run). See also SANITARY PIPEWORK.

BRANCH FLUE (SHUNT FLUE)

A system of flueing in which the CONVENTIONAL FLUES from more than one dwelling join to run up in a main common flue. See also SHARED FLUE.

(See illustration opposite)

BRANCH FORMING TOOL

A tool used to assist the forming of branch joints on light gauge copper tube. There are several types, some form a saddle onto the end of a pipe which in turn is brazed over a hole cut into the main pipe; another type pulls up the sides of a hole cut in the main pipe into which the branch pipe is inserted. Should the second method be

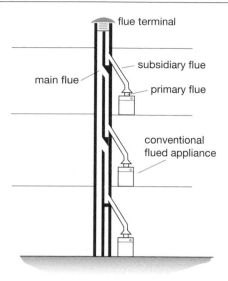

Branch flue

chosen, care must be taken to ensure that the branch pipe inserted does not protrude into the main pipe blocking its flow. By forming lugs on the end of the pipe with dimple pliers this can be prevented. Joints made with these tools are to be made water tight with HARD SOLDERING only due to the small area in which the pipe surfaces meet.

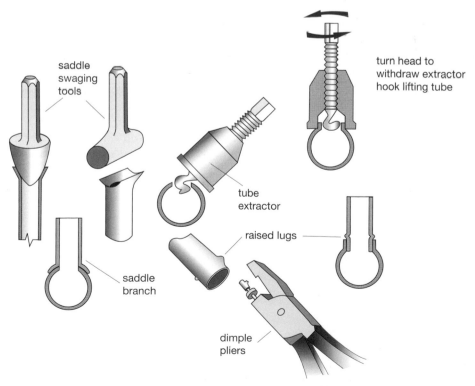

Branch forming tools

BRANCH VENTILATING PIPE

A VENTILATING PIPE which connects to a BRANCH DISCHARGE PIPE. The branch ventilating pipe should connect to the branch discharge pipe within 300 mm from the trap and should not connect to the discharge stack below the spillover level of the highest fitting served. Where such a pipe run would be unsightly a loop vent, as shown, could prove an alternative solution. The minimum size for a branch ventilating pipe serving a single appliance should be 25 mm, but, where the branch run is longer than 15 m, or contains more than five bends, or serves more than one appliance the minimum pipe size should be 32 mm. See also SANITARY PIPEWORK.

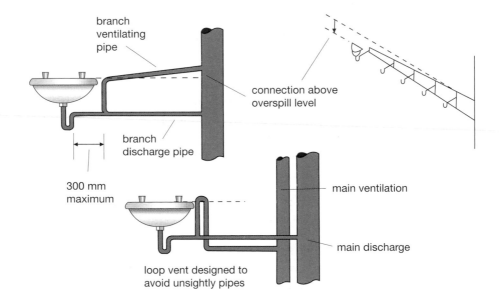

Branch ventilating pipes

BRASS

An ALLOY of copper and zinc which is used in the production of many types of valves and fittings.

BRAZED JOINT

A type of HARD SOLDERING in which the two parts being united are being joined together with molten brass or similar ALLOY. To braze a joint successfully relies on penetrating small gaps between the close fitting surfaces of the two metals, thus it is in fact a CAPILLARY JOINT unlike BRONZE WELDING which is distinguished by the local building up of the welding material and is often confused with brazing. The joint is often referred to as a hard soldered joint because it is much stronger than a soft soldered joint and for this reason smaller jointing surfaces are required between the metals. For example, the depth of a socket joint when soft soldered is about 15 mm whereas the same joint if brazed need only be 5–8 mm. To achieve this stronger joint a much higher temperature is required, 600–800°C depending upon the brazing alloy used. Whatever heating fuel is used, in most cases oxyacetylene, it is essential not to use a small concentrated flame but a large spreading flame which will give a larger heated area of the joint. See also SILVER SOLDERING.

BREAK CISTERN
Sometimes called a break tank because it has a bolted-on cover. But due to the fact it has a FLOAT OPERATED VALVE and is open to the atmosphere via a filter it is in fact a cistern. It is used on boosted cold water supplies to act as a reservoir of stored drinking water which can be pumped up in high rise buildings. It is required to overcome the problems of a reduction of mains water pressure caused by pumping directly from the water main. For example, see BOOSTED SYSTEM OF COLD WATER SUPPLY. See also BREAK PRESSURE CISTERN.

BREAK CORNER   See EXTERNAL AND INTERNAL CORNERS.

BREAK PRESSURE CISTERN
A cistern fitted in high rise buildings to overcome the problem of excessive pressures. The head of water in any system should be limited to 30 metres, therefore for buildings over this height cisterns may be required at about every tenth floor. These are called break pressure cisterns. For example, see BOOSTED SYSTEM OF COLD WATER SUPPLY. See also BREAK CISTERN.

BREECH FITTING (BREECHES JUNCTION)
A pipe fitting designed so that two pipes running parallel can turn in and unite as one thus in effect it is similar to a Y branch. See BRANCH for example.

BRITISH BOARD OF AGRÉMENT
The British Board of Agrément was set up in 1966, and is presently sponsored by the Department of the Environment as an independent body to assess building materials, products, components and processes. The board supplies an Agrément Certificate which provides architects, specifiers and contractors, etc. with information that will ensure the correct use of the product and help avoid unnecessary costly and dangerous failures.

BRITISH STANDARDS (BS)
A standard specifying quality or dimensions of a product, laid down by the British Standards Institution (BSI). They ensure that items are designed, manufactured and installed to a standard specification. For example, a hot storage vessel will be made to a standard which will ensure that it will withstand a certain pressure. If an item is manufactured to a certain specification as laid down by the BSI it is given a BS number and stamped with the British Standard kitemark. Some examples are:

BS 1212: Specification for float operated valves
BS 1010: Specification for draw-off taps and above-ground stopvalves

BSI also produces Codes of Practice specifying what is considered good practice in a trade described. Today in order to comply with European Standards many are now given a BS EN number, examples include:

BS EN 752: Drain and Sewer Systems
BS EN 12056: Gravity Drainage

BRITISH THERMAL UNIT (BThU/BTU; THERM)

One BThU can be identified as the amount of heat required to raise one pound of water one degree fahrenheit. Since the introduction of the metric system in this country BThU, pound or the fahrenheit temperature scale are no longer used as units of measurement. See METRIC SYSTEM.

BROADSTONE BALLVALVE

A ballvalve rarely seen, but it was a patent ballvalve designed to enable rewashing and servicing as necessary without turning off the water supply. See also SUPATAP.

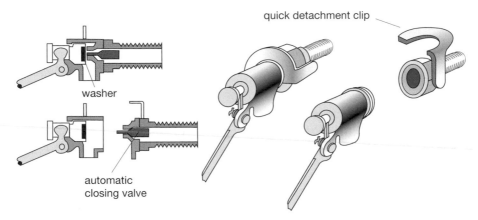

Broadstone ballvalve

BRONZE

An ALLOY of copper and tin which is used in the production of many types of valves and fittings. See also GUNMETAL.

BRONZE WELDING

A method used to join various metals, it is referred to as a welded joint but in fact no melting of the parent metal takes place therefore it is a type of HARD SOLDERED joint. To make a bronze welded joint a space is required between the two surfaces to be joined, this can be achieved by grinding thick material to form a 'veed' edge or in the case of pipe a space could be left or a bell joint formed. The joint should be thoroughly cleaned to remove any OXIDE coating then heat is applied to the joint with a slightly OXIDISING FLAME, working the flame from side to side or in small circles over the surface. An oxidising flame is chosen because the excess oxygen in the flame is required to mix with the zinc in the filler rod, which boils and changes to a vapour below the melting temperature of copper. When the oxygen mixes with the zinc a zinc oxide is formed which melts at a much higher temperature (pure zinc oxide melts at 1975°C), failure to observe this will result in the zinc volatilising (changing to a vapour) and bubbling up through the weld, leaving a series of blow-holes. The fluxed filler rod is introduced to the joint and the flame is lifted to stroke the rod which, upon melting, should run and adhere to the metal being joined. The process is continued slowly progressing along the joint, adopting the LEFTWARD WELDING TECHNIQUE, forming a series of characteristic weld ripples. The filler rods mainly used for bronze welding consist of approximately 60% copper and 40%

zinc with a small amount of tin and silicon to act as deoxidisers and assist in its flowing characteristics.

One of the advantages of bronze welding is the temperature at which the filler rod melts which is around 850–950°C, this minimises the distortion which would otherwise take place should the joint be FUSION WELDED requiring a temperature of at least the melting point of the parent metal. Also bronze welding can be used to join dissimilar metals such as copper to iron. A special flux is used to bronze weld which consists of BORAX and SILICON, this is either made into a paste by mixing with water and applied directly to the joint, or a heated filler rod is dipped into the powdered flux which in turn sticks to the rod. Special flux impregnated filler rods can be purchased at a little extra cost. Like all joints requiring a flux, ensure that any excess flux is removed on completion of the weld.

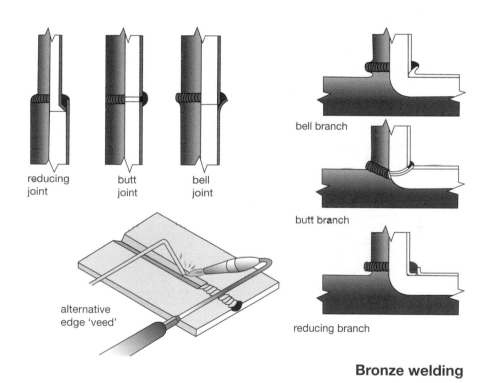

Bronze welding

BRS BALLVALVE (GARSTON BALLVALVE)
Abbreviation for Building Research Station ballvalve. This was the first design of the diaphragm type ballvalve and it was at the Building Research Station at Garston, Herts, where this type of valve was developed. See DIAPHRAGM BALLVALVE.

BSPT
Abbreviation for British Standard pipe thread, which is the type of thread used in this country on all plumbing pipework. It is worth noting that when this country changed over to the metric system LOW CARBON STEEL was one material that did not become metricated; and is still sold in the old imperial measurements due to the BSPT sizes used, i.e. $3/8$ in. $1/2$ in. $3/4$ in. 1 in. $1^{1}/4$ in. $1^{1}/2$ in. 2 in., etc.

BThU (BTU)
Abbreviation for BRITISH THERMAL UNIT. See METRIC SYSTEM.

BUBBLE TOP   See CYLINDER AIR GAP.

BUCKET PUMP
A type of JACK or LIFT PUMP. These pumps incorporate a non-return valve which is displaced by the action of pumping, this non-return valve is called the bucket or clack valve. For example, see LIFT PUMP.

BUCKET SINK   See CLEANER'S SINK.

BUFFER CLIP
A pipe clip used on flushpipes. The clip incorporates a rubber buffer which prevents damage to the pipe when the WC seat is lifted because it rests on the buffer.

BUILDER'S OPENING
The recess or opening at the base of a chimney in which a fire is burnt or a gas fire installed.

BUILDING PAPER
A fibre and bitumen reinforced paper which can be used as an underlay between the sheet weathering material and the roof structure. See UNDERLAY.

BUILDING REGULATIONS
The Building Regulations provide a uniform system of building control for England and Wales, including Inner London. There is a separate set of Building Regulations for Scotland. The regulations are statutory and must be observed. They are designed for the purpose of:

(1) Securing the welfare, health and safety and convenience of people in or around buildings.
(2) Conserving fuel.
(3) Preventing waste, undue consumption, misuse or contamination of water.
(4) Regulating the design and uniformity of construction.

   The Building Regulations contain no technical detail. This is found in a series of Approved Documents, covering subjects such as Drainage and Waste Disposal, Heat Producing Appliances, Hygiene, etc., as well as in other non-statutory documents such as British Standards and Codes of Practice.

BUILDING TERMINOLOGY
See sketch opposite for examples of various positions around a building and the terms used to identify them.

(See illustration opposite)

BULBOUS OUTGO WC PAN   See SIPHONIC WC PAN.

B

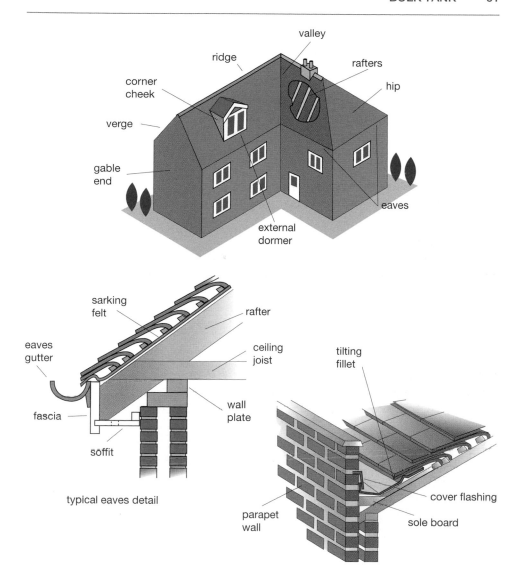

Building terminology

BULK TANK

A tank designed for the storage of liquefied petroleum gas (LPG). The size of the tank depends upon the maximum hourly gas consumption. The larger the tank the greater the vapour off-take will be. This is because the liquid stored within the tank needs to extract heat from the surrounding air to allow the liquid to boil and change to its gaseous state. Larger tanks have a greater wetted surface area. Tanks are never filled to more than 87% capacity to allow for expansion due to temperature changes and to allow room for the vapour to form. The pressure within a tank varies with the temperature (e.g. at 15°C the pressure is 6.8 bar) therefore two regulators are used to reduce this high pressure to 37 mbar, i.e. that required for a domestic supply. The first stage regulator reduces the varying high tank pressure to a constant medium pressure of 0.75 bar and the second stage regulator further reduces it to 37 mbar. Within the supply line an over pressure shut-off and an under pressure shut-off valve (OPSO/UPSO) should be incorporated to prevent excessive pressures entering the domestic installation and to prevent gas entering the building where insufficient pressure is detected, such as may result from an open end. A pressure relief valve will be incorporated into the tank, set at about 14 bar, which is designed to open should the pressure increase substantially, such as in a fire. The siting of the tank should be, as illustrated, on a concrete base, away from the house, boundaries and any source of ignition. The tank is fitted with a contents gauge and liquid level indicator designed to tell how much liquid is within the tank. A shut-off valve is also incorporated to permit the tank to be isolated.

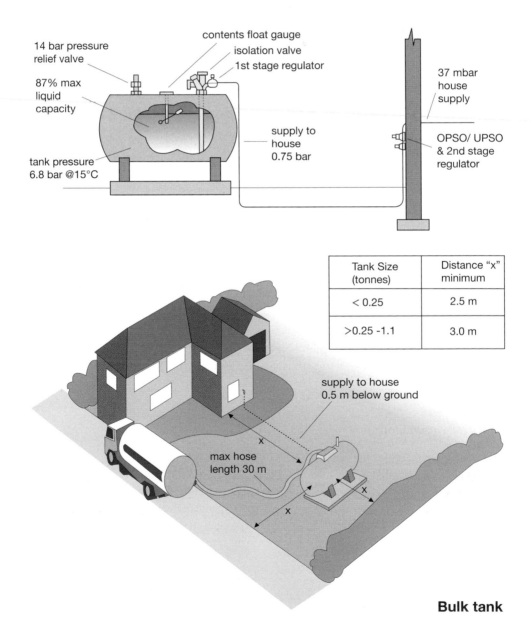

Tank Size (tonnes)	Distance "x" minimum
< 0.25	2.5 m
>0.25 -1.1	3.0 m

Bulk tank

BUMPER TOOL

A SOCKET FORMING TOOL.

BUNSEN FLAME   See AERATED FLAME.

BURLINGTON BELL FLUSHING CISTERN   See FLUSHING CISTERN.

BURLOC FIXING TOOL

An old fashioned tool used to secure a short piece of copper pipe or cap and lining, etc. into the internal surface of lead pipe prior to wiping a joint. For example, see WIPED SOLDER JOINT.

BURNER

The part of an appliance where the fuel is burnt to produce heat. For example, see GAS BURNER and VAPORISING BURNER.

BURNER PRESSURE   See OPERATING PRESSURE.

BURNING IN

The first run of a lead welded joint, it is the fusing of the two mating edges. Burning in is carried out on welds which require a build up of welding material to give support to the finished job. For example, see ANGLE WELDED JOINT.

BURNING VELOCITY

The speed at which a flame spreads through a gas/air mixture. NATURAL GAS has a slower burning velocity than TOWN GAS which used to be supplied in Britain. See also GAS BURNER.

BUSH

A pipe fitting designed to connect a female thread to a male thread of smaller size. It is basically a thread within a thread.

BUTANE GAS

A liquefied petroleum gas which is extensively used by plumbers to fuel their blowlamps. It is supplied in steel containers usually coloured blue or grey. See LIQUEFIED PETROLEUM GAS.

BUTLER SINK

A 'Belfast' or 'London pattern' sink, see SINK.

BUTT WELDED JOINT

A method of joining sheet or pipe by FUSION WELDING or BRONZE WELDING. It is carried out by butting two edges together and welding between the joint; note, some methods of butt welding require a gap to be left thus ensuring full penetration to the joint. For examples see LEAD WELDING and BRONZE WELDING.

BUTTERFLY VALVE

A valve that uses a centralised pivoting gate to close a pipeline. It has a large port opening and therefore allows high rates of flow. Where complete shut off is needed a valve with a soft NEOPRENE lined seating is generally used.

BUTTON BURNING

Lead welding which is carried out on an incline. It is sometimes termed button burning because the lead is deposited up the incline in the form of overlapping buttons or beads. See also INCLINED LEAD WELDING.

BYELAWS

These were the old Water Regulations that were in force prior to the current WATER SUPPLY REGULATIONS, which should be sought for further study. Byelaws were a set of requirements to be observed by the local water authority, which set them. They varied from authority to authority, which gave rise to confusion. Modern Regulations are national and the same rules apply throughout the country.

BY-PASS

The arrangement of pipework which directs the flow of gas or liquid around the normal route of flow. For example, see GAS METER.

CALOR GAS
A trade name for a well-known producer of butane and propane LIQUEFIED PETROLEUM GAS.

CALORIFIC VALUE (HEAT OF COMBUSTION)
The specific amount of heat produced by a specific amount of fuel, in joules, when completely consumed. The calorific value of fuels differs from one another, for example the amount of heat produced by burning 1 kg of wood would be less than that produced by burning 1 kg of coal. Natural gas has an even higher calorific value. The calorific value of gases is usually expressed as the amount of heat in joules they contain per cubic metre (m^3) while for solid and liquid fuels the calorific value is expressed in joules per kilogram.

Some calorific values

Fuel	Calorific value
Anthracite	32 MJ/kg
Coal	30 MJ/kg
Coke	25 MJ/kg
Wood	19 MJ/kg
Domestic grade oil	45 MJ/kg
Liquefied petroleum gas	30 MJ/kg
Natural gas	38 MJ/m^3
Town gas	19 MJ/m^3
Electricity	3.6 MJ/kW

CALORIFIER
The term calorifier generally refers to a type of INDIRECT CYLINDER heated by steam or high temperature hot water used mostly for non-domestic purposes, but see also STEAM HEATED CALORIFIER.

CAME (CAM OR CALM/FRETTED LEAD)
The H-shaped bar of lead which is used to make leaded lights (leaded windows) or stained glass windows. The glass is fixed between the lead grooves.

CANOPY
An overhanging cover fixed above doorways to afford shelter or above ovens to collect the fumes and steam.

CANTILEVER BRACKET
A bracket which is built into the wall holding it secure, on top of which sits a storage vessel or some other appliance such as a wash basin which requires support.

CAP
A fitting used to fix over the end of a pipe blanking it off.

CAP AND LINING
A copper alloy fitting having a straight tail piece (the lining) which is wiped into a lead pipe, the other end has a union nut (cap) for fixing onto a male iron thread. For example, see UNION CONNECTOR.

CAP FLASHING
The metallic or non-metallic covering which is sometimes used to cover the top of chimney stacks or parapet walls, thus affording some protection against the weather.

CAPACITOR (CONDENSER)
A device designed to store a charge of electricity. Capacitors are often used to give a boost to small electric motors that run on a single-phase supply. They provide the initial impetus to get the motor running.

CAPACITY
The amount of liquid a pipe, vessel or other receptacle will hold in litres, either when full or up to a specified mark known as the water line.

CAPILLARY ATTRACTION (CAPILLARITY)
A well-known phenomenon, which many people find hard to believe, in which liquids can rise upwards between two surfaces. To demonstrate this phenomenon put a clear drinking straw into a glass of water; if the straw is squeezed together the water will rise even higher. Capillary attraction is caused by the liquid adhering to the side walls of the straw and then, due to the COHESION of the water molecules (the need to hold together), pulling itself up to the level of adherence, the water then adheres again to the sides of the straw and again the water pulls itself up to the level of adherence. The process continues until the weight of the water molecules is too great and pulling itself up to the level of adhesion proves too much. Thus it can be stated that capillary attraction is caused by the adhesion and cohesion qualities of liquids. The closer two surfaces are together the higher the liquid will rise. Also should the surface be greasy or oily the liquid will not readily adhere. Capillary attraction can be put to use in plumbing works, such as with capillary solder joints, hence the need for cleanliness, to get good adhesion. But it can also cause problems and damage such as water passing between the laps in sheet weathering material; to overcome problems such as these, see DAMP PROOF COURSE and ANTI-CAPILLARY GROOVE.

(See illustration opposite)

CAPILLARY GROOVE (GAP)   See ANTI-CAPILLARY GROOVE.

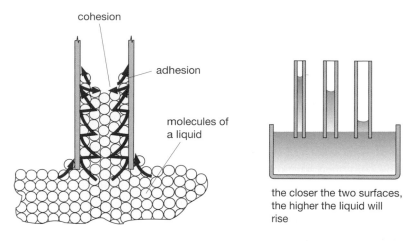

the closer the two surfaces,
the higher the liquid will
rise

Capillary attraction

CAPILLARY JOINT

A joint used extensively on copper and stainless steel tubing; the joint is made by melting solder which fills up into the fitting by CAPILLARY ATTRACTION. There are two types of capillary joints, those which are SOFT SOLDERED and those which are HARD SOLDERED. If capillary joints are used on mains supply the solder used must be lead-free. The method of making a soft soldered joint is to first thoroughly clean the pipe end and internal surface of the fitting with wire wool; emery paper should not be used as this leaves a dirty film on the surface unseen to the human eye. Once cleaned, apply a smear of flux to the pipe and joint then assemble. Now heat up the joint with a blowlamp until solder which is applied melts and runs round and fills the socket. Do not melt on too much solder as it flows into the pipe possibly blocking it, remove the heat and allow the joint to cool. Upon completion flush out the pipe line and clean the external surface of the pipe of excessive flux. See also SOCKET FORMING TOOL.

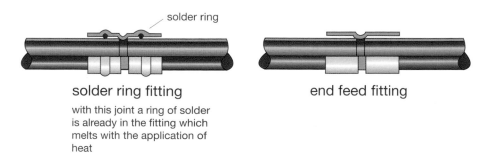

solder ring fitting

with this joint a ring of solder
is already in the fitting which
melts with the application of
heat

end feed fitting

Capillary joints

CAPPING PIECE

The top piece of material which covers the BATTEN ROLL. See also CAP FLASHING.

CAPSTAN HEAD
A type of head used on taps, for example, see BIB TAP.

CAPTIVE PLUG　　See ANTI-THEFT WASTE PLUG.

CARBON
A non-metallic chemical element which is found in three common forms; diamonds, graphite and non-crystalline carbon. Carbon black is a nearly pure form of carbon while charcoal, coal, coke and soot, etc. are other carbon-rich substances. Carbon is often found combined with other elements such as with oxygen to give carbon dioxide in the air or dissolved in water; it also occurs as a carbonate in rocks such as limestone and chalk or as a hydrocarbon (a compound of only hydrogen and carbon) in coal, natural gas and petroleum.

CARBON DIOXIDE (CARBONIC ACID GAS)
A colourless gas found in the atmosphere having the chemical symbol of CO_2. When carbon dioxide is pressurised it undergoes a physical change and changes to a solid. In this form plumbers use it to assist in freezing pipes to make a quick repair (see FREEZING EQUIPMENT), thus avoiding the need to drain down the whole system. Carbon dioxide is absorbed into water, e.g. when it rains, tending to make the water more corrosive and is then often known as a weak carbonic acid.

CARBON DIOXIDE INDICATOR
A device used to find out the amount of carbon dioxide (CO_2) contained in the products of combustion within a flue. CO_2 is added to and mixed with a special gas absorbing fluid, this causes the fluid to rise to a new level in the container, from this new position one can read from the scale the amount of CO_2 it contains. The CO_2 indicator works as follows:

(1) The fluid is poured into the special container which consists of a special non-return valve at its inlet and an enlarged base which is fitted with a diaphragm washer.
(2) The gauge on the container is adjusted so that zero is in line with the liquid level.
(3) A probe is put into the flue pipe and its hand pump is depressed several times to eliminate any air. The plunger is then connected to the inlet of the container.
(4) The plunger is pressed down, which in turn opens the non-return valve; now the pump is depressed 18 times and during the 18th bulb deflation the plunger is removed from the container, thus closing the non-return valve trapping the CO_2 inside the container.
(5) The container is now inverted several times which causes the CO_2 to be absorbed by the liquid. Because the CO_2 has been absorbed it is no longer a gas in the container and thus a partial vacuum is formed, which in turn causes the diaphragm to flex up and lift the fluid up the centre bore of the container, raising its height.
(6) The new fluid liquid level is read from the scale, which represents the percentage of CO_2 gas at the point of sampling.

These percentages could now be compared with the flue temperature to give the combustion efficiency of the appliance. See also COMBUSTION EFFICIENCY TESTING.

Generally accepted percentages of CO_2 for good combustion	
Fuel	CO_2 Percentage
Light oil	11–12.5
Heavy oil	11–13
Natural gas	9–10
Propane gas	9–11
Anthracite coal	12–14

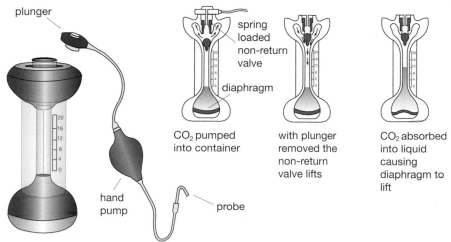

Carbon dioxide indicator

CARBON MONOXIDE
A highly toxic, colourless and odourless gas which can result from incomplete combustion of fuels. Carbon monoxide is produced due to the incomplete conversion of carbon or carbon containing fuels to carbon dioxide. Upon inhalation of carbon monoxide, the red blood cells of our body will accept these in preference to oxygen, thus often proving fatal. The plumber upon completion of any work involving the combustion of fuel must ensure that the appliance is left safe and that no fuel gases are discharged where they could cause harm.

CARBONISING FLAME (CARBURISING FLAME)
A flame which has an excess of acetylene at the oxy-acetylene welding blowpipe. This tends to show as a whitish feather which surrounds the bright inner cone of the flame (see OXY-ACETYLENE FLAME). A very slight carbonising flame is used to prevent OXIDATION of the weld when welding metals such as aluminium, stainless steel, magnesium and nickel. See also LINDEWELDING, where a heavy carbonising flame is used to weld steel piping.

CARCASS
Term used to indicate the pipe runs laid in a building, with no appliances connected, during the 'first fix', prior to the floors being laid and walls plastered.

CARDWIRE
A type of flat wire brush.

CASCADE INTERCEPTING TRAP
A type of intercepting trap designed with its inlet INVERT LEVEL higher than its outlet invert level causing the liquid sewage to cascade from its inlet pipe down to the water level. For example, see INTERCEPTING TRAP.

CASE HARDENING (CARBURISING)
A process undertaken in which mild steel has carbon introduced into its outer layers to form a 'case' of up to 0.15 mm thick. Case hardening is carried out by heating the steel, which is surrounded by a substance from which it can absorb carbon.

CAST IRON
The main composition of cast iron is a mixture of iron with 2–4% carbon. Cast iron has a high resistance to corrosion due to its high carbon content and is used for drainage and sanitation pipework, various appliances, inspection covers and a vast range of products. See also, SAND CAST and SPUN CAST IRON.

CAST IRON PIPE AND FITTINGS
A material used extensively for the design of water mains and above and below ground drainage systems. Traditionally cast iron was jointed by CAULKING joints but nowadays more and more proprietary jointing methods are being used, these allow for movement in the pipe work and are much quicker to install.

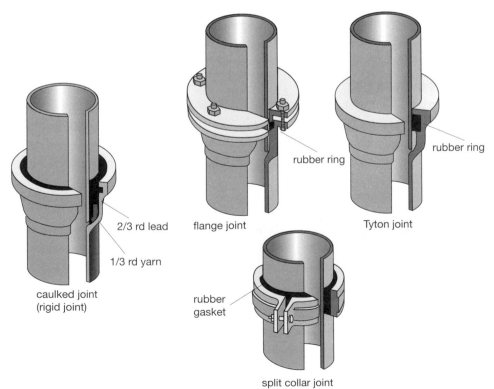

rubber ring

rubber ring

2/3 rd lead

flange joint

Tyton joint

1/3 rd yarn

caulked joint
(rigid joint)

rubber
gasket

split collar joint

Methods of jointing cast iron pipe

CAST IRON WELDING

Cast iron is best welded using oxy-acetylene equipment, with the blowpipe set to a NEUTRAL FLAME. Too much oxygen will result in a porous weld. For best results, the work should be done with a filler rod which has a high silicon content and the work should proceed allowing for a slow uniform cooling of the work. Prior to welding, the work should be pre-heated to avoid expansion stresses which will cause cracking of the metal. To achieve good penetration the edges of the material should be 'veed' to an angle of 90°; material over 9 mm thick should be veed on both sides. See also LOW CARBON STEEL WELDING.

CAST LEAD

Sheet lead which has been produced by pouring molten lead into a large sand filled tray designed for this purpose. Years ago all lead was produced in this way and the work was done on the building site. Lead produced in this way tends not to be as uniform in its thickness as milled lead. See also MACHINE CAST LEAD SHEET and SAND CASTING.

CASTING SAND   See SAND CASTING.

CATALYTIC COMBUSTION (FLAMELESS COMBUSTION)

The flameless reaction that occurs when using a catalytic burner. The burner consists of a flat pad of refractory material fibre, that has been impregnated with a very fine platinum powder. When the pad is heated to the required operating temperature and a suitable gas/air mixture passed through the pad a flameless chemical reaction occurs, just as if the gasses were burning. To establish the temperature of the pad initially and to bring about the chemical reaction an electric element is sometimes embedded in the surface; alternatively the gas burns on the surface until the pad reaches the required temperature at which 'flameless combustion' can occur.

CATCH CLOTH (GLOVE CLOTH/BLANKET)

A large WIPING CLOTH used to catch the molten solder which is poured onto a large diameter lead pipe when making a wiped soldered joint with a melting pot and ladle. The catch cloth is large enough to insert the hand inside the back fold, thus giving better protection from the molten solder.

CATCH POT TRAP

A BOTTLE TRAP designed so that the lower part is deep enough to retain the waste water, which can be removed for examination.

CATCHMENT SPACE

The void at the base of a chimney to allow the accumulation of soot, dead birds, etc. that may fall. This space should periodically be inspected and all debris removed.

CATCHPIT (CESSPOOL)

A box-shaped receiver constructed in a roof or gutter to collect rainwater which then passes into a rainwater pipe connected to it. A catchpit is generally used for internal rainwater pipes and if chosen it is advisable to fit an overflow to the side because, should the catchpit become blocked, flooding of the roof could result, eventually leading to water getting into the building. For example, see GUTTER.

CATHODE

A metal which would destroy an anode in ELECTROLYTIC CORROSION. Copper would be a cathode to aluminium whereas it would be an anode to silver, see ELECTROMOTIVE SERIES.

CATHODIC PROTECTION (GALVANIC OR SACRIFICIAL PROTECTION)

A means of using a sacrificial metal to save the destruction of metals by ELECTROLYTIC CORROSION. The sacrificial metal (or anode) would be a metal such as magnesium, which is very low down the ELECTROMOTIVE SERIES, this metal would be destroyed before any other thus giving a longer life to such metals as zinc when mixed with copper. The sacrificial metal in a plumbing system could be a solid block of magnesium placed inside the feed and storage cistern, or could be identified as iron which has been GALVANISED.

CAULK

Term used to mean 'to stop up' or pack any voids or crevices by driving in lead, GASKIN, soft rope such as asbestos cement cord or any other suitable material. See also CAULKED JOINTS.

CAULKED JOINTS (CAULKING)

A method of jointing cast iron pipes. The spigot end of one pipe is placed in the socket end of the other then the joint is one third filled with YARN which has to be well compacted with a YARNING IRON, once this is completed it is followed by pouring molten lead into the socket filling it to the top of the joint. The lead is left to cool and solidify and in so doing contract. Now the lead must be consolidated into the

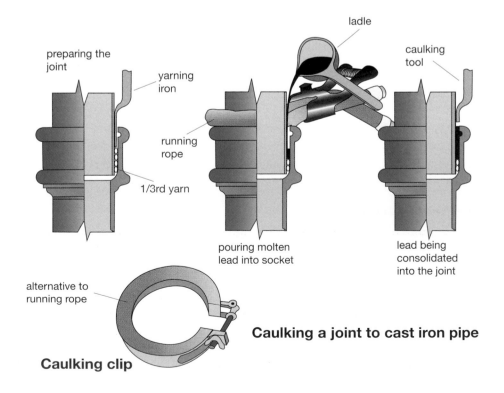

preparing the joint

yarning iron

running rope

1/3rd yarn

ladle

caulking tool

pouring molten lead into socket

lead being consolidated into the joint

alternative to running rope

Caulking a joint to cast iron pipe

Caulking clip

joint with a CAULKING TOOL. In wet conditions the molten lead would 'spit', so LEAD WOOL is used instead, this being known as a cold caulking method. For jointing pipes in the horizontal position, either a RUNNING ROPE or a CAULKING CLIP is used to assist in pouring the molten lead and retaining it in the joint.

CAULKING CLIP
A tool used to assist the running of molten lead into the socket when making a horizontal CAULKED JOINT on cast iron pipe. The wing nut is placed at the top to form a pouring hole. See also RUNNING ROPE.

CAULKING TOOL (CAULKING IRON)
A tool used for consolidating lead into the socket of a cast iron pipe. A caulking tool is similar in shape to a YARNING IRON although the blade is somewhat shorter and thicker. For example, see CAULKED JOINT.

CAVITATION
As water rushes through a pipe it forms air bubbles; as the water passes through valve seatings it sometimes experiences a sharp drop in pressure which causes the air bubbles to collapse with explosive force, this is cavitation: the formation and collapse of air bubbles. Over a period of time cavitation can lead to the EROSION of the valve seatings which appear pitted or scored due to this continued explosive impingement upon their surface. The scoring caused on brass by cavitation is sometimes confused with DEZINCIFICATION due to its appearance.

CE MARK
A symbol indicated on a product to indicate that it complies with European Standards.

CELSIUS
The temperature scale used in Great Britain since the introduction of the metric system. The word Celsius replaced the word centigrade. See also TEMPERATURE.

CENTIGRADE   See CELSIUS.

CENTRAL HEATING
A term used to mean that the heat used to warm a building is taken from a central source. There are two basic types of central heating, that which heats water, HOT WATER HEATING SYSTEMS and that which heats air, WARM AIR HEATING SYSTEMS.

CENTRALISED HOT WATER SUPPLY
A method of hot water supply where the hot water is heated and stored centrally. The water is heated either directly or indirectly, but the principle is the same for both. Water is heated in the boiler and circulates around the PRIMARY FLOW and return pipes by CONVECTION CURRENTS or by use of a pump to the hot storage vessel, here the water is stored directly or passed through a HEAT EXCHANGER to indirectly heat the water in the hot store vessel. Once the water in the storage vessel has been heated it can simply be drawn off for use. It must be noted in the sketch over that allowance must be made for expansion of the water by means of a vent pipe, this also serves as a means of allowing air into the system keeping the pressure equal

to the surrounding atmosphere. Also in the sketch will be seen a cold feed cistern, this is positioned as high as possible to give a suitable pressure at the draw-off points. Both the feed cistern and vent pipe can be omitted as in unvented domestic hot water systems, which are under mains pressure. See also ELECTRIC WATER HEATING and TEMPERATURE CONTROL FOR DOMESTIC HOT WATER, DIRECT and INDIRECT SYSTEMS OF HOT WATER SUPPLY.

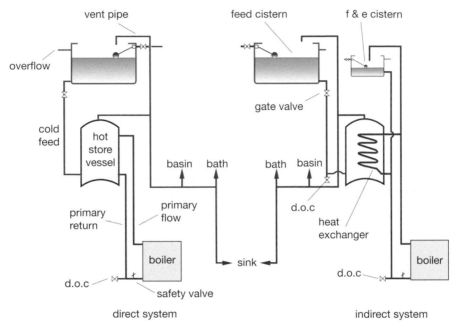

Centralised systems of domestic hot water supply

CENTRE GUTTER   See VALLEY GUTTER.

CENTRE ROLL   See KING ROLL.

CENTRIFUGAL FAN
A fan in which the impeller blades are parallel to the axis of rotation of the motor, unlike those of an AXIAL FLOW FAN. The rotation of the impeller causes the air to exit the fan at right angles to the direction of entry as shown in the illustration.

(See illustration opposite)

CENTRIFUGAL PUMP
A pump in which the pressure is created by centrifugal force caused by a rotating impeller. This type of pump would be used in domestic plumbing when the use of a pump is required to assist circulation. If the pressure developed by the pump is too great it could give rise to an excessive noise level due to the fact that the water velocity is too high. Most modern pumps are made so that their outlet pressure can be regulated. Pumps fitted in plumbing systems should have a valve fitted either side to allow for removal of the pump without draining down the whole system. Before fitting a pump to a new system the system should be flushed through to remove any

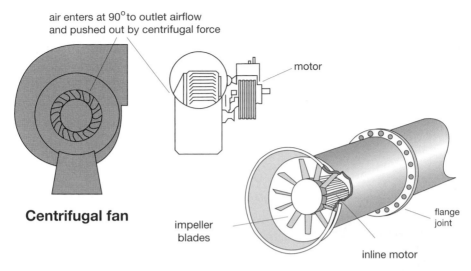

air enters at 90° to outlet airflow and pushed out by centrifugal force

motor

Centrifugal fan

impeller blades

flange joint

inline motor

Axial flow fan

traces of grit in the pipework. Also the pump once fitted should be PURGED of air, as these pumps rely on water for lubrication. A pump fitted into a system which is open to the atmosphere must be fitted into the system at a 'NEUTRAL' POINT, meaning that it is not fixed in a position where it could pump water over into the feed and expansion cistern, or where it might draw air in from a vent or radiator valve. For example, see PUMP.

CERAMIC WARE
Various forms of pottery such as earthenware, fireclay, stoneware or vitreous china.

CESSPOOL (CESSPIT)
A watertight container used under ground for the collection and storage of foul water and crude sewage. A cesspool might be used in areas where the main drainage has not been connected to a property. The contents from a cesspool should periodically be removed for proper disposal. See also SEPTIC TANK.

CESSPOOL   See CATCHPIT.

CFCs   See CHLOROFLUOROCARBONS.

CHAIN CUTTER   See LINK PIPE CUTTER.

CHAIN VICE   See PIPE VICE.

CHAIN WRENCH   See WRENCH.

CHAIR
A concealed frame which supports wall hung sanitary appliances (CORBEL APPLIANCES), therefore enabling the floor to be kept clear. The chair should be designed so that the weight of the appliance is transferred to the floor.

CHALK LINE
A length of string which has been coated in chalk dust, and is used to mark a straight line onto a surface. The chalk line is held down taught between the proposed line's start and finish point then the line is plucked or struck, and an accurate chalk mark is made.

CHANGE COLLAR
A collar or coupling designed specially to join the spigot of a metric sized pipe to that of an imperial sized pipe. The term is also used when joining one material to another material, i.e. plastic to clayware.

CHANGE OF STATE   See PHYSICAL CHANGE and CHEMICAL CHANGE.

CHANGE-OVER VALVE   See REGULATOR.

CHANNEL
An open waterway being semi-circular in cross-section. A channel would be found in an INSPECTION CHAMBER.

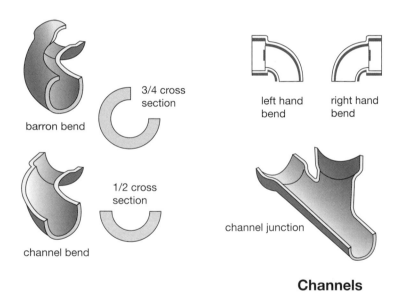

barron bend

3/4 cross section

left hand bend

right hand bend

1/2 cross section

channel bend

channel junction

Channels

CHANNEL BEND
A fitting used in an INSPECTION CHAMBER where a branch joins the main drain. A channel bend may be half or three-quarters cross-section, although if it is $^3/4$ cross-section it is often known as a barron bend. When installing a channel bend, the INVERT of the bend must be above the main channel unlike a CHANNEL JUNCTION.

CHANNEL JUNCTION
A purpose made drainage fitting which is used to join a branch into a main drain run. The channel junction ensures that the INVERT of the branch and main channel are at the same level unlike that of a CHANNEL BEND.

CHARLES' LAW

The relationship between volume and temperature. Providing the pressure remains constant, the volume of a gas is directly proportional to its absolute temperature. For example, a customer has a gas meter fitted in an outside meter box. This allows gas to flow into a warm room where it is consumed by an appliance. Because of this rise in temperature the gas increases in volume, therefore the customer gets more than they had paid for! See also BOYLE'S LAW.

CHASE

A recess formed in walls and floors to receive various services, i.e. pipes and cables.

CHASE WEDGE

A boxwood tool used in sheet roofwork to chase in angles, rolls and drips, although it is sometimes used to set in the crease at a corner prior to BOSSING.

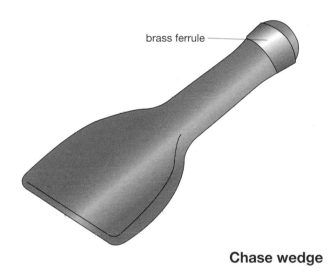

brass ferrule

Chase wedge

CHASER DIE AND STOCKS   See STOCKS AND DIES.

CHATTERING GOVERNOR

The noise caused by a governor/regulator banging open and closed as a result of surges in gas pressure created by appliances rapidly closing. The problem is often the result of the breather hole being too large. By carefully closing this hole, reducing it in size, the noise can often be dampened.

CHECK VALVE

A backflow prevention device used to allow water to flow in one direction only. When there is no flow of water in the pipeline or backflow the valve closes. For example, see NON-RETURN VALVE.

CHEEK

The vertical side of a DORMER.

CHEMICAL CHANGE

A chemical change happens when a chemical reaction causes an entirely new substance to be formed with its own properties and characteristics. For example two volumes of hydrogen and one volume of oxygen combined constitute water (H_2O); the new substance formed is permanent in that it cannot be changed back to its constituent elements by normal means, i.e. freezing, boiling, pressurising or straining, etc. The elements hydrogen and oxygen have not been mixed like sand in water, they have undergone a chemical reaction. It is not possible to produce water by mixing 2 volumes of hydrogen and 1 volume of oxygen, the two would always be individual elements; but, should the hydrogen gas ignite it would burn fiercely and water would be a by-product. To convert water back to the individual gases, hydrogen and oxygen, an electrical current is passed through the water. See also PHYSICAL CHANGE.

CHEMICAL CLOSET

A portable or fixed WC pan which contains a liquid with disinfectant and deodorant properties. It is used where there is no drainage or water supply which can flush the WC pan by normal means. Chemical closets require the periodical removal of their contents.

CHEMICAL FORMULA

A 'shorthand' used to represent the composition of a molecule of a substance. For example carbon dioxide is CO_2 and water is H_2O. See also CHEMICAL SYMBOL.

CHEMICAL PIPE CLEANING

A method sometimes employed to remove lime scale deposits in discharge pipes and stacks. When carrying out any work of this nature protective clothing and eye shields should be worn and all work upon completion should be thoroughly flushed and washed down. A descaling fluid containing 15–30% inhibited hydrochloric acid and 20–40% phosphoric acid is poured into the pipes in small quantities at predetermined points, or applied via a drip feed into the pipe at a rate of about 4 litres every 20 minutes. When carrying out any de-scaling, all windows should be opened to ensure good ventilation. Most acid descaling fluid will attack linseed oil-bound putty therefore prolonged contact with these jointing materials should be avoided.

If the discharge pipework is only lined or blocked with grease or soap residues the pipework should be flushed with very hot water which has 1 kg of soda crystals dissolved in every 9 litres. *Note* soda crystals should not be confused with caustic soda.

CHEMICAL SYMBOL

A 'shorthand' used to represent elements and consists of a capital letter or a combination of a capital letter and small letter. If the letters are followed by a small number then this shows how many atoms are present. Often the shorthand used is derived from the Latin name of the element. For example, lead is written Pb, the Latin name being plumbum. See also CHEMICAL FORMULA.

Some chemical symbols

Aluminium	Al
Carbon	C
Copper	Cu
Hydrogen	H
Iron	Fe
Lead	Pb
Oxygen	O
Tin	Sn
Zinc	Zn

CHEMICAL TEST   See SCENT TEST.

CHEMICALS TO TREAT SCALE

Chemicals which are added to water to reduce scale deposits. 'Calgon' is a well-known brand name of a liquid type of scale reducer. See also WATER CONDITIONERS.

CHEZY FORMULA

A formula used to find the velocity of flow and gradient for a drainage pipe or channel.

Chezy's formula is expressed as

$$V = C\sqrt{MI}$$

V = velocity in metres per second
C = 55 (C is a constant figure used being an average coefficient of
    friction caused by water passing through the pipe)
M = hydraulic mean depth (HMD)
$= \dfrac{\text{wetted cross sectional area of liquid}}{\text{wetted perimeter}}$
I = inclination (for example 1 in 40 = $^{1}/_{40}$)

The method used to find the hydraulic mean depth is as stated above and should be used when calculating the velocity or gradient of a rectangular channel such as boxed shaped eaves guttering, etc. It is also used for round pipes and gutters, but, due to the difficulty which can be experienced trying to calculate the cross sectional area of pipes flowing at one quarter or one third their depth of flow, an easier calculation can be used using the following table.

HMD for various depth of flows

Depth of flow	Value of HMD
$^{1}/_{4}$	diameter × 0.15
$^{1}/_{3}$	diameter × 0.19
$^{1}/_{2}$	diameter × 0.25
$^{2}/_{3}$	diameter × 0.29
$^{3}/_{4}$	diameter × 0.3
full bore (same as $^{1}/_{2}$)	diameter × 0.25

Therefore, for example, with a 100 mm pipe flowing full bore the HMD would be 0.1 m × 0.25 = 0.025.

Example of Chezy's formula put to use
Find the velocity of water flowing through a 100 mm diameter pipe flowing half full bore, when the gradient is 1 in 40.

$$V = C\sqrt{MI}$$
$$V = 55\sqrt{0.1 \times 0.25 \times \tfrac{1}{40}}$$
$$= 1.375 \text{ metres per second}$$

Should the gradient of the pipe be required the formula could be transposed to form $I = V^2/MC^2$

The answer just found 1.375 m/s can also be expressed in litres per second, if required, to do so the following calculation is carried out

$$Q = VA \times 1000$$

Q = volume of flow in l/s
V = velocity of flow in m/s
A = the cross-sectional area of pipe in m^2

Example
$Q = VA \times 1000$
$Q = 1.375 \times D^2 \times 0.7854 \times 1000$
$\quad = 1.375 \times 0.1^2 \times 0.7854 \times 1000$
$\quad = 10.8$ litres per second.
(*note* $D^2 \times 0.7854$ is one method of finding the area of a circle)

See also IMPERMEABLE FACTOR, TRANSPOSITION OF FORMULA, DISCHARGE UNITS and GRADIENT.

CHILLING
A situation in which a flame has been reduced in temperature below its ignition temperature. Natural gas, for example, needs to reach about 704°C in order to be completely consumed. If, however, it touches a heat exchanger, although hot in the ordinary sense, possibly 80–90°C, it is far lower than is required for complete combustion and as a result will not be completely consumed.

CHIMNEY
Any structure or part of a structure which contains a vertical flue.

CHIMNEY BACK (CHIMNEY GUTTER)
A BACKGUTTER.

CHIMNEY BREAST
The projection beyond the wall line that contains the flue and BUILDER'S OPENING.

CHIMNEY FLASHINGS (CHIMNEY WEATHERINGS)
The roof weathering which prevents rain water penetrating the building where a chimney stack passes through the roof structure. Chimney flashings consist of a

FRONT APRON, SOAKERS, STEP FLASHING, BACKGUTTER and a COVER FLASHING. The soakers could be omitted should a STEP AND COVER FLASHING be used. The material used for chimney flashings can be any of the standard sheet weathering materials, although lead tends to be the most popular due to its ease of application.

C

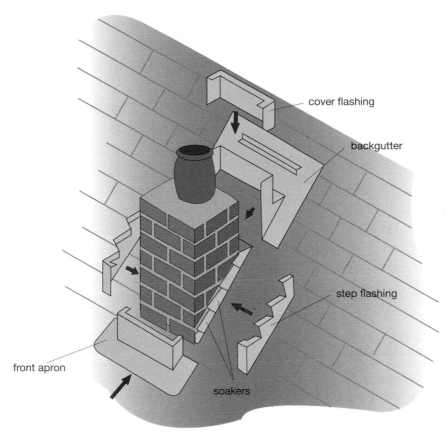

cover flashing

backgutter

step flashing

front apron

soakers

Chimney flashings

CHIMNEY PLATE (NOTICE PLATE)
A permanent plate or label located at a prominent position, for example, on the gas meter, water supply stopcock or next to a chimney. It gives details of the flue location/hearth and the category of the flue (i.e. what can be fitted to it), the type and size and its installation date.

CHISELS
An assortment of tools which are used to remove wood, brick or metal. Chisels designed to cut wood are different from those used to cut brick or metal. They are sharpened to a very sharp cutting edge and are usually fitted with a wooden or plastic handle. Chisels used to cut metal or brickwork are often referred to as cold chisels, because they can cut mild steel when it is cold. As a point of safety when the end of the cold chisel becomes burred over (mushroomed) due to excessive use the excess metal should be ground off to prevent it breaking away possibly flying into the eye.

(See illustration over.)

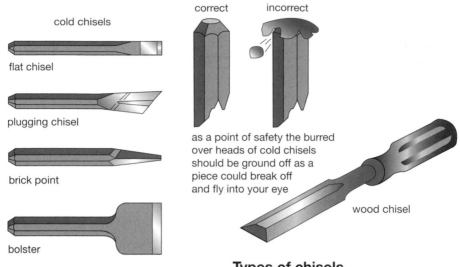

Types of chisels

CHLORINATION

A process of treating water with chlorine, usually in the form of gas. Chlorine is added to water to destroy any harmful bacteria which may be present in the water. Ammonia is sometimes added with the chlorine, this tends to remove the distinctive taste from the water. Should ammonia be added the process is known as a chloramine process. See also STERILISATION OF INSTALLATIONS WITHIN BUILDINGS.

CHLOROFLUOROCARBONS (CFCS)

Materials once extensively used as REFRIGERANTS and for the production of foamed pipe insulation. However, as they damage the ozone layer within the atmosphere, they have been banned from use.

CHROMIUM

A hard white metal which will take a high polish. Chromium is used in ALLOYS to increase their strength and corrosion resistance.

CHUTE

Term used to mean that a valley gutter discharges directly through the parapet wall, or over the edge of a building into a hopper head. For example, see GUTTER.

CIRCUIT

The formation of pipes and fittings in a hot water system in which water can circulate either by CONVECTION CURRENTS or pump assisted. For examples, see PRIMARY and SECONDARY CIRCULATION.

CIRCUIT BREAKER

A device that is designed to protect electrical equipment from becoming overloaded and damaged. There are several types of circuit breaker, which work by means of a switch opening as a result of the increased heat or magnetism that occurs in a conductor during overload. Circuit breakers, unlike fuses, simply trip a switch that can be remade.

CIRCUIT PROTECTIVE CONDUCTOR

The name given to the earth wire/conductor found within any cable, such as a 2.5 twin and earth. Within flex this will be covered in a green and yellow sheath, whereas in cable this wire will be found bare and un-insulated.

CIRCULATING HEAD (CIRCULATING HEIGHT)

The vertical distance between the centre-line of the boiler to the centre-line of the hot storage vessel, see CIRCULATING PRESSURE.

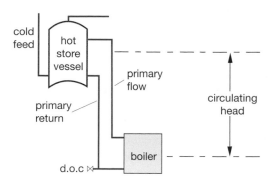

Circulating head

CIRCULATING PRESSURE (CIRCULATING FORCE)

The movement of water caused by CONVECTION CURRENTS. The greater the distance of the CIRCULATING HEAD in the PRIMARY CIRCUIT the greater will be the circulating pressure of water, and therefore circulation will be better. For example faced with the choice of placing the hot storage vessel directly above the boiler or well above it in a room upstairs, the room upstairs should be chosen, this would give a shorter heat-up period of the water, or give one the choice of reducing the primary circuit pipe size.

CIRCULATING PUMP

A pump, usually of the centrifugal type, used to assist in the circulation of water around a circuit, such as pumping water to radiators in a central heating system.

CIRCULATOR   See GAS CIRCULATOR.

CIRCUMFERENCE

The distance around the outside of a circle with every point being the same distance from the centre.

CISTERMISER

The trade name of a patented valve which is used to control the flow of water to an automatic flushing cistern. See HYDRAULICALLY OPERATED VALVE.

CISTERN

A vessel open at the top to the atmosphere, used to hold a supply of cold water. A cistern should have a lid to prevent the entry of dirt and vermin. See also FLUSHING CISTERN, STORAGE CISTERN and FEED CISTERN.

CISTERN HEAD
Term used in some parts of the West Country to mean the HOPPER HEAD for a rain-water pipe.

CLACK VALVE (BUCKET VALVE)
A type of NON-RETURN VALVE fitted into LIFT PUMPS.

CLADDING
Term used to mean the non-load bearing covering fixed to walls and roofs of buildings to keep the weather out.

CLARKS PROCESS (LIME SODA PROCESS)
A water-softening process carried out by some water authorities in which small quantities of soda lime are added to temporary hard water in order to reduce its hardness. Soda lime is a mixture of calcium hydroxide and sodium hydroxide which is capable of absorbing the carbon dioxide (CO_2) present in the bicarbonate found in temporary hard water. Thus once the carbon dioxide has been removed from the bicarbonate, the calcium carbonate is no longer held in suspension within the water and therefore settles in the bottom of the tank or is filtered from the water. See TEMPORARY HARD WATER.

CLAW HAMMER   See HAMMER.

CLAYWARE PIPES
Pipes used for below ground drainage systems. The pipes are made with or without sockets depending upon the method of jointing which is employed. Modern practice favours unsocketed pipes which rely on polypropylene jointing pieces, this allows for movement in the pipeline due to ground settlement and expansion. Socketed joints relying upon cement/mortar fixings are nowadays generally limited to connections or repairs to existing drains.

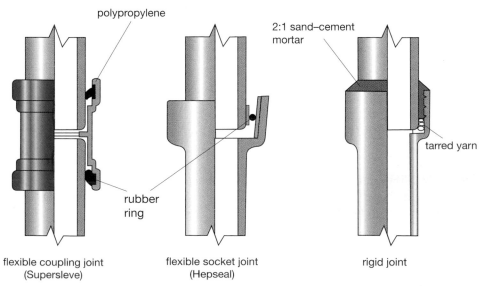

flexible coupling joint   flexible socket joint   rigid joint
(Supersleve)   (Hepseal)

Joints to clayware pipes

CLEANER'S SINK (BUCKET SINK)

A small sink fitted at low level with a GALVANISED or brass hinged grating on which a bucket could be rested while being filled. Cleaner's sinks are usually made from glazed earthenware.

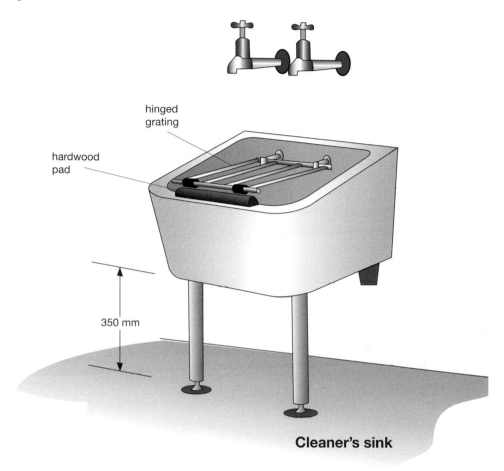

Cleaner's sink

CLEANING EYE

An ACCESS COVER.

CLEAT   See FIXING CLIP.

CLENCHER STOPPER (LEVER LOCKING STOPPER)

The stopper found in an intercepting trap. It is designed with a locking arm which can be removed by pulling a chain from a fixed point high in the inspection chamber. It is opened or removed should rodding of the drain beyond the intercepting trap be necessary. For example, see INTERCEPTING TRAP.

CLOAKING PIECE (COVER PIECE)

A shield positioned between urinals in order to give privacy to the user. The term is also used to represent any covering piece hiding the joint between adjacent wash basins.

CLOAKROOM

A term used to mean a toilet or a WC compartment. See also SANITARY ACCOMMODATION.

CLOSE COUPLED WC SUITE (COMBINATION SUITE)

A WC suite which does not require a flush pipe, the flushing cistern is connected directly to the WC pan. See WC SUITE.

CLOSE COUPLING

A term sometimes used to indicate that the cold feed and vent pipe connections to an open vented FULLY PUMPED SYSTEM have been terminated within close proximity to one another (maximum 150 mm apart) thereby ensuring the NEUTRAL POINT of the system is such that air is not drawn into the system due to the operation of the pump.

CLOSE CUT HIP or VALLEY (MITRED OR SOAKERED HIP OR VALLEY)

A HIP or VALLEY in which the tiles or slates have been cut to meet at the joint along the line of the hip or valley. A soaker is cut to fit beneath the tiles as they pass round the corner thus making a weathered joint.

CLOSED CIRCUIT

A system of pipework where the same water is circulated in which none is drawn off, such as in a hot water heating system.

CLOSED FLUE SYSTEM

The name given to a flue system that does not have any openings into it, such as a draught diverter or flue break. Within the gas industry this term is now falling out of favour; in modern BS terminology the flue system is classified as B2.

CLOSED SYSTEM   See SEALED SYSTEM and RODDING POINT SYSTEM.

CLOSED WELT   See STIFFENER.

CLOSURE PLATE

A sheet metal plate that is secured to the BUILDER'S OPENING in order to allow a gas fire to be installed. The closure plate ensures that the correct flue draught is maintained and prevents products entering the room, it also provides access to the catchment space for cleaning and test purposes.

CO$_2$ WELDING   See METAL ARC GAS SHIELDED WELDING.

COAL GAS

A gas produced by heating coal to approximately 1000°C. See TOWN GAS.

COCK

A valve through which liquid or gas flows, it can be adjusted to regulate the velocity of flow. The term cock is often used to describe those valves fitted with a parallel or tapered plug. See PLUG COCK.

CO/CO$_2$ RATIO

A figure calculated to indicate the optimum efficiency of a fuel when completely consumed. This is found by comparing the carbon monoxide (CO) against the carbon dioxide (CO$_2$) using the following formula:

$$\frac{CO(ppm)}{CO_2 \times 10\ 000}$$

C

Example

Assume the following readings were taken from a sample of flue gases: CO = 25 ppm and CO$_2$ = 6%.

Therefore $\dfrac{25}{6 \times 10\ 000} = 0.00042$ CO/CO$_2$ ratio

Note the ratio must not exceed 0.02 for appliances burning natural gas to ensure safe operation.

CODE OF PRACTICE   See BRITISH STANDARDS.

COEFFICIENT OF THERMAL EXPANSION

The coefficient of thermal expansion is the amount 1 mm of material expands when heated 1°C. The amount a material expands when heated can be measured by simple calculation as shown over. The coefficients of thermal expansion rates are given in the following table.

Coefficients of thermal expansion

Material	Approximate coefficient per °C
Aluminium	0.000026
Cast iron	0.000011
Copper	0.000016
Invar	0.0000009
Lead	0.000029
Mild steel	0.000011
Plastic	0.00018
Tin	0.000021
Zinc	0.000029

The calculation used to find the amount a material expands in length (linear expansion) is expressed as

length (in mm) × temperature change × coefficient

Example

Find the amount a 9 m long plastic discharge stack will expand due to a temperature rise of 24°C.

= 9000 mm × 24 × 0.00018 = 38.9 mm

This emphasises the need for EXPANSION JOINTS to plastic piping. To find the amount of expansion over an area (superficial expansion) the calculation is expressed as

area (in mm^2) × temperature change × coefficient × 2

(the answer being expressed in mm^2).

To find the amount of expansion per volume (cubical expansion) the calculation is expressed as

$$\text{volume (in mm}^3) \times \text{temperature change} \times \text{coefficient} \times 3$$

(the answer being expressed in mm^3).

COHESION
The pulling force which holds the MOLECULES of liquids and solids together. The molecules of a gas have no cohesion.

COLD CAULKING
A method of jointing cast iron pipes with either LEAD WOOL or ASBESTOS-CEMENT CORD (PC4).

COLD CHISEL   See CHISELS.

COLD FEED
A pipe conveying water from a cistern to a vented or unvented hot water supply.

COLD FEED CISTERN   See FEED CISTERN.

COLD WATER SUPPLY
Throughout the United Kingdom POTABLE WATER is provided by the local water authorities to individual premises and the various industries. When a supply of water is required the water authority will supply water to a point just outside the property boundary line, from this point the supply pipe should be run into the dwelling ensuring that precautions to protect the pipe from frost damage and aggressive soil are taken. It is also advisable to allow for movements in the ground. Notice that any pipe passing through or under a building must be ducted, this allows for its removal should the need arise. A consumer's STOP TAP should be fitted where the

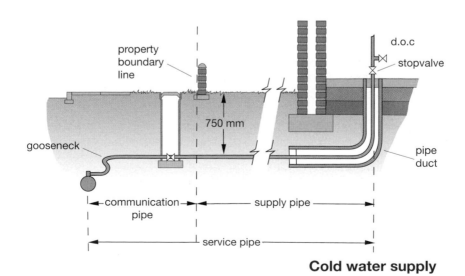

Cold water supply

supply pipe enters the building and should be fixed as low as possible with a DRAIN OFF COCK immediately above it. From this point the supply could be either of the direct or indirect system of cold water supply. See SERVICE PIPE, see also BOOSTED SYSTEM OF COLD WATER SUPPLY.

COLD WORKING
The working of metal in a cold state which can lead to the material becoming WORK HARDENED. Some metals can be better worked and shaped if heated.

COLLAPSE OF HOT STORAGE CYLINDERS
A situation where the hot water storage cylinder collapses due to the pressure inside the cylinder becoming less than that of atmospheric pressure. The condition is brought about by the cold feed and vent pipe serving the cylinder becoming blocked due to rust, scale or the pipe being frozen in winter. When these two pipes are blocked the system is closed to the atmosphere, thus if water is drawn off from the taps no air or water can get into the cylinder to fill the space where the hot water has been removed, and a partial vacuum will be formed, resulting in the sides of the cylinder being pushed in by atmospheric pressure. The cold feed and vent pipe being blocked can also result in a partially collapsed cylinder caused by hot water in the cylinder cooling and thus contracting.

COLLAR
A pipe fitting used to join the spigot ends of two pipes, it is basically a socket backing onto another socket. See also COUPLING.

COLLAR BOSS
A multi-branch fitting designed to allow bath and shower waste discharge pipes to connect to a discharge stack at any point around its circumference. The collar boss is designed to ensure that there is no cross flow of water from one discharge pipe to another. The biggest disadvantage with the collar boss is one of size, it tends to be bulky. For example, see BRANCH DISCHARGE PIPE.

COLOUR TEST
A test carried out in below ground drainage systems to trace the flow of the drain or sewer pipe, or for locating any leaks. A common liquid which is poured into the drain for this purpose is fluorescein, which turns the water to a bright green colour. See also TRACING EXISTING DRAINS.

COLUMN RADIATOR   See RADIATOR.

COMBINATION BOILER (COMBI BOILER)
A specially designed boiler which is used to heat up the domestic hot water instantly as and when it is required and also to serve a SEALED SYSTEM of hot water central heating. Using this type of boiler reduces the installation costs, in that, no feed and storage vessels are required for the supply of water, also by omitting the storage of domestic hot water a saving can be made in not heating the water unnecessarily. These boilers basically operate as follows:

(1) Should the central heating system call for heat the pump is energised which starts the water flowing. As the water passes through a VENTURI a pressure differential occurs in the deficiency valve causing the gas valve to open.

(2) Gas flows through the main burner and is ignited by the pilot flame.

(3) The water is rapidly heated in the low water content HEAT EXCHANGER and can only circulate around the boiler, through the DHW heat exchanger. The expansion of the water being taken up in the CH SEALED EXPANSION VESSEL.

(4) As and when the temperature of the water reaches 55–60°C, the thermostatic element expands causing the hot water system valve to close and the heating system valve to open, allowing water to flow around the heating circuit.

(5) The closing of the hot water system valve also causes a rod to rise and activate a micro-switch. This notifies the boiler control box that higher temperatures can be achieved which are manually determined by the setting of the flow temperature selector, on the control panel, ranging from 60–90°C.

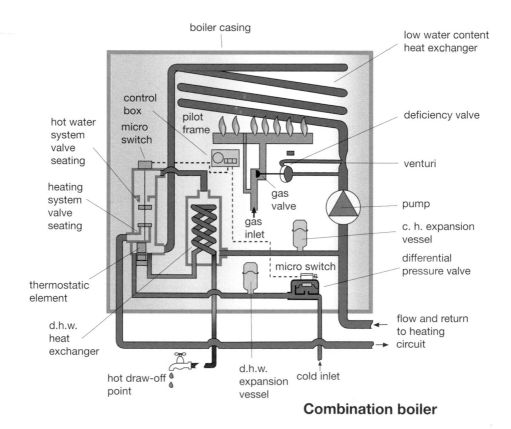

Combination boiler

Should the domestic water be required the following operation takes place:

(1) When a hot draw-off point is opened, water flows through the differential pressure valve which causes the diaphragm to lift and actuate a micro-switch, energising the pump.

(2) The main burner ignites and the boiler functions as in (1), (2) and (3) above.

(3) As the cold water passes over the thermostatic element it keeps it cool, ensuring it does not expand, resulting in the heating system valve opening. Therefore the central heating hot water only circulates through the DHW heat exchanger and around the boiler.

(4) As the cold water passes over the DHW heat exchanger it is rapidly heated to be discharged out through the hot tap.

See sketch. *Note* the domestic hot water takes precedence at all times over the central heating.

COMBINATION SUITE   See CLOSE COUPLED WC SUITE.

COMBINATION TAP   See MIXER TAP.

COMBINATION UNIT (COMBINATION TANK)
A compact unit in which the hot storage vessel and cold feed cistern are already 'plumbed' together, therefore it is only necessary to run the cold supply to the unit and pick up the hot DISTRIBUTION PIPE and run the overflow.

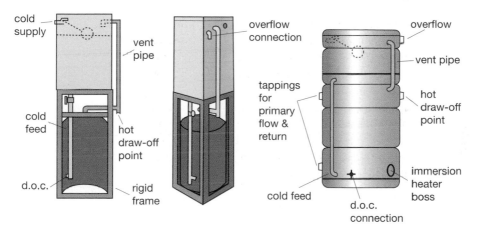

Combination units

COMBINATION WASTE   See WASTE FITTING.

COMBINED HEAT AND POWER (CHP)
A method used to generate usable heat and power, usually electric, in a single process. In its simplest form, it uses a turbine to turn an alternator, allowing the electricity generated to be used wholly within the premises where it was generated or it may be put back into the national grid.

COMBINED PRIMARY STORAGE UNIT (CPSU)
A heating unit that is designed to supply domestic hot water instantaneously plus it maintains the ability to provide central heating water as and when required. The CPSU basically operates similar to a THERMAL STORAGE SYSTEM. The difference being that the boiler and hot storage vessel are contained within the one unit in a CPSU whereas these are two separate components in a thermal storage system.

COMBINED SYSTEM OF DRAINAGE

A design of below ground drainage in which one pipe is used to convey FOUL, WASTE and SURFACE WATER. When this system is used all points in the system open to the atmosphere must be trapped, the only exception to this would be VENTILATING PIPES and FRESH AIR INLETS. The advantages of this system over SEPARATE and PARTIALLY SEPARATE SYSTEMS is one of cost, being cheaper and easier to install. Also during periods of rain the whole system gets a good flush through. The disadvantage of this system is also one of cost, it may be cheaper to install but it is dearer to run; with this system all water (foul and surface) has to be treated at the sewerage works. Also at times of heavy rainfalls the drain, if not adequately sized, could be prone to SURCHARGING, which makes water course pollution very probable. See also BELOW GROUND DRAINAGE.

COMBINED TEMPERATURE AND PRESSURE RELIEF VALVE

See TEMPERATURE RELIEF and PRESSURE RELIEF VALVE. A combined valve is capable of performing both functions, thus reducing cost and the need for two tappings.

COMBUSTION

The burning of fuel. When installing any appliances which burn fuel to produce heat it is essential that good combustion can take place within the appliance, failure to do so could result in CARBON MONOXIDE being produced. See STOICHIOMETRIC MIXTURE.

COMBUSTION ANALYSIS

A term used to indicate that sample flue gas products have been taken for the purpose of determining whether the appliance is operating efficiently and effectively. See COMBUSTION EFFICIENCY TESTING and CO/CO$_2$ RATIO.

COMBUSTION CHAMBER

The position in a boiler or burner where the fuel is ignited and burnt.

COMBUSTION EFFICIENCY TESTING

In burning any fuel some of the heat is lost up the chimney and by other miscellaneous heat losses. A boiler can only be expected to operate at approximately 80% efficiency and to ensure that a fuel burns as efficiently as possible several tests can be carried out in the flue pipe of the appliance.

  These consist of:

(1) The temperature of flue gases.
(2) The carbon dioxide (CO$_2$) content within the flue.
(3) The solids within the flue gases.

One of the most important tests is to find out the temperature of the flue gases; this is simply carried out by inserting a special thermostat into a test hole in the flue pipe. To ensure maximum efficiency the flue temperature should be kept as low as possible. The need to find the CO$_2$ content gives an indication of how much air is used to support the combustion of the fuel. Too much air will result in wasted heat due to the warming of the air; too little air will result in incomplete combustion. If the CO$_2$

percentage is high it means the air supplied to the burner is small. To find the CO_2 content a CARBON DIOXIDE INDICATOR is used; and this entry must be sought for further study. Once the flue temperature and CO_2 percentage have been found, it is possible to calculate the combustion efficiency of the appliance. This is simply found by taking the flue gas temperature, deducting from it the ambient or boiler room temperature and referring to the chart as shown.

C

Example
Find the percentage of combustion efficiency of an oil fired boiler in which the CO_2 content is 10% and the flue temperature is 260°C, the temperature in the boiler room is 60°C.

$$\text{Answer} = 260 - 60 = 200°C \text{ net temperature}$$
$$CO_2 \text{ in flue gases} = 10\%$$

By referring to the chart we find that the combustion efficiency = 82.9%.

Combustion efficiency: domestic fuel oil

CO_2	Net temperature °C						
%	150	200	250	300	350	400	450
15	88	86.2	84.4	82.6	80.8	79.1	77.6
14	87.6	85.6	83.9	81.9	80.1	78.1	76.5
13	87.2	85	83.1	81	79.1	77.2	75.3
12	86.8	84.5	82.1	80.2	78	76	73.9
11	86	83.7	81.2	79	76.9	74.5	72.1
10	85.4	82.9	80.3	77.8	75.3	72.9	70.1
9	84.5	81.6	79	76.2	73.1	70.6	68
8	83.5	80.5	77.2	74	71	68	64.9
7	82	78.9	75.4	72	68.1	65	61
6	80.5	76.7	72.5	68.5	64.1	60.1	50.9

(See illustration over.)

To find out if any solids are contained in the flue gases a special tester is used. This consists of a suction pump in which a special filter paper is held. After warming the pump to avoid condensation the filter pump probe is inserted into the flue and a sample of flue gas is withdrawn by operating the pump. The filter paper is then compared with the smoke scale supplied with the pump. Too much smoke could indicate that not enough air is being supplied to the fuel to support combustion and thus fully burn the carbon in the fuel. When carrying out a smoke test one should aim to get as low a reading as possible and in any case no higher than number three on the smoke scale. Therefore, in the example given where the combustion efficiency was 82.9%, if the smoke reading was three or less, satisfactory combustion should be taking place. A fourth test can be carried out in which the flue draught is measured using a special instrument. A probe is inserted into the flue pipe and the draught registers on a gauge in pascals (Pa) or N/m^2; the draught should be as specified by the manufacturer of the appliance. Many of the tests described above can be achieved using an electronic FLUE GAS ANALYSER.

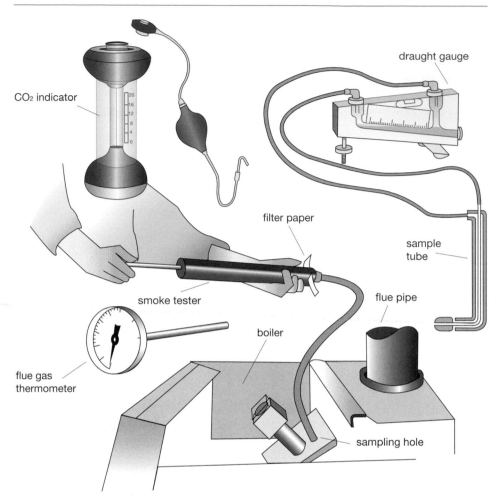

CO₂ indicator

draught gauge

filter paper

sample tube

smoke tester

flue pipe

boiler

flue gas thermometer

sampling hole

Combustion efficiency testing

COMMISSIONING
Term used to represent the initial checks and adjustments carried out to a system or appliance in order to confirm that it is working correctly.

COMMON DISCHARGE PIPE (COMMON WASTE)
A discharge pipe serving a number of waste appliances, such as wash basins or urinals. The pipe has a running trap fitted at its end close to the discharge stack. This method of connecting appliances into an above ground drainage system is no longer permitted, due to objectionable odours entering the building. See also FLOAT, which differs in that each sanitary appliance is fitted with its own individual trap.

COMMON FLUE SYSTEM (MODULAR FLUE)
A flue system that serves two or more appliances that are installed within the same room. See also BRANCH FLUE and SHARED FLUE.

COMMUNICATION PIPE   See SERVICE PIPE.

COMPARTMENT VENTILATION

Where a gas or oil burning appliance is installed within a cupboard or small compartment the space will require ventilation for the purpose of keeping the compartment cool. Two ventilation grilles will be required, one at high level and one at low level. The size of the vents is indicated in the following table.

Minimum free air vent requirements cm^2/kW of appliance net input

Appliance	Vent to	Lowest vent	Highest vent
Room sealed	Inside	10	10
Room sealed	Outside	5	5
Open flued	Inside	20	10
Open flued	Outside	10	5

Note both vents should communicate within the same room or both be on the same external wall. Should the air be supplied via ducts these must be independent of each other. When appliances have been installed with MODULATING BURNERS the air vent should be sized according to the maximum input rating. The door of the compartment should have a label affixed to warn against blocking the vents. See also VENTILATION GRILLE.

COMPENSATING VALVE

A modulating three port mixing valve, installed in a central heating circuit, which is designed to save on fuel consumption. Heating systems are usually designed on a $-1°C$ outside temperature. However, where it is not this cold the system is somewhat oversized. This valve is used in conjunction with an external temperature sensor which monitors the outside environment and adjusts the central heating flow temperature to suit, allowing water to by-pass the boiler so cooler flow temperatures can be circulated around the system to maintain the desired temperature. See also OPTIMISER.

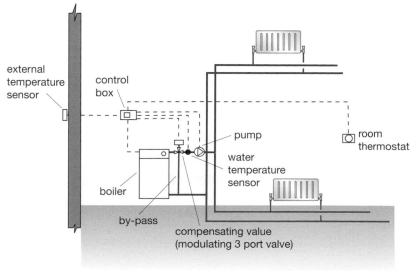

Compensating valve

COMPENSATOR

A control unit that takes into consideration the outside air temperature in order to reduce the amount of heat needed to be generated by a boiler. Thus on a very cold day the boiler generates a large amount of heat; however when the air temperature outside is warmer, the boiler will not be required to produce as much heat and an ACTUATOR located on a three port valve allows water within the system to run at a cooler temperature.

COMPOSITE VALVE

A valve that serves more than one function. One example would be the COMBINED TEMPERATURE AND PRESSURE RELIEF VALVE as found installed on an unvented domestic hot water heating system.

COMPOUND

A substance in which the MOLECULES contain more than one kind of ATOM. For example, water (H_2O) is a compound containing two hydrogen atoms and one oxygen atom.

COMPRESSION (BACK PRESSURE)   See TRAP SEAL LOSS.

COMPRESSION JOINT

A joint made by clamping the pipe to a fitting. There are two main types, either manipulative or non-manipulative. The difference is that with the manipulative compression joint the end of the tube has to be manipulated to form the joint whereas with the non-manipulative compression joint the nut is simply slipped on followed by the compression ring (olive). It must be noted that joints on copper pipe laid under ground must be of the manipulative type. Compression joints have been used for many years on copper pipework, but more and more in recent years they have been used on various other materials. For compression joints, used on plastic supply pipes a copper sleeve is inserted inside the pipe to prevent it from collapsing. This is not required on plastic pipes used for waste supplies, working under much

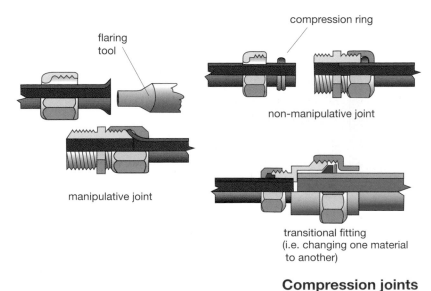

flaring tool

compression ring

non-manipulative joint

manipulative joint

transitional fitting
(i.e. changing one material
to another)

Compression joints

lower pressures. Nowadays compression joints are used to join to lead piping as WIPED SOLDER JOINTS are no longer preferred and also a compression joint is used for LOW CARBON STEEL tube. When jointing different materials the joint is often referred to as a transitional fitting.

COMPRESSION RING (OLIVE)
A wedding ring shaped piece of brass, copper or plastic used to assist the making of compression joints.

COMPRESSOR   See AIR COMPRESSOR.

CONCENTRIC FLUE
A TWIN WALLED FLUE system in which one pipe/void passes through the length of another and maintains an ANNULAR SPACE. Generally the flue products pass through the central flue and air is drawn into the annulus or remaining outer pipe void.

CONCENTRIC REDUCER   See REDUCER.

CONCENTRIC TAPER   See TAPER.

CONCRETE PIPES
Pipes used for SURFACE WATER drains and sewers, they can be used for FOUL WATER drains only where acceptable by the local authority. Precautions should be taken when laying pipes in acidic soils to prevent their destruction by the soils. The method of jointing can be either by sand and cement giving a rigid joint or by use of a rubber jointing ring which has the advantage of allowing for movement and being easily assembled. See also CLAYWARE PIPES.

CONDENSATE
The 'liquid' water resulting from condensation of water vapour or steam.

CONDENSATE DRAIN
The plastic pipe run from a CONDENSING BOILER with the intention of removing the water formed as part of the combustion process from the boiler. The condensing drain should be run from the appliance to connect to the house drain or suitably located soakaway. Where connection is to be made to a FOUL WATER DRAIN, the pipe would need to be trapped to ensure smells and possible methane gases rising from the sewer do not enter the appliance.

CONDENSATE PIPE
(1) A pipe which carries water due to condensation in a steam heating system. Whenever possible this water should be carried back to the boiler, for economic running costs. Failure to do so results in cold fresh water being needed to supply the boiler to make up the loss which has two drawbacks. Firstly cold water at approximately 10°C has to be heated whereas the condensate water could well be as high as 82°C, thus fuel consumption is reduced. Secondly in HARD WATER districts more treatment to TEMPORARY HARD WATER is required. See also STEAM HEATING.

(2) A pipe which is fitted to a point at the lower end of a flue pipe in order to remove any water which may be present in the pipe due to condensation caused by hot gases rising up a cold flue. See OPEN FLUE.

CONDENSATE RECEIVER (CONDENSATE POCKET)

A low point in a gas installation, a steam heated system or a flue pipe in which water can collect due to condensation within the pipe. From the condensate receiver the water can be drained off naturally or pumped as necessary. NATURAL GAS does not contain any moisture as TOWN GAS did therefore it could be said that a condensate receiver is not required, but, should water get into the pipe due to a leakage the condensate receiver proves useful to remove this. Thus when running a gas service pipe underground it should be made to fall to a low point where water can be removed if the need arises.

CONDENSATION

The PHYSICAL CHANGE of a vapour into its liquid which is brought about by either cooling the vapour or subjecting the vapour to a pressure (forcing the MOLECULES closer together). One of the most common causes of condensation can be seen as the deposition of a liquid, such as water from its vapour, generally onto a surface which is cooler than that of the vapour itself. As steam cools from its gaseous state it condensates back into a liquid. Air contains water vapour, the hotter the air becomes the more water it will hold, should the air temperature be reduced, e.g. due to contact with a cold object, or its pressure increased, the water is given up. See also DEW POINT.

CONDENSATION GUTTER

A small gutter fitted internally behind a SKYLIGHT, designed with a small condensate pipe which passes out through the building structure to discharge any water which collects. The condensation gutter is designed to remove any water which forms on the internal surface of the glass due to condensation.

CONDENSER

A device found within refrigerators, chillers and air conditioning units that is designed to remove the heat from the REFRIGERANT in order to convert the vapour back into its liquid form. See GAS REFRIGERATOR.

CONDENSING BOILER

A comparatively new design of boiler which has an increased efficiency over the more traditional types. The efficiency of a typical 'non-condensing' boiler is around 67–77% whereas with condensing boilers a figure of over 87% can be expected. This increased efficiency is due to the extraction of heat from the otherwise wasted flue gases. Most boilers have a single combustion chamber enclosed by the waterways of the HEAT EXCHANGER, through which the hot gases can pass and be eventually expelled from the boiler into the flue at a temperature as high as 250°C. Condensing boilers, on the other hand, are designed to first allow the heat to rise upwards through a primary heat exchanger; then at the top of the heater the flue gases are divided and turned 180° to pass back down over two secondary heat exchangers. These reduce the flue gas temperature to about 55°C, this reduction of temperature causes the water vapour, formed during the combustion of the fuel, to be condensed out thus releasing the LATENT HEAT which would otherwise escape with the flue gases. As the flue gas temperature is reduced, the droplets of water formed fall by gravity to collect at the base of the flue manifold. From here a condensate drain allows the liquid to run off and the remaining gases are expelled to the outside environment through a fan assisted BALANCED FLUE. The condensation produced within the appliance should be drained as necessary into the waste discharge pipework or externally into a purpose made SOAKAWAY.

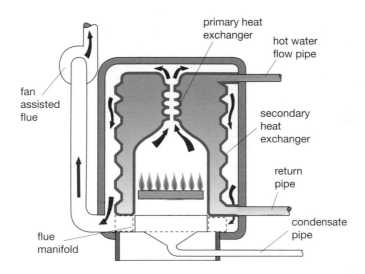

Condensing boiler

CONDUCTION

The TRANSFERENCE OF HEAT through or along a solid. If you hold a metal rod and heat up one end with a blowlamp the other end would soon become warm, this is because the heat is being transferred through the metal. Heat travels through all materials but the speed at which the heat passes through varies. The faster the heat travels the better the material is at conduction.

Thermal Conductivity of Materials

Copper	good conductors
Aluminium	↑
Iron	
Glass	
Brick	
Water	
Wood	↓
Still air	bad conductors

CONDUCTOR

A material that allows heat or electricity to pass through it with little difficulty. All substances are composed of atoms. Atoms have electrons revolving around a central nucleus. The electrons that are some distance from the central nucleus are free to move to another atom, and a random movement of these electrons occurs. Not all substances have free electrons. However, those that do, for example most metals, will conduct heat and electricity readily. Materials that do not have any free electrons, such as plastics or wood, will not allow the easy transference of heat or electricity and are known as INSULATORS.

CONDUIT

Plastic or steel pipe used to pass electric cables through, thereby providing suitable protection.

CONICAL ROLL

An EXPANSION JOINT used on copper and aluminium roof coverings. It is a type of batten roll. See COPPER SHEET AND ROOF COVERINGS.

CONNECTION TO EXISTING DRAINS

When joining into an existing drain run several methods can be adopted, in most cases the most suitable method is to connect into an existing INSPECTION CHAMBER or construct a new one. Sometimes a direct connection has to be made, for example where a house drain is to join a public sewer. The method of joining into the pipe would depend upon the material but in many cases either a junction block or saddle junction could be used, which must be connected into the top half of the drain. If the pipe which is to be cut into is less than 225 mm in diameter it is best to insert a new branch junction because a hole cut in such a small pipe would weaken it.

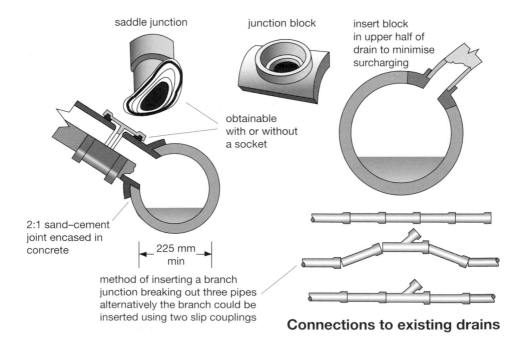

saddle junction

junction block

insert block in upper half of drain to minimise surcharging

obtainable with or without a socket

2:1 sand–cement joint encased in concrete

225 mm min

method of inserting a branch junction breaking out three pipes alternatively the branch could be inserted using two slip couplings

Connections to existing drains

CONNECTOR

Any of several types of connectors for example, FLUSHPIPE CONNECTOR, PAN CONNECTOR or LONGSCREW, etc.

CONSERVANCY 'DRAINAGE' SYSTEM

A method of drainage in which the waste matter is collected and retained in a suitable vessel and disposed of as and when necessary, for example, see CHEMICAL CLOSET.

CONSTANT OIL LEVEL CONTROL

A device which controls the feed of oil to a VAPORISING BURNER. For a vaporising burner to function efficiently a constant level of oil is required. Should the oil flow directly from the storage tank its pressure and velocity would vary as the oil level in the tank drops. The constant oil level control serves four functions.

(1) To maintain a constant level of oil, irrespective of the oil level in the storage tank.
(2) To precisely control the flow of oil to the burner.
(3) To provide a secondary means of shutting off the flow of oil to the burner (the first being an isolation valve located close to the storage tank).
(4) To provide a trip to shut down the flow of oil to the burner should the needle valve fail to close.

There are various designs of oil level control but in general they consist of either a single or double float.

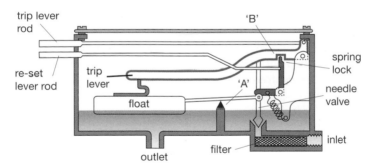

Single float constant oil level control

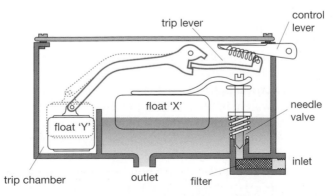

Double float constant oil level control

The double float type is no longer manufactured but many are still to be found in service. The single float constant oil level control operates as follows:

(1) Oil flows through a filter past the needle valve and slowly lifts the float which in turn causes its lever arm to pivot at A which applies a pressure onto the needle valve closing off the supply. As the oil is used and the level lowers so does the float allowing more oil to enter.

(2) Should the needle valve fail to close due to, for example, excessive head pressure, a faulty seating or grit between the valve and seating the oil level continues to raise the float until it bears against the trip lever, causing it to lift.

(3) As the trip lever is lifted the spring lock is disengaged from the slot in the trip lever B to apply a sufficient force to drive the needle valve home.

(4) The control cannot then be used until the oil level has been lowered and the re-set lever has been pushed in, lifting the spring lock back into position.

The double float constant oil level control operates as follows:

(1) To start the oil flowing into the control, the control lever is pressed down; conversely to shut it off the lever is lifted.

(2) As oil enters the inlet it slowly lifts float X which in turn brings the needle valve down to close off the supply at the required level. As the oil level drops more oil is allowed to enter the chamber.

(3) The second trip chamber is only a fail safe device and is not part of the efficient running of the control. If float X does not shut off the oil supply, due to grit blocking the inlet, etc., the oil would continue to rise and discharge into the trip chamber, where it would raise float 'Y' which would, when it reached a predetermined level allow the trip lever to drop and with the assistance of the spring force the needle valve shut. The control cannot then be used until attention has been given to it to rectify the problem and reset the trip mechanism, which is done by draining the trip chamber and pressing down the control lever.

CONSTANT VOLUME GOVERNOR   See GAS GOVERNOR.

CONSUMER UNIT
The electrical inlet fuse board and circuit breaker located where the electricity supply enters the property.

CONTINUITY BONDING (CROSS-BONDING)
Continuity bonding refers to a situation where certain metal parts within a building such as pipes or steel beams, etc. have been connected together to ensure they all remain at an equal electrical potential. In the case of metalwork that enters the building, such as a gas or water supply pipe, a connection is made from the metalwork to the electrical earthing system, so that should a fault arise within the electrical system the electrical fault goes down to earth via a suitable route. This is referred to as the 'main equipotential bond'. In specific locations within the building such as bathrooms, additional 'supplementary bonding' is used; this may or may not be connected to the electrical earthing system, but in any case will again ensure that all

the metalwork is at the same potential. When removing any pipework a new cross-bond must be set up to ensure the continuity of bonding is maintained, failure to do so could result in someone receiving an electric shock. All plumbers should carry with them a temporary bonding wire for use in repair work involving the temporary removal of pipework, this is for their own protection as much as anything else. See TEMPORARY BONDING WIRE.

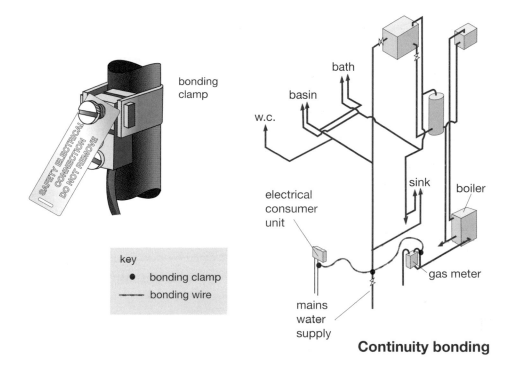

Continuity bonding

CONTINUOUS STEP FLASHING
A COVER FLASHING which rises up the side of a roof against an ABUTMENT, stepping at each rise in the brick course. There are various names given to a continuous step flashing such as cockney flashing, dog tooth flashing and herringbone flashing. For example, see STEP FLASHING.

CONTRACTION
To get smaller; see EXPANSION.

CONVECTION CURRENTS
When water or air is heated it expands and occupies more space thus becoming lighter per unit area. Because it is lighter it is pushed upwards by the cooler denser water/air which falls to replace that rising. When the water/air at the lower level becomes hotter it expands and in turn is thus pushed upwards. This process continues over and over, circulating around the room, in the case of air, or around a vessel in the case of water. See also TRANSFERENCE OF HEAT.

(See illustration over.)

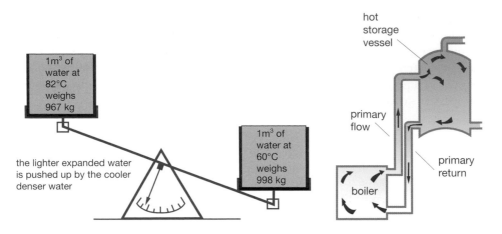

1m³ of water at 82°C weighs 967 kg

the lighter expanded water is pushed up by the cooler denser water

1m³ of water at 60°C weighs 998 kg

hot storage vessel

primary flow

primary return

boiler

Convection currents

CONVECTOR HEATER

A HEAT EMITTER designed with a small area of heating surface which is surrounded by a series of fins through which air flows. Due to the fact that the heating surface is surrounded by fins it is impossible to touch the pipe, therefore this type of heater is ideal for SEALED SYSTEMS of heating. There are two basic types of convector heaters, those which rely on the natural air movement, caused by CONVECTION CURRENTS, and those which are fan assisted. The term convector heater is also given to some types of gas and electric fires.

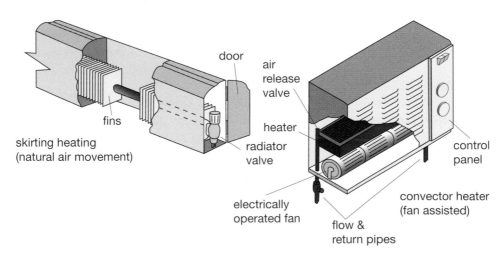

door

air release valve

fins

heater

skirting heating (natural air movement)

radiator valve

control panel

electrically operated fan

flow & return pipes

convector heater (fan assisted)

Convector heaters

CONVENTIONAL BOILER   See REGULAR BOILER and SYSTEM BOILER.

CONVENTIONAL FLUE   See OPEN FLUE.

COOK'S HOLE CUTTER   See HOLE CUTTERS AND HOLE SAWS.

COOLING TOWER

A device that cools the hot water that carries waste heat from a condenser. They are used in locations where there is no large body of water available, such as a river, to cool the water via a heat exchanger. The cooling tower works by allowing the heat to be drawn from the water by evaporation. Referring to the illustration, it can be seen that the water falls by gravity and collects at the base of the tower before returning to the condenser to draw more heat.

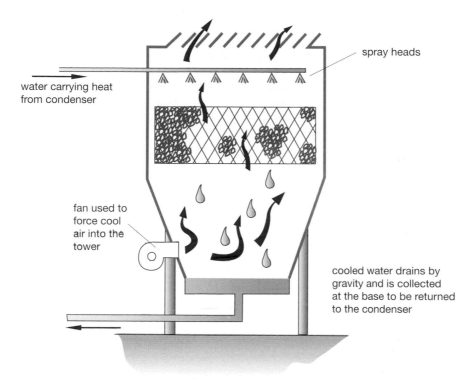

spray heads

water carrying heat from condenser

fan used to force cool air into the tower

cooled water drains by gravity and is collected at the base to be returned to the condenser

Cooling tower

COOPER'S PLUNGER   See SUCTION PLUNGER.

COPING

A concrete or stone protection which overhangs the top of a wall, such as a PARAPET wall.

COPPER

A NON-FERROUS METAL produced for many uses, but its most common form will be found in wire, pipe and sheet.

Chemical symbol	Cu
Colour	Reddish-brown
Melting point	1083°C
Boiling point	2595°C
Coefficient of linear expansion	0.000016/°C
Density	8900 kg/m^3

The chemical symbol Cu comes from its Latin name, cuprum. See also, COPPER PIPE
AND FITTINGS and COPPER SHEET AND ROOF COVERINGS.

COPPER BIT (COPPER BOLT)   See SOLDERING IRON.

COPPER PIPE AND FITTINGS

Seamless round copper tube has been used for water, gas and heating applications
for many years in a range of sizes 6–267 mm diameter. The size selected denotes the
external pipe diameter. Today it is manufactured to BS EN 1057 and supplied in
straight lengths. For pipe sizes up to 54 mm it can also be supplied as a continuous
coil up to 25 m in length. The TEMPER of the tube can be specified as a soft or half
hard condition. The methods used to join the pipe could either be by a manipulative
or non-manipulative compression joint, capillary fitting or bronze weld. Only manipulative compression joints are permitted to be used under ground, which must be
either gunmetal or of a brass resistant to DEZINCIFICATION. Capillary fittings, if
used on mains supply, must contain a lead free solder, preventing the contamination
of the water. A comprehensive range of both capillary and compression fittings are
available and trade catalogues, etc. should be examined, but one vital piece of
knowledge is required to order 'tee' pieces, and this is to order as shown.

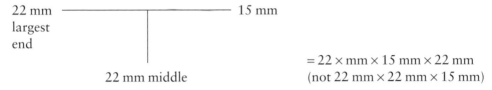

22 mm
largest
end

15 mm

22 mm middle

$= 22 \times mm \times 15\ mm \times 22\ mm$
(not 22 mm × 22 mm × 15 mm)

i.e. end × end × middle. If the order is not made in this fashion the wrong fitting
could be supplied.

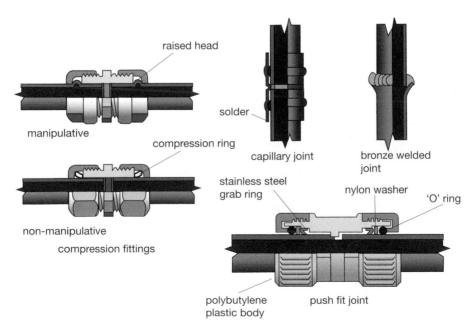

raised head

manipulative

compression ring

non-manipulative

compression fittings

solder

capillary joint

stainless steel
grab ring

polybutylene
plastic body

push fit joint

bronze welded
joint

nylon washer

'O' ring

Copper pipe fittings

COPPER SHEET AND ROOF COVERINGS

Sheet copper can be obtained in almost any size, although the standard sizes usually specified are 1830 mm × 610 mm which should be fully ANNEALED for general roof work. Sheet copper comes in a range of thickness and the thickness used would depend upon circumstances and cost.

Recommended thickness of sheet copper

Location	Thickness
Cover flashings	0.45 mm
Chimney flashings, canopies, gutters	0.60 mm
Flat roofs, gutters.	0.70 mm

Sheet copper remains unaffected by changes in temperature and will not CREEP but will become WORK HARDENED if COLD WORKED too much. The method of jointing sheets of copper together varies depending on the circumstances and specifications, but in general, for joints which run across the flow of water (transverse joints) the method of jointing is either by drips, double lock welts or single lock welts.

Recommended transverse joint to be used on pitched roofs

Jointing method	Degrees of roof pitch
Drips	up to 6
Double lock welt (sealed)	6–20
Double lock welt	20–45
Single lock welt	over 45

(See illustration over.)

For roofs with a pitch of between 6–20° if a double lock welt is used it must be sealed, either with boiled linseed oil or a non-setting mastic. For joints which run in the same direction as the flow of water standing seams or batten rolls are used. In the case of roofs where foot traffic is expected, standing seams should not be used to prevent any flattening of the joint if trodden on. Large roofs are divided into a series of smaller areas called bays and the size of a bay will depend upon the thickness of the sheet copper. The roof is divided into bays to overcome the problems caused by expansion and contraction and the problems created by wind pressure. See also DOUBLE LOCK WELTS, SINGLE LOCK WELTS, STANDING SEAM, LONG STRIP ROOFING and SHEET FIXINGS.

COPPER TUBE CUTTER   See PIPE CUTTER.

COPPER WELDING

Copper can be welded with oxy-acetylene equipment if required, but to do so a larger welding tip is required to allow for the high conductivity of the copper. As a general guide, a welding tip of one to two sizes larger is required than for welding low carbon steel of a similar thickness. The gases should be adjusted to a NEUTRAL FLAME and the inner cone should be held about 9–10 mm from the molten pool of

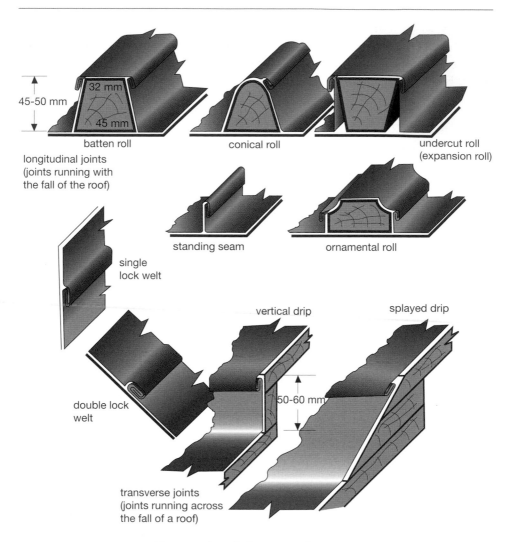

32 mm
45 mm
45-50 mm

batten roll

longitudinal joints
(joints running with
the fall of the roof)

conical roll

undercut roll
(expansion roll)

standing seam

ornamental roll

single
lock welt

vertical drip

splayed drip

double lock
welt

50-60 mm

transverse joints
(joints running across
the fall of a roof)

Expansion joints used on copper roof coverings

copper, this avoids the copper absorbing the gases which can cause porosity to the
finished weld. The filler rod used should contain a deoxidising agent such as phosphorus. To weld copper successfully, the TUNGSTEN INERT GAS WELDING approach
is often adopted which uses a gas shield to protect the surface from OXIDATION, also
by using electricity as the heat source, absorption of the burning gases by copper is
avoided. Upon completion of a welded joint, when the molten metal has solidified, a
light hammering to the red hot weld tends to increase its strength.

COPPERISED LEAD
A type of cast lead in which any copper element within the lead is evenly distributed
throughout the thickness of the material, unlike milled lead where the lighter copper
elements float to the surface of the cooling billets of lead prior to rolling or milling.
See MACHINE CAST LEAD SHEET.

C

CORBEL APPLIANCES
Corbel appliances are SANITARY APPLIANCES which have been fixed directly into a wall and no floor support is evident, although the appliance could be sitting on a concealed chair.

CORGI
The Council for Registered Gas Installers. In order to undertake any gas installation work, other than DIY, one must be registered with this national organisation. Failure to do so would mean that you are breaking the law. CORGI has been set up to ensure, as far as practically possible, all gas installation work is safe in terms of gas soundness and effective flue and ventilation provision.

CORNICE
A projecting moulding situated at the top of an outside wall. It is designed to throw the drips of rainwater away from the wall. The cornice is sometimes used by the plumber to support a gutter.

CORROSION
The destruction of metals resulting from a chemical reaction on the surface by the action of air, water or chemicals. When metal comes into contact with air for example a reaction takes place and often powdery deposits are formed on the surface of the metal. If copper and iron are joined together and placed in an ELECTROLYTE a simple cell would be formed and corrosion would be speeded up. There are two basic types, see ATMOSPHERIC and ELECTROLYTIC CORROSION. See also EROSION.

CORROSION INHIBITOR
A chemical which is put into hot water heating systems in order to reduce corrosion. Impurities in the water, such as flux residues, are also neutralised and rendered harmless. Some corrosion inhibitors also have pump lubricating properties and will destroy bacterial growth. A corrosion inhibitor should be added to all new systems as soon as possible and in any case within 14 days of filling. When adding any type of chemical to a system one must observe the maker's recommendations.

COSHH
The abbreviation for Control of Substances Hazardous to Health. This is the legal framework requiring all employers to assess their work situations where employees may be exposed to harm caused by the use of solids, liquids, gases, dust, fumes and vapours, etc. The law requires action to be put in place to remove or reduce the risk of illness or injury from such substances.

COTTER PIN   See SPLIT PIN.

COULOMB
A unit quantity (or charge) of electricity. Heat quantity is measured in joules; a quantity of water is measured in litres. The amount of electricity passing any point along a conductor in one second with a current of 1 ampere flowing is 1 coulomb.

COUNTER TOP BASIN   See VANITY BASIN.

COUPLING
A pipe fitting used to join two pieces of pipe in a straight line. See also COLLAR.

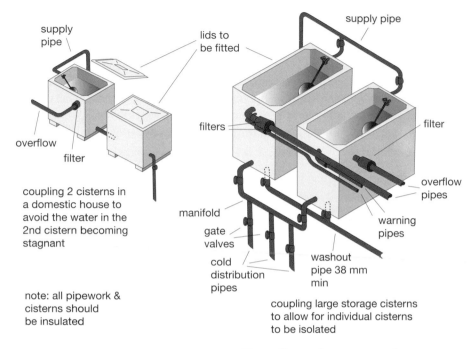

supply pipe

lids to be fitted

supply pipe

overflow

filter

filters

filter

coupling 2 cisterns in a domestic house to avoid the water in the 2nd cistern becoming stagnant

manifold

gate valves

cold distribution pipes

overflow pipes

warning pipes

washout pipe 38 mm min

note: all pipework & cisterns should be insulated

coupling large storage cisterns to allow for individual cisterns to be isolated

Coupling of storage cisterns

COUPLING OF STORAGE CISTERNS
Sometimes it is desirable to have two or more cisterns coupled together instead of one large cistern. By coupling cisterns together one can always be isolated and drained down if required. On a smaller scale, in a house, sometimes space limits the size of a storage cistern, two smaller cisterns can be joined together giving the required capacity. There are different methods used for the above examples. If means are required to isolate one cistern, an ISOLATING VALVE is fitted at each point in or out of the cistern, which can be shut off. For the domestic house it would be uncommon for one cistern to be isolated, therefore the mains supply is usually taken into one cistern and the cold distribution or feed pipe is taken out of the other, by designing it in this way the water in the second cistern would not become stagnant. See also WASHOUT PIPE.

COVER FLASHING (SIDE FLASHING)
The sheet weathering where a roofing material meets an ABUTMENT, designed to prevent the entry of water into the building. All cover flashing should have a minimum cover of at least 75 mm. See also GUNSTOCK LAP.

(See illustration opposite)

CRADLE
A support designed to fit the underside and hold an appliance such as a bath or storage cistern in position. See also CHAIR.

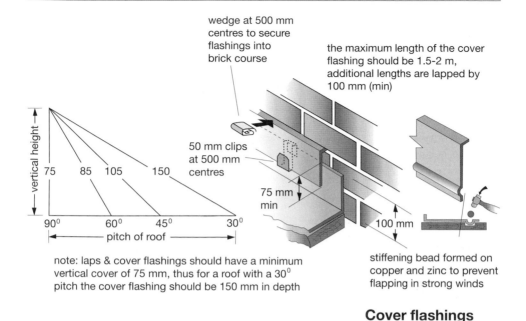

wedge at 500 mm centres to secure flashings into brick course

the maximum length of the cover flashing should be 1.5-2 m, additional lengths are lapped by 100 mm (min)

50 mm clips at 500 mm centres

75 mm min

100 mm

vertical height

75   85   105   150

90°   60°   45°   30°

pitch of roof

note: laps & cover flashings should have a minimum vertical cover of 75 mm, thus for a roof with a 30° pitch the cover flashing should be 150 mm in depth

stiffening bead formed on copper and zinc to prevent flapping in strong winds

Cover flashings

CRANK SET   See PASS-OVER OFFSET.

CREEP

The amount, if any, a metal will stretch. When a metal is heated or is under a constant load it will become longer, when the heat or load is removed the metal should return to its original size. If it has suffered any permanent increase in size, i.e. stretched, it could be said 'this is the amount of creep'.

CROSS-BONDING   See CONTINUITY BONDING.

CROSS-CONNECTION

A situation where wholesome drinking water is contaminated by a cross-connection with an unwholesome supply. Such an example could occur where a bath shower mixer tap is used, taking the cold water from the mains supply pipe and the hot water from a low pressure cistern-fed system without the use of a BACKFLOW PREVENTION DEVICE.

CROSS-FITTING

A double branch designed to connect two branch pipes or channels from opposite sides.

CROSS-FLOW

Water flowing down a branch discharge pipe enters the discharge stack and shoots across the stack and enters another branch pipe. To prevent this from occurring, branch pipes should be staggered as they enter the main stack. Where a WC branch connects to the discharge stack the area opposite that pipe, from its centre line down to a distance of 200 mm, should be restricted. For examples, see BRANCH DISCHARGE PIPE. The term cross-flow is also used to identify the flow from one side of a hot/cold water mixer fitting to the other, possibly leading to the contamination of the water supply.

CROSS-PEIN HAMMER   See HAMMER.

CROSS-SECTIONAL SKETCH
A sketch which shows the internal, rather than the external view of a component/object. Many of the illustrations in this book are cross-sectional sketches, for one example, see BALLVALVE.

CROSS-VENT (YOKE VENT)
A pipe joining the main discharge stack to the main ventilating pipe. The purpose of cross-venting is to ensure that a pressure does not build up inside the main discharge pipe as the water falls down the stack. For example, see SANITARY PIPEWORK.

CROSS-WELT
A type of EXPANSION JOINT used to join metallic sheet materials across the flow of water. See SINGLE LOCK WELT and DOUBLE LOCK WELT.

CROWN
The highest point on the inside of a trap outlet. See also SOFFIT.

CROW'S FOOT
A name used in some parts of the country to mean a BASIN SPANNER.

CROYDON BALLVALVE
A BALLVALVE with the piston moving in a vertical plain, it is because of this design that it is no longer made. This vertical action has a mechanical disadvantage and thus tends to have a jerky and sluggish action. The water level to this type of ballvalve is adjusted by means of bending the lever arm up or down. See also EQUILIBRIUM BALLVALVES.

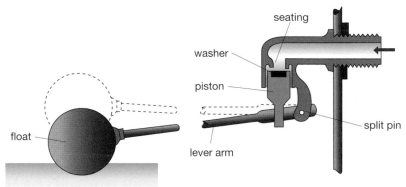

Croydon ballvalve

CRUTCH HEAD
A type of head used on taps, for example, see BIB TAP.

CUBICAL EXPANSION   See COEFFICIENT OF THERMAL EXPANSION.

CUP JOINT

A TAFT or BELL JOINT as used on lead or copper pipe.

CUPROSOLVENT

The ability of water to dissolve copper. Almost all waters are cuprosolvent, but only rarely does cuprosolvency reach a level harmful to health. See also SOLVENCY.

CURB ROOF

A MANSARD ROOF. The curb refers to the point where the roof pitch changes.

CURRENT

The flow of free electrons passing round a circuit. This is measured in amperes. See ALTERNATING CURRENT. See also CONDUCTOR.

CURTILAGE

The land that surrounds a series of buildings and therefore forming the property boundary.

CUTAWAY WC SEAT   See OPEN FRONT WC SEAT.

CUTTING BLOWPIPE (CUTTING TORCH)

An oxy-acetylene blowpipe which is designed to cut iron or mild steel. The iron or steel is heated to a bright red colour then, by pushing a lever on the handle, a jet of oxygen is blown onto the heated area. Immediately OXIDATION of the metal occurs and changes it to a magnetic oxide (magnetic oxide has a lower melting temperature than iron or mild steel). The magnetic oxide melts and is blown away by the stream of oxygen. For example, see WELDING BLOWPIPE.

CUTTING LUBRICANTS

A fluid used to prolong the life of cutting tools such as dies and drill bits. Soluble oil tends to be the most versatile lubricant although special cutting pastes are available. Paraffin is sometimes used to assist cutting copper and aluminium.

C.V.   See CALORIFIC VALUE.

CYCLING

A situation where a boiler, after having warmed a body of water for heating purposes, continually switches on and off in response to a boiler thermostat. Cycling of a boiler in this way is a waste of fuel and the result of a poor system design. In fact, almost as much heat is lost up the flue as is given to heating the water. Today Building Regulations impose requirements that prohibit system designs that allow unnecessary cycling.

CYLINDER

A cylinder in the shape of a pipe. The term cylinder is used more times than not to mean a vessel used for the storage of hot water. See HOT STORAGE VESSEL and INDIRECT CYLINDER.

CYLINDER AIR GAP (BUBBLE TOP)
The void within an unvented domestic hot water cylinder to allow the expansion of the hot water, due to it being heated.

CYLINDER COLLAPSE   See COLLAPSE OF HOT STORAGE CYLINDERS.

CYLINDER TANK SYSTEM   See SUPPLEMENTARY STORAGE SYSTEM.

CYLINDER THERMOSTAT   See THERMOSTAT and TEMPERATURE CONTROL FOR DOMESTIC HOT WATER SUPPLY.

CYLTROL VALVE   See THERMOSTATIC CONTROL VALVE.

DAMP PROOF COURSE

An IMPERVIOUS layer of material which prevents water penetrating into the top and bottom of a building's brickwork by CAPILLARY ATTRACTION. A layer of material such as lead, slate, plastic or engineering bricks is incorporated within the brickwork where damp may be a problem. To ensure earth does not build up beyond the height of the damp proof course at low level it should be located at least 150 mm above ground level. When joining damp proof course materials the joints should be lapped, with at least 100 mm overlay.

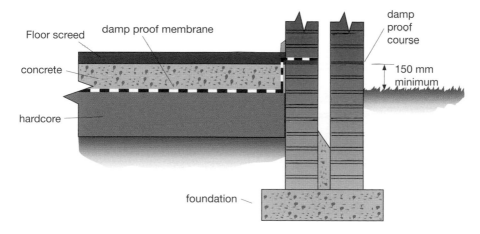

Damp proof to a solid ground floor

DAMPER

A hinged flap which permits fresh air into a vessel or system of duct work. SOLID FUEL boilers are often fitted with a damper at low level which can be adjusted to regulate the amount of air supplied to the fuel, this affects the speed and amount of fuel it will burn.

DAMPER REGULATOR

A device which is sometimes fitted onto independent solid fuel boilers to control the amount a DAMPER is open or closed. The damper regulator is filled with a liquid held inside a bulb, as the liquid heats up it expands and compresses a metal bellows which in turn pushes on a lever connected to the damper, slowly closing it. Nowadays it is more common that the air used to increase the speed at which the fuel is burnt is supplied by a forced draught unit, e.g. fan assisted.

DATA BADGE (DATA PLATE or IDENTITY BADGE)
A permanently fixed plate located on appliances such as boilers, cookers, fires, etc. that gives information relative to its operation, such as burner pressures, heat inputs and fuel type.

DATUM   See SITE DATUM.

DATUM POINT   See NEUTRAL POINT.

DC
Abbreviation for direct current, see ALTERNATING CURRENT and IONS.

DEAD LEG
(1) A section of cold water pipework through which water cannot flow and be drawn off from a valve fitted to an appliance. The section of pipe can only be emptied from a DRAIN OFF COCK.
(2) Term used to mean that cold water has to be drawn off from a hot tap before hot water is drawn off. It is caused by the hot storage vessel being fitted a long way away, thus you have an excessively long hot DISTRIBUTION PIPE. To overcome the problem a secondary circuit could be designed into the hot pipework or one could install a LOCALISED HOT WATER SUPPLY. For example, see SECONDARY CIRCUIT.

DEAD SOFT TEMPER
The TEMPER or softness of sheet copper. Copper in this state should be used for copper sheet weatherings.

DEAD-WEIGHT TYPE SAFETY VALVE
A type of SAFETY VALVE in which the pressure needed to hold the valve shut is supplied by lead or cast iron weights. Should the pressure within a hot water system be greater than that supplied by the weights it would lift the valve from its seating. For example, see PRESSURE RELIEF VALVE.

DE-AERATOR (AIR SEPARATOR)
A fitting designed to overcome the problems of pumping water in the PRIMARY FLOW AND RETURN circuit where water is sometimes pumped over the VENT PIPE or air is drawn in through it. The de-aerator is designed to make the point where it is fitted the NEUTRAL POINT within the system and thus a position where the pump has little or no influence to the flowing water. The de-aerator is also designed so that as flowing water passes through it, it is made to turbulate and thus air bubbles within the water can form and rise out of the system. See also AIR PURGER.

(See illustration opposite)

DE-AERATOR ('TIGERLOOP')
A device which allows oil to be drawn up from a sub-gravity oil storage tank to a pressure jet burner, without the need for a return pipe back to the storage tank, as would be required in a two pipe oil supply. The de-aerator removes the air from the line in a single one pipe lift. It must be fitted outside the building, to allow any oil

D

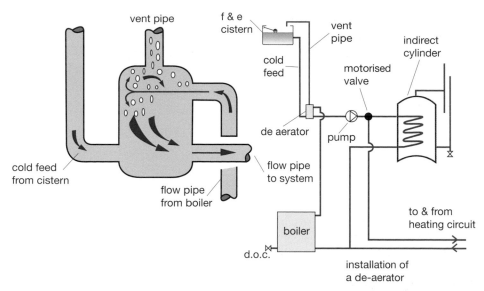

De-aerator (air separator)

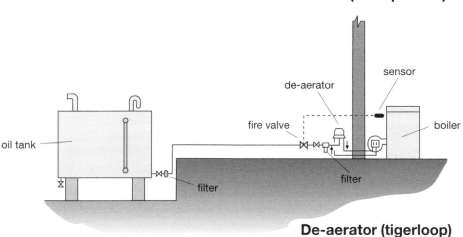

De-aerator (tigerloop)

vapour to disperse, with a two pipe loop to allow oil to circulate from the oil pump in the burner.

DE-BURRING REAMER   See REAMER.

DECKING
Term used to mean the roof timbers which support the sheet weathering material. In most cases the decking will consist of either chip-board, compressed straw slabs, blockboard, plywood or tongue and groove boarding. On top of the decking an underlay should be positioned prior to the fixing of any metallic roof covering.

DECORATIVE FUEL EFFECT FIRE (DFE GAS FIRE)
A gas appliance designed to simulate the effect of burning SOLID FUEL. These appliances are primarily for decorative purposes and allow for yellow flame patterns,

which would usually be regarded as totally unacceptable in a gas-burning appliance. The products of combustion pass unrestricted from the fire bed to the flue and the fires are therefore highly inefficient.

Where the appliance is installed, a ventilation grille of 100 cm² minimum is usually required.

DEEP SEAL TRAP
A trap having a TRAP SEAL of 75 mm or more.

DEEP WELL   See WATER CYCLE.

DEHUMIDIFICATION
A process in which the moisture within the air is removed. One type of dehumidifier works by passing the air over a cooling coil within an air-handling unit, which has the effect of reducing the temperature to below the DEW POINT. The water condenses and changes from a vapour to a liquid, which is simply drained away or collected for reuse.

DE-LAMINATED JOINT
A method of jointing 'NURALITE' sheet. The de-laminated joint consists of heating the sheet edge and parting it between its laminated layers of material. The joint is opened to a depth of 25 mm and a no. 1 welding block of bitumen is then coated to the two de-laminated surfaces of the joint. The whole joint is re-heated and the laminated edges are re-closed clamping the second sheet between its surfaces and the whole joint is consolidated firmly together, removing any excess bitumen. This method of jointing is not to be used to join sheets of the material together but is only used when inserting a gusset, such as might be needed when forming an external corner. An example of a completed de-laminated joint will be found under the entry of NURALITE.

DELAYED ACTION BALLVALVE
A device which is fitted into a storage cistern on systems which may be supplied with a pump or where the water is supplied from an AUTOMATIC PNEUMATIC CYLINDER. The delayed action ballvalve is designed to hold the ballvalve closed even though the water level in the cistern may be falling. By fitting such a device, the pump does not have to be continually switched on and off and in the case of a pneumatic cylinder the air pressure within the cylinder is conserved. The delayed action ballvalve shown over works as follows:

(1) Sketch A. The ballvalve is closed due to the water level in the float canister holding the float up.
(2) Sketch B. The water level falls within the cistern, but the level inside the float canister remains high due to the lower float valve being held closed by the lower float.
(3) Sketch C. As the water level within the cistern drops to a low enough level the lower float drops and opens the lower float valve letting the water out of the float canister, thus the ballvalve float drops letting more water into the cistern. The cistern upon finally refilling shuts off as in sketch A.

D

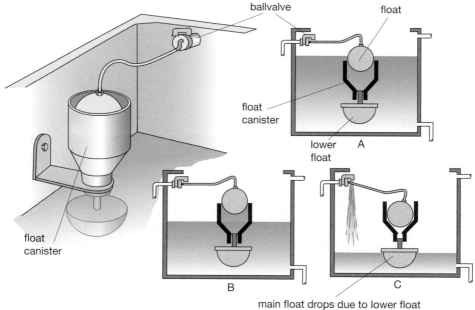

ballvalve

float

float
canister

lower   A
float

float
canister

B

C

main float drops due to lower float
valve opening thus allowing in water

Delayed action ballvalve

DELUGE SYSTEM

A sprinkler system designed to supply the maximum amount of water in the minimum amount of time. The deluge system is comprised of open SPRINKLER HEADS which are supplied with water as soon as it is required, unlike those sprinkler systems which are fitted with heat fusing elements. Therefore time can be saved by not having to wait for the sprinkler fuse head to blow. The deluge system is put into operation by automatic detectors which open the flow control valve. This system is particularly useful in high risk areas such as where flammable liquids are stored. See SPRINKLER SYSTEM.

DENSITY

The mass of molecules per volume in a particular substance. One cubic metre of water at 4°C weighs 1000 kg whereas a cubic metre of water at 82°C weighs only 967 kg, thus it can be seen that the water at the lower temperature is heavier, this is due to there being more molecules, the molecules are closer or 'more dense' (more compact).

Water when cooled below 4°C begins to expand again, this is a strange phenomenon which is peculiar to water therefore it will be seen that water is at its maximum density at 4°C. See also SPECIFIC GRAVITY.

DENZO TAPE   See ANTI-CORROSIVE BANDAGE.

DE-OXIDATION (REDUCTION)

A process in which the oxygen has been excluded from a substance. The oxy-acetylene flame burns with an outer reducing zone of carbon monoxide and hydrogen.

This zone burns and protects the metal from oxidising or absorbing oxygen from the air. Due to the fact that OXIDISATION is prevented many welding processes such as lead welding can be carried out without the need of a FLUX. See OXY-ACETYLENE FLAME for example.

DEPTH OF SEAL   See TRAP.

DESCALING
A process usually associated with instantaneous water heaters in which scale build up within the heat exchanger is removed. First the heat exchanger is removed and connected via two plastic tubes to a tank, which is filled with a proprietary descalant. A pump is switched on to force the liquid through the heat exchanger slowly causing the scale to dissolve. After the scale has dissolved, the heater must be thoroughly flushed through. When using the acid solution, it may be necessary to dilute it. Should this be the case, the acid must be added to the water rather than the water added to the acid to prevent the water boiling and spitting out with force.

DESIGN PRESSURE
The pressure to which something, e.g. a pipe or vessel, is designed to operate safely and not leak.

DE-STRATIFICATION FAN
A fan, mounted at a high level inside a tall room or building that has an in built thermostat to monitor the air temperature. Should this rise to about 28°C a fan is put into operation to blow and push the warm air back down to the occupied space, so avoiding unnecessary heat losses through the roof.

DETRITUS
Any inorganic material such as grit which accumulates in rainwater and drainage fittings.

DETRITUS GULLY   See GRIT GULLY.

DEW POINT
The temperature at which any water vapour present in the atmosphere is saturated. Upon cooling to approximately 55°C, usually via contact with a cooler surface, the water vapour condenses to its liquid form.

DEZINCIFICATION OF BRASS
A condition where all the zinc in brass has been destroyed leaving the fitting porous and brittle. This condition is brought about by the aggressive water in soil surrounding the fitting, providing an ELECTROLYTE which accelerates ELECTROLYTIC CORROSION and brings about the destruction of the ANODE (the zinc). For this reason water regulations require that brass should not be used underground, instead gunmetal or a dezincification resistant brass must be used. See also MERINGUE DEZINCIFICATION.

DFE FIRE   See DECORATIVE FUEL EFFECT FIRE.

DIAMETER

A straight line, starting at one point on the CIRCUMFERENCE of a circle, passing through the centre and terminating at the circumference on the opposite side.

DIAPHRAGM BALLVALVE

A BALLVALVE designed with the advantage of having fewer working parts than the PORTSMOUTH and CROYDON BALLVALVES previously available, it is also designed to make the adjustment of the water level easier by means of an adjusting screw on the lever arm. There are several types of diaphragm ballvalve available, the older earlier type having its outlet at the bottom of the valve was more prone to BACK-SIPHONAGE. See also EQUILIBRIUM BALLVALVE.

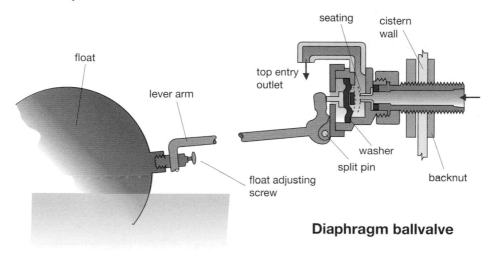

Diaphragm ballvalve

DIAPHRAGM PUMP

A pump where the water is drawn into a chamber and discharged through a non-return valve. The water is drawn into the chamber by flexing a diaphragm.

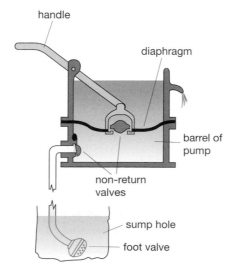

Diaphragm pump

DIAPHRAGM WASHER

Any one of a collection of large washers used to operate anything from a pump to a gas governor. Inside a plunger type of flushing cistern the diaphragm washer used nowadays is usually made of thin polythene sheet and is usually made by the plumber due to the difficulty which can be experienced in trying to buy one of the correct size.

DIE   See STOCKS AND DIES.

DIFFERENTIAL

The difference, usually in pressure, between two opposing forces.

DIFFUSER

A grille which directs the discharge of hot air or fresh air into a room. See WARM AIR HEATING.

DILUTED CHLORINE (SODIUM HYPOCHLORITE)

A substance that is used in the process of disinfecting drinking water systems.

DILUTING RECEIVER

A vessel which holds a quantity of water and is fixed under laboratory sinks. As the contents of the sink, which might contain say acid, drain away they mix with the water and are diluted before being discharged down the waste pipe.

DILUTION

Term used to mean that air is added to the flue gases in a flue pipe, generally to cut down the amount of condensation within the flue. The air used for dilution is usually admitted to the flue pipe via the DOWN DRAUGHT DIVERTER or a condensate pocket which has a pipe open to the atmosphere.

DIMINISHING COUPLING   See REDUCER.

DIMPLE PLIERS

A special pair of pliers designed to give a raised dimple to the end of a pipe, thus preventing it from passing too far into a formed branch joint on copper tube, blocking the flow. For example, see BRANCH FORMING TOOL.

DIP PARTITION TRAP

A type of bottle trap, see TRAP.

DIP PIPE

The name given to a cold feed pipe or return pipe which enters into the top of a hot storage vessel and extends down internally to the bottom of the vessel, to discharge any cooler water from the pipe low in the vessel so as not to mix with the hotter water, which is at a higher level due to the hot water rising.

DIP TUBE TRAP

A type of bottle trap, see TRAP.

D

DIPPED COLD FEED

Some plumbers 'dip' the cold feed pipe a short distance as it enters into a hot storage vessel. The reason for doing this is to avoid ONE PIPE CIRCULATION taking place in the cold feed pipe. This is an error on the part of the plumber because the dip often becomes a trap for sludge and other deposits which would otherwise collect in the bottom of the hot storage vessel and thus can lead to a blockage.

DIRECT CURRENT   See ALTERNATING CURRENT and IONS.

DIRECT CYLINDER

A cylinder used in the direct system of hot water supply. It differs from an INDIRECT CYLINDER because no HEAT EXCHANGER is fitted inside.

DIRECT SYSTEM OF COLD WATER SUPPLY

A system of cold water supply in which all the cold water in the house is on mains supply. The size of the supply pipe for the average house need only be 15 mm unless the pressure supplied is not very good in which case 22 mm may be required. When connecting appliances, and for all draw-off points it is essential that precautions against BACK-SIPHONAGE are taken. See also INDIRECT COLD WATER SUPPLY.

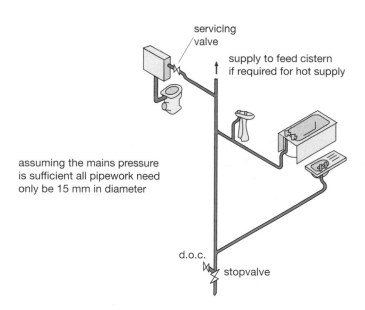

Direct system of cold water supply

DIRECT SYSTEM OF HOT WATER SUPPLY (CENTRALISED)

A method of CENTRALISED HOT WATER SUPPLY in which all the water drawn off at the appliances has passed through the boiler. See also, INDIRECT SYSTEM OF HOT WATER SUPPLY and LOCALISED HOT WATER.

(See illustration over.)

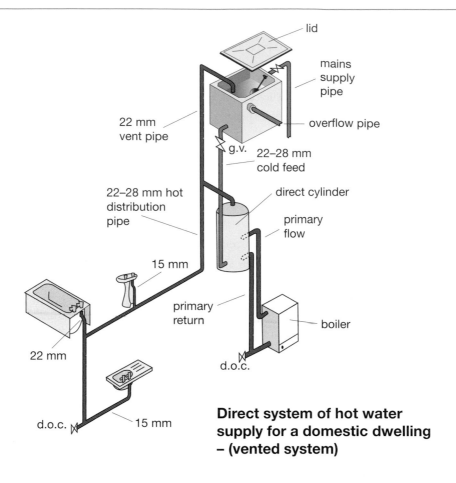

Direct system of hot water supply for a domestic dwelling – (vented system)

DISC FEED
A means of regulating the flow rate of water to an AUTOMATIC FLUSHING CISTERN. A thin disc with a hole in it is fixed into a fitting in the feed pipe serving the cistern, thus the water cannot get through too fast. A water authority seal is sometimes fixed to these fittings to prevent people tampering with the flow rate.

DISC GLOBE VALVE
A valve used on high pressure pipework, it is similar to an ordinary stopcock although more robust. There are two basic types of globe valve, those with fibre washers and those with a metal washer thus giving a metal to metal seating. Those with the metal washers are often used on heating systems and may be either straight or angular.

(See illustration over.)

DISC VALVE
A valve with a small hole in it to slow up the velocity of water. See DISC FEED.

DISCHARGE PIPE
A pipe which conveys surplus water or liquid sewage to the house drains. See ABOVE GROUND DRAINAGE.

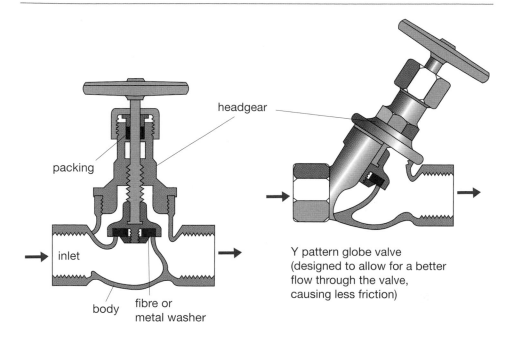

packing

headgear

inlet

body

fibre or
metal washer

Y pattern globe valve
(designed to allow for a better
flow through the valve,
causing less friction)

Disc globe valve

DISCHARGE POINT

The position where any water is discharged and beyond the concern of the person discharging it. Therefore the term discharge point varies in different circumstances. The house holder might class the sink waste or drain as a discharge point. The plumber may classify a discharge point as a drain, a soakaway or a nearby drainage ditch; the water authority could again have a new definition being a river or sewage works or even the sea.

DISCHARGE STACK

A vertical DISCHARGE PIPE.

DISCHARGE UNIT

A figure that can be read from a table that represents the average amount of water that would flow through a DISCHARGE PIPE or drain. By adding all the required discharge units (DUs) together the required waste water flow rate in litres per second (l/s) can be found using the following formula:

$$K\sqrt{DU} = \text{waste water flow l/s}$$

where K is a frequency factor that depends upon the building type, e.g. domestic dwellings 0.5, schools and hospitals 0.7, public buildings 1.0.

Once the flow rate has been determined, the pipe size can simply be determined from a further table. This book is of insufficient size to list all DUs, K factors and pipe sizes and only a small sample is given to illustrate its use. See BS EN 12056 for further information.

Sanitary appliance	DU	Max. water flow rate (l/s)	Discharge pipe size (mm)
6–7.5 litre WC	2		
9 litre WC	2.5	0.5	60
Basin or bidet	0.5	1.5	70
Bath or sink	0.8	2.0	80
Urinal with flush valve	0.5	2.7	90
Urinal with cistern	0.6	4.0	100
Shower	0.6	5.8	125
		9.5	150

Example

Find the diameter of a discharge stack serving a school with 20 7.5 litre WCs, 25 washbasins, 6 urinals and 4 sinks.

$$WCs = 20 \times 2.0 = 40$$

$$Basins = 25 \times 0.5 = 12.5$$

$$Urinals = 6 \times 0.6 = 3.6$$

$$Sinks = 4 \times 0.8 = \underline{3.2}$$

$$Total = \underline{59.3 \text{ DUs}}$$

Therefore:

$$0.7 \times \sqrt{59.3} = \underline{5.4 \text{ litre/s}}$$

By referral to the table, a 125 mm discharge stack would be needed. In addition to the pipe size, the incline of a horizontal drain can also be determined using another table, as found in the BS. Alternatively a formula, such as the CHEZY FORMULA could be used.

DISCONNECTING JOINT

A pipe fitting which can easily be disconnected from a pipe run. A disconnecting joint could be a LONGSCREW, a UNION CONNECTOR, and any compression or pushfit joint.

DISCONNECTING TRAP (DISCONNECTOR)   See INTERCEPTING TRAP.

DISINFECTION

A procedure carried out to hot and cold water systems in commercial and industrial premises in order to kill off any harmful bacteria. Disinfection is carried out by filling the system with a measured quantity of chlorinated water at 50 ppm. For this sodium hypochlorite is generally used. Disinfection is carried out typically in the following cases:

- New installations and extensions
- Where contamination is suspected
- In systems that have not been in constant use.

DISSOLVED ACETYLENE (DA)

Acetylene gas cannot be compressed and pressurised without the risk of it exploding, therefore it is mixed with ACETONE, which stabilises it. It is in this form that it is known as dissolved acetylene. See ACETYLENE GAS.

DISTILLED WATER
Water that is completely pure and contains no gases or solids in suspension. Water that has been evaporated and condensed would be distilled.

DISTORTION
A term used in the welding process to indicate that the metal being welded is becoming twisted and deformed due to the heat causing expansion of the metal. Distortion can be reduced by using the correct size of welding tip and not welding at too slow a rate along the weld, causing over heating. To overcome the problems of distortion it is best to give several tacks along the weld run prior to welding and, if necessary and possible, clamp the work in position.

DISTRIBUTION MAIN (SERVICE MAIN)
The water mains pipe from the water authorities works to the various housing and industrial estates. See also MAINS, TRUNK MAINS and SERVICE PIPE.

DISTRIBUTION PIPE
Any pipe supplying cold water from a storage cistern to the various draw-off points or hot water from a hot storage vessel, which is supplied with water from a feed cistern. For examples see INDIRECT COLD and INDIRECT HOT WATER SUPPLIES.

DIVERSITY FACTOR
Where there are a large number of appliances it is unlikely that they will all be used at the same time. The diversity factor is a factor sometimes used in the design of pipework to determine the maximum rate of flow to allow for, and is found by dividing the possible maximum rate of flow into the probable rate of flow.

$$\text{Diversity factor} = \frac{\text{probable rate of flow}}{\text{possible total rate of flow}}$$

See also LOADING UNITS.

DIVERTER   See DOWN DRAUGHT DIVERTER.

DIVERTER VALVE (THREE WAY VALVE/THREE PORT VALVE). See MOTORISED VALVE and COMPENSATING VALVE.

DIVIDERS (COMPASSES)
Tool with two steel points which pivot at their top end where they are joined. Dividers prove handy when developing sheet and pipework, they are also used in conjunction with a SCRIBING PLATE to mark out branch joints on lead pipes.

DOG EAR CORNER (BULLS LUG CORNER)
A method of forming a corner on the harder sheet materials such as copper and aluminium or on non-metallic material such as NURALITE. Forming the dog ear corner is a very simple process which can be assisted by using a forming block or a pair of dog earing pliers as shown in the diagram. The ear formed can be finished internally or externally depending on the side of material which is on show.

(See illustration over.)

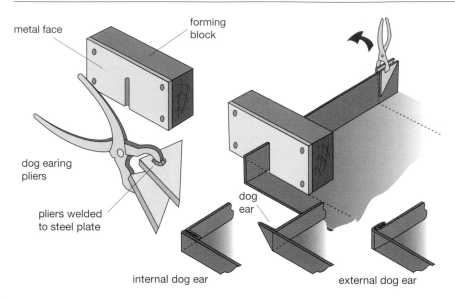

Dog-eared corners

DOG EARING PLIERS

A pair of pliers which are home made and designed to assist in the forming of a DOG EAR CORNER.

DOMICAL GRATINGS

A domed-shaped grating over the outlet of a waste fitting in a urinal or roof outlet. Designed to allow the liquid to escape down the discharge pipe should any debris block its passage, the water level rises and flows over the obstruction.

DORMER

A window opening formed in a roof slope. The dormer could be internal or external, see BUILDING TERMINOLOGY.

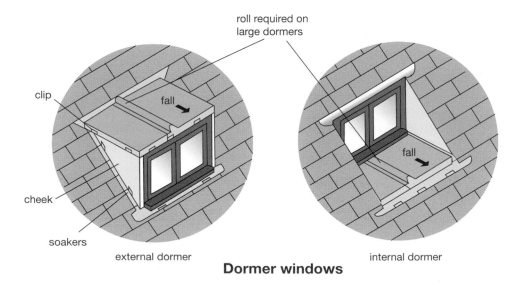

Dormer windows

DOT MOULD   See LEAD DOT.

DOUBLE BRANCH
A fitting designed to connect two branch pipes or channels from opposite sides. See BRANCH for example.

DOUBLE CHECKVALVE ASSEMBLY
Two CHECKVALVES fitted in a pipeline, with a DRAIN OFF COCK fitted between them. It is used as a means of backflow prevention or to prevent BACK-SIPHONAGE. The reason a drain off cock is fitted between the two checkvalves is to remove the water which would otherwise not be drained from the pipe, as may be required in winter, because the valves would be closed. A double checkvalve assembly could be made up by the plumber or a double checkvalve fitting could be purchased.

DOUBLE ENTRY RADIATOR VALVE
A radiator valve designed specially for micro-bore central heating systems. The double entry valve permits the plumber to connect both the flow and return pipes into the radiator at the same end. The valve is designed with a long insert tube which passes right through the bottom of the radiator to pick up the return flow of water. For example, see MICRO-BORE.

DOUBLE FEED INDIRECT CYLINDER
An indirect cylinder which has separate cold feed connections for the primary circuit heat exchanger and the secondary waters. See also SINGLE FEED CYLINDER.

DOUBLE LOCK WELT
An EXPANSION JOINT used on metallic roof coverings for joints which run across the fall of the roof (transverse joints). Double lock welts should not be used on roof pitches lower than 20° unless the joint is sealed with a non-setting mastic or boiled linseed oil, in which case they can be used on roof pitches as low as 6°. The dimensions for double lock welts are open to debate and in most cases the plumber makes them a size to suit the forming blocks or irons used to assist making the joint. Sometimes the size of the welt is specified by the client in which case these specified sizes should be observed. The sizes given in the diagram are for copper and aluminium welts which have been found suitable in the past. For sheet lead, a much larger welt will be required. See also SINGLE LOCK WELT and COPPER SHEET AND ROOF CO VERINGS.

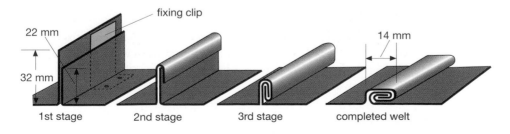

22 mm   fixing clip   14 mm

32 mm

1st stage   2nd stage   3rd stage   completed welt

Double lock welt

DOUCHE SPRAY   See ASCENDING SPRAY.

DOWN DRAUGHT
A draught which blows down flue pipes possibly causing VITIATED AIR to affect the correct combustion of the fuel. A down draught is caused by the bad siting of flue terminals, such as below the eaves of the building. For example of correct positions see FLUE. See also DOWN DRAUGHT DIVERTER, SPILLAGE and DRAUGHT STABILISER.

DOWN DRAUGHT DIVERTER
A device fitted in OPEN FLUE pipes directly above gas boilers. The purpose of a down draught diverter is to divert any excess down draught from entering the boiler and causing it to affect the combustion of fuel. This is illustrated in the sketch, as the draught blows down it hits the baffle and is expelled through the opening in the flue pipe. Should the updraught in the flue also be excessive the down draught diverter also prevents the flames from being lifted off the burner jets.

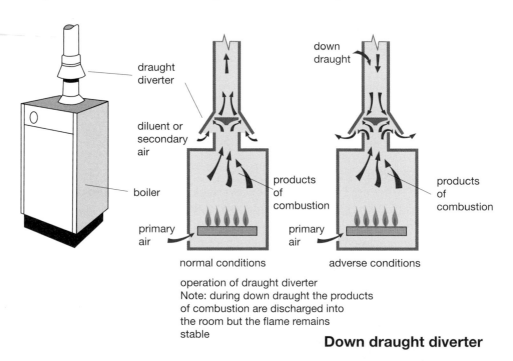

draught diverter

diluent or secondary air

boiler

primary air

down draught

products of combustion

primary air

products of combustion

normal conditions

adverse conditions

operation of draught diverter
Note: during down draught the products of combustion are discharged into the room but the flame remains stable

Down draught diverter

DOWN PIPE (DOWN SPOUT/DOWN COMER)   See RAINWATER PIPE.

DOWNSTREAM
A term sometimes used to indicate that part of a water or gas pipeline that is further along the pipe from a specific point away from the direction to which the fluid has been travelling. Conversely, UPSTREAM would be further along the pipe towards to incoming supply or source of the fluid.

DRAEGER ANALYSER
A small manually operated gas analyser that is designed to sample the concentration of gas within an environment. The principle of operation is the same as that used by

the police when using a breathalyser. Basically, a sample of air or flue gas is sucked through a special glass tube filled with chemicals that react to the presence of a particular gas. Tubes of different chemicals can be selected to detect various gases such as carbon dioxide and carbon monoxide. For example, see CARBON DIOXIDE INDICATOR.

DRAIN
A pipe for 'draining' away surplus water or liquid sewage from a building. See BELOW GROUND DRAINAGE. See also SEWER.

DRAIN AUGER
A flexible spring with a handle at one end which is available in a range of lengths and sizes. Its purpose is to pass down discharge pipes in order to remove blockages. The drain auger is sometimes called a snake. See also SNAKENTAINER.

DRAIN CHUTE
A drainage fitting designed to provide better access for rodding of drains in deep INSPECTION CHAMBERS.

DRAIN CLEANING
Occasionally discharge pipes and drains become blocked due to various causes and it is for the plumber to overcome the problem of unblocking the pipe with the least

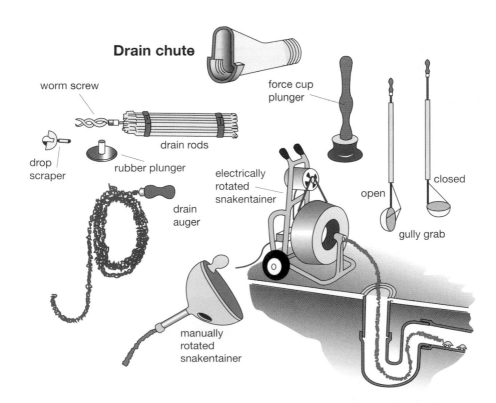

Drain cleaning

inconvenience. The methods can vary in different circumstances from using drain rods to a suction plunger or even the use of chemicals may be needed. Drain rods are passed down the pipe until the obstruction is met then by giving a few blows the blockage is dislodged and with the pressure caused by the built up water it is often washed away. A variation to this is to pass a spring or DRAIN AUGER down the pipe which is rotated, thus dislodging the blockage, the spring could either be hand held or machine operated. Possibly one of the methods used most often to remove blockages is to use a plunger; the reason why the plunger proves so successful is because it creates a pressure on the blockage of at least that of atmospheric pressure (101.3 kN/m^2). As the plunger is withdrawn, it leaves a void behind thus a partial vacuum is formed. When plunging sink wastes it is essential that the overflow pipe, if connected to the waste, is blocked up otherwise air will travel down this and relieve the negative pressure within the pipe, see SUCTION PLUNGER. See also CHEMICAL PIPE CLEANING and INSPECTION OF DRAINS.

DRAIN OFF COCK (DOC OR DRAIN VALVE)
A valve which is primarily used to drain down hot and cold services to empty the pipes or vessels of their water. A DOC should be fitted at all low points where water could not normally be drained. When draining any system it is advisable to inspect the washer in case it needs replacing. See also PLUG COCK.

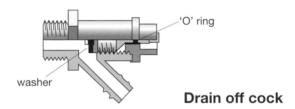

'O' ring

washer

Drain off cock

DRAIN PLUG
An expandable device used to seal off the ends of drain pipes, normally for testing purposes. See also, INFLATABLE STOPPER.

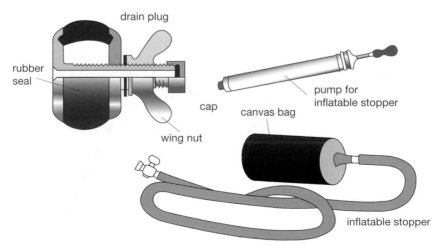

drain plug

rubber seal

cap

pump for inflatable stopper

canvas bag

wing nut

inflatable stopper

Drain plugs and stoppers

DRAIN RODS   See DRAIN CLEARING.

DRAIN TESTING   See TESTING OF DRAINS.

DRAUGHT BREAK
Any opening into the flue way of an open flue system such as a down DRAUGHT DIVERTER or DRAUGHT STABILISER.

DRAUGHT DIVERTER   See DOWN DRAUGHT DIVERTER.

DRAUGHT FLAP   See ANTI-DRAUGHT FLAP.

DRAUGHT GAUGE
See INCLINED MANOMETER or COMBUSTION EFFICIENCY TESTING.

DRAUGHT STABILISER
A hinged flap usually fitted to the base of a chimney of oil fired and solid fuel boilers. It is fitted in the same room as the appliance and its purpose is to ensure that the chimney maintains the required draught condition. Should the draught within the chimney be too high, caused for example by high winds, the flap will open permitting air to enter directly into the chimney, thus reducing the draught through the boiler. The draught stabiliser is also forced open should the pressure inside the chimney become excessive, when the pressure and draught within the chimney become stable, the flap automatically closes.

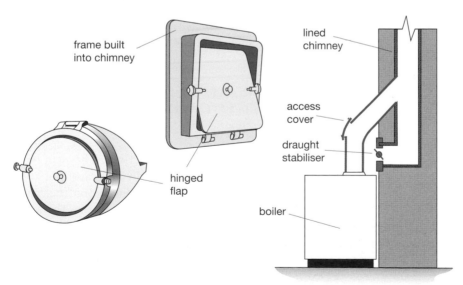

Draught stabiliser

DRAW KNIFE   See LEAD KNIFE.

DRAW OFF POINTS
A valve with an outlet from where water is taken, e.g. hot and cold taps to a sink.

DRENCHER SPRINKLER SYSTEM

A sprinkler system designed to discharge water over the external surfaces of buildings, to prevent the spread of fire from adjacent properties. The system is designed with pipework fitted on the outside of the building with nozzles (drenchers) at suitable positions. The nozzles could be either automatically operated and be fitted with heat sensitive SPRINKLER HEADS or the nozzles could be open discharge points which are controlled by opening a valve. See SPRINKLER SYSTEM.

DRESSER   See FLAT DRESSER.

DRESSING

Term sometimes used to mean BOSSING.

DRIFT   See FLARING TOOL.

DRINKING FOUNTAIN

An appliance designed to give a small discharge of drinking water from a nozzle to which the mouth can be positioned to catch the liquid. Drinking fountains should have the control of water supplied via a NON-CONCUSSIVE VALVE to ensure that the supply is not left running.

DRINKING WATER CISTERN   See BOOSTED SYSTEM OF COLD WATER SUPPLY.

DRIP PLATE

A sheet of 2 mm thick steel which can be laid between two sheets of lead to assist the process of BOSSING in awkward positions, such as when lead is being pushed into a corner. The drip plate assists the lead riding over the lower piece of lead.

DRIP SINK

A small sink designed primarily to catch small drips from, for example, a tap. The term drip sink is sometimes used to mean a CLEANER'S SINK.

DRIPS

A method of allowing for expansion on metallic roof coverings. They can be used for non-metallic coverings (Nuralite) although this is usually for reasons of appearance only. There are different types of drips, each depending upon the material used and circumstances. Drips are expansion joints which run across the fall of a roof and are used on roofs with a fall of usually less than 15° pitch. The depth of all drips should be at least 50 mm, drips formed on lead roofs can be reduced to 40 mm if an ANTI-CAPILLARY GROOVE is incorporated, but drips of this size are not to be recommended. For examples see COPPER, LEAD and ZINC SHEET AND ROOF COVERINGS.

DRIVING STICK

A cane or piece of straight timber used to drive a MANDREL or BOBBINS and FOLLOWERS through lead pipe.

DROP CONNECTION   See BACKDROP CONNECTION.

DROP END (STOPEND OUTLET)

An eaves gutter fitting which incorporates both the stopend of a gutter and outlet to the rainwater pipe.

DROP FAN COCK

A plug cock which has a finger plate or fan incorporated. When the valve is open the plate is held upright, when the cock is turned off the plate falls. The cock cannot be opened unless the finger plate is deliberately held up, this prevents the cock accidentally being turned on. For example, see PLUG COCK.

DROP TEST

Name sometimes given to a TIGHTNESS TEST, which should be referred to for further study.

DROP VALVE

A name sometimes used when referring to the valve mechanism found within a VALVED FLUSHING CISTERN.

DRY PIPE SPRINKLER SYSTEM

A sprinkler system used in unheated buildings where the temperatures may fall below 0°C therefore possibly causing freezing. The system is charged with compressed air and the water is held back behind a special differential valve. Should a SPRINKLER HEAD open the air is released, thus the differential valve opens and allows water to enter the pipework to discharge from the sprinklers which have burst open due to excessive heat. See SPRINKLER SYSTEM.

DRY RISER

An empty pipe which runs up vertically inside the building with a 64 mm fire brigade hydrant at each floor and at roof level. An inlet is provided at ground level for the fire brigade to connect to the nearest hydrant. Dry risers are incorporated into high buildings to save time running canvas hoses up the staircases should the building be on fire. The size of the dry riser should be 100 mm in diameter for buildings up to 45 metres in height and for buildings 45–60 m in height the diameter should be increased to 150 mm. A 150 mm pipe is also required for any building should two 64 mm valves be fitted at each floor. For buildings over 60 metres (20 storeys), high dry risers should not be installed and a WET RISER would be required.

(See illustration over.)

DRY WELL

An underground chamber which houses sewage lifting equipment and its associated pipework. See PNEUMATIC EJECTOR.

DRYING COIL

An arrangement of pipework to form a coil or similar design, to give off a small amount of heat in an airing cupboard, thus acting as a clothes drier. The drying coil should be connected into the PRIMARY FLOW AND RETURN pipes to the hot storage vessel this ensures use in the summer months when the central heating is not being used.

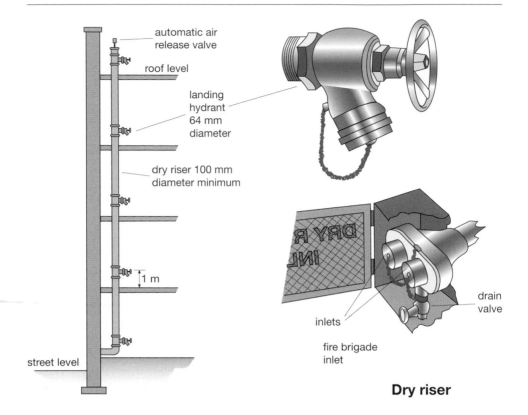

automatic air release valve

roof level

landing hydrant 64 mm diameter

dry riser 100 mm diameter minimum

1 m

street level

drain valve

inlets

fire brigade inlet

Dry riser

DUAL FLUSH

A siphon fitted inside a FLUSHING CISTERN to give the user a choice of two different volumes of flushes. When the lever arm is operated and released, flushing commences and continues until air is admitted half-way down the siphon, thus breaking SIPHONIC ACTION. To obtain a full flush, the handle has to be held down, this, in effect, blocks up the air hole preventing air from entering the siphon.

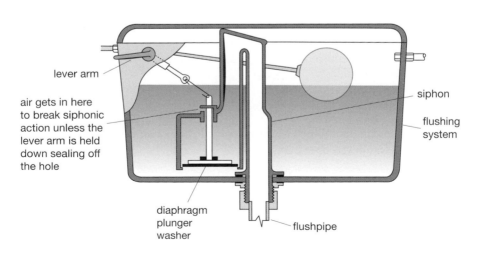

lever arm

air gets in here to break siphonic action unless the lever arm is held down sealing off the hole

siphon

flushing system

diaphragm plunger washer

flushpipe

Dual flush

DUAL FUEL BURNER

A special forced draught burner found on some commercial appliances in which two fuels are used, such as oil and gas. These appliances are multi-fuelled to enable the user the opportunity to choose the fuel with the best price or to overcome the problem of fuel shortages. Generally the system is supplied by two fuel lines and arranged so that as one line opens the other closes.

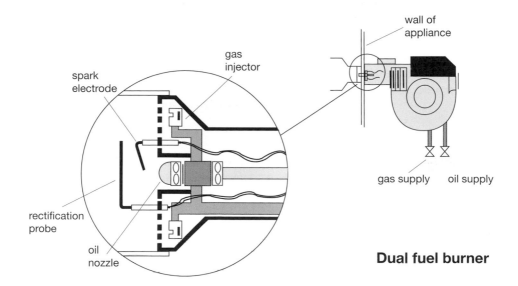

Dual fuel burner

DUALSTREAM

The trade name of a patented system that provides a system of constant pressure and flow. The system utilises a vessel containing a water filled rubber bladder. Where the supply flow rate is poor, opening too many taps at the same time would result in the starvation of water at one or two outlets. However, with the 'Dualstream' system when all the outlets within the system are closed water slowly fills the bladder within the vessel and creates a pressure equal to that of the maximum supply main. Thus when the taps are opened this stored volume of water is released back into the pipe work, allowing for a maintained flow through all outlets which continues until the bladder is empty. The length of time depends upon the size of vessel used. This system is therefore highly suitable where pressure or flow rate problems are a constant nuisance.

(See illustration over.)

DUCK-FOOT BEND   See REST BEND.

DUCT

An enclosure designed to accommodate one or more services. For example, see COLD WATER SUPPLY. A duct is also used to convey the forced or natural air flow from a means of ventilation or warm air central heating system.

DUCTED AIR HEATER   See WARM AIR CENTRAL HEATING.

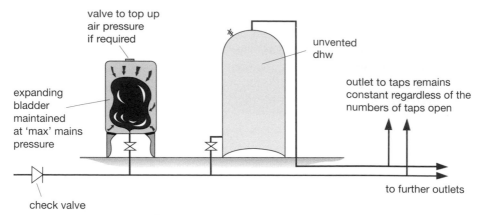

valve to top up
air pressure
if required

unvented
dhw

outlet to taps remains
constant regardless of the
numbers of taps open

expanding
bladder
maintained
at 'max' mains
pressure

to further outlets

check valve

Dualstream constant pressure & flow system

DUCTILITY

The ability of a metal to withstand distortion without fracturing. A good example of a ductile metal would be copper which can be drawn out to a fine wire.

DUMMY

A tool which has a soldered blob about the size of an egg on the end of a Malacca cane handle. There are two types of dummy; the short handed dummy which is used to assist in the bossing process (see BOSSING) and the long handed dummy which is passed down large diameter lead pipes and used to knock out the kink formed when bending lead pipe. The long handed dummy is used prior to passing BOBBINS through the pipe thus restoring its bore.

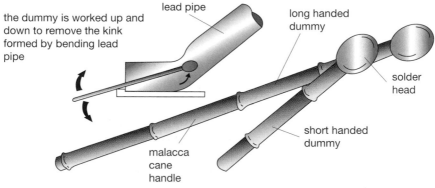

the dummy is worked up and
down to remove the kink
formed by bending lead
pipe

lead pipe

long handed
dummy

solder
head

malacca
cane
handle

short handed
dummy

The dummy

DURABILITY

The ability of a material to resist wear and tear; also a material which is long lasting and non-perishable.

EARTHENWARE   See CERAMIC WARE.

EARTHING (EARTH CONTINUITY)
Term used to mean that an appliance or pipe has been connected to the earthing point of the building's electric system. See CONTINUITY BONDING.

EASTERN CLOSET   See SQUATTING WC PAN.

EASY BEND   See LONG RADIUS BEND.

EASY CLEAN COVER
A smooth cover, usually chromium plated, which fits over the head and gland nut of BIB TAPS, PILLAR TAPS, etc. It is designed to provide the tap with a more pleasing appearance and make cleaning easier.

EAVES
The lowest, usually overhanging part of a sloping roof where the fascia board and gutter are located. See BUILDING TERMINOLOGY.

EAVES GUTTER
The guttering fixed at the lower edge of a pitched roof, there are many designs of eaves gutter the most common being half round, square line and 'O Gee'. Probably the most common material used today for guttering is plastic although it is also made in cast iron, galvanised pressed steel and asbestos cement. The method of securing eaves guttering to the building depends upon the building design but in general FASCIA BRACKETS are used. When installing any guttering material allowances for expansion must be taken into account especially with plastic guttering materials as the finished gutter will be exposed to constant temperature changes. Occasionally putty is used to join the non-plastic materials, but it is not to be recommended because it dries hard and therefore there is no allowance for thermal movement, instead plumbers mait should be used, this being a non-setting putty. Plastic materials should be installed as the manufacturers' instructions state, failure to do so would soon result in leaky, buckling work, due to the high expansion rates of plastic. The fall or gradient of a gutter should be approximately 1 in 600. This fall would scarcely be noticeable to the human eye, for example over 10 metres the fall would only be ($10\,000 \div 600 = 16.6$) $16\,^1/_2$ mm. Once the fall has been determined the brackets should be fixed at about 1 metre intervals as shown in the diagram. Notice how the highest bracket is fixed as high as possible. Once the gutter has

been completed, it is connected to the drain via a rain water pipe. See also ZINC EAVES GUTTERS.

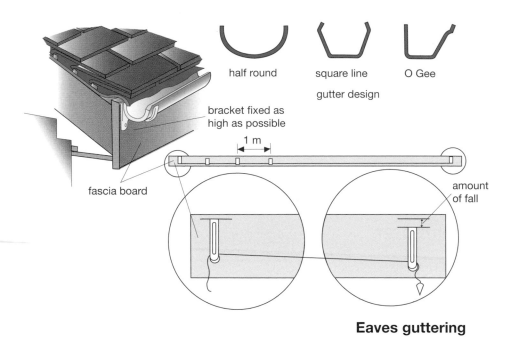

half round   square line   O Gee

gutter design

bracket fixed as high as possible

1 m

fascia board

amount of fall

Eaves guttering

ECCENTRIC REDUCER   See REDUCER.

EFFICIENCY
A term used as an expression of an appliance's capability of producing the desired results in terms of performance and economy.

EFFLUENT
The outflowing liquid resulting from the purification treatment of sewage. The effluent must be of a condition and standard which will not cause a nuisance or be prejudicial to health. See also TRADE EFFLUENT.

EIGHTH BEND
A bend which turns through 45°. See BEND.

EJECTOR PUMP
A pump designed to eject liquid sewage from a SUMP or DRY WELL. See PNEUMATIC EJECTOR.

ELASTICITY
The ability of a material to return to its original shape after being distorted. Rubber is an excellent example of a material which is elastic. Metals are basically not elastic, although some steels, etc. can be made into spring shapes which possess certain characteristics of elasticity. In metallic roof coverings and pipes the elasticity is generally referred to as 'spring back'; a sample piece of material is bent and then how

much it tends to return to its original position, if at all is noted. Sheet metals such as commercial quality zinc and hard copper show a tendency to spring back, but lead and soft copper do not. When bending low carbon steel pipe it is generally over pulled to allow for a certain amount of spring back. See also YIELD POINT.

ELASTOMERIC
The elastic or 'rubber-like' properties of a specific material.

ELBOW
A pipe fitting giving a sharp 90° bend in the pipeline. See also BEND and TWIN ELBOW.

ELBOW ACTION TAP (SURGEONS TAP)
A tap designed to be fully opened or closed by turning the operating handle through only 90°.

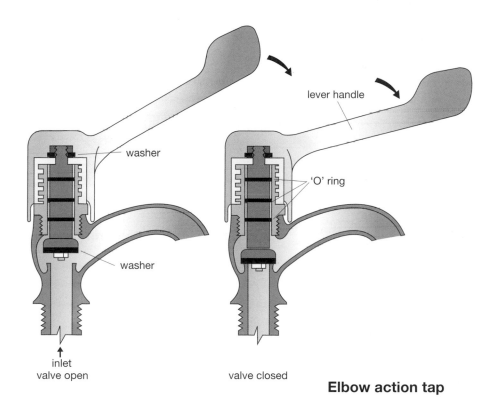

Elbow action tap

ELECTRIC ARC WELDING   See ARC WELDING.

ELECTRIC BOILER
A specially designed heater containing a copper tube that is surrounded by a heating element. These boilers are used in a similar way as any boiler, with a flow and return connection. The method of heating the water is simply achieved by way of electricity instead of, say, gas or oil. The output required for the system is regulated by the number of these heaters connected in series to heat the water. See also ELECTRIC STORAGE BOILER.

ELECTRIC RESISTANCE SOLDERING MACHINE

A special soldering machine designed to assist making capillary solder joints up to 54 mm on copper tube without the use of a naked flame. The machine consists of a portable transformer to produce a current of between 90 and 190 amps and a special pair of heating tongs with two carbon electrodes which are available with rounded or V shaped heads or as a straight rod (designed to assist reaching joints difficult to reach). The machine, although expensive to purchase, can prove useful when making solder joints in close proximity to decorated surfaces and, due to the reduced heat conduction along a pipe, joints can be made on pre-insulated pipe without the need to remove excessive amounts of insulation.

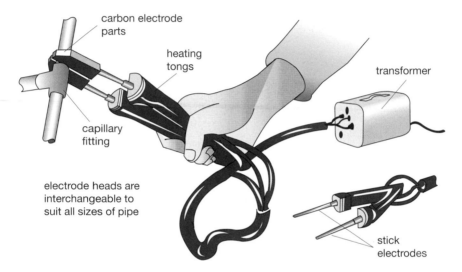

Electric resistance soldering machine

ELECTRIC STORAGE BOILER

A 'dry core' boiler designed to store cheap rate electricity at night in special storage blocks. Electricity is passed through the heating elements which pass down through the special interlocking Feolite refractory bricks which make up the dry core of the boiler. During the day when heat is required a fan blows air up through the blocks and down the hot draught tube to force hot air onto the water filled HEAT EXCHANGER and it is from here that the water is pumped around a hot water heating system. The air is re-circulated around the boiler in a CLOSED CIRCUIT to cut down on the heat losses. See sketch.

(See illustration opposite)

ELECTRIC STORAGE HEATER

A design of convector heater in which an electrical element passes through an arrangement of dense blocks that have a high SPECIFIC HEAT capacity. The blocks are usually heated overnight using cheap rate electricity. The amount of heat given off by the heater during the day depends upon its use during the day and the quantity of heat that was stored the previous night.

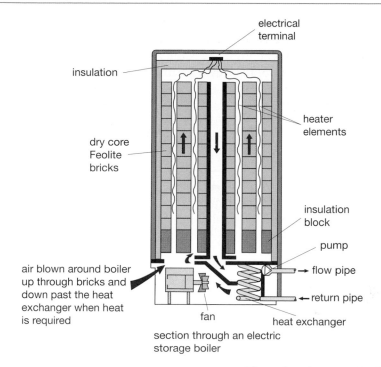

electrical
terminal

insulation

heater
elements

dry core
Feolite
bricks

insulation
block

pump

air blown around boiler
up through bricks and
down past the heat
exchanger when heat
is required

flow pipe

return pipe

fan

heat exchanger

section through an electric
storage boiler

Electric storage boiler

ELECTRIC WATER CONDITIONERS   See WATER CONDITIONERS.

ELECTRIC WATER HEATING

Electricity is sometimes used to heat quantities of water for domestic use. This generally consists of either INSTANTANEOUS WATER HEATERS in which the water is heated only as and when it is required, these heaters are fixed locally to where the water is required, or the water can be heated by installing a purpose made storage vessel or by fitting an IMMERSION HEATER into a hot storage vessel, at some central point within the building. When using an immersion heater it is essential that the heater element reaches low down near the bottom of the storage vessel because it will only heat the water above this point, as hot water rises (see CONVECTION CURRENTS). The heater should be at least 50 mm from the base of the storage vessel to prevent convection currents disturbing any sediment. Sometimes two heater elements are used, one fitted at low level and one fitted much nearer to the top. The higher immersion heater is switched on if only enough hot water is required to fill a sink, etc, whereas should a lot of hot water be required, e.g. to fill a bath, the lower heater is switched on. The power supply to the element is controlled by a thermostat which should be set to cut off the supply of power at a temperature of about 60°C in HARD WATER and about 65°C in SOFT WATER areas. Most hot storage vessels are fitted with a 2 1/4 in BSP thread for connections of immersion heaters; should this be missing an immersion heater boss can be fitted.

The electrical power supply to an immersion heater must come directly from the consumer unit, with no other connections on the line and it must have its own

15 amp fuse supplied with a 2.5 twin and earth sheathed copper cable run from the consumer unit to terminate close to the hot storage vessel with a double pole switch, it is from here that a 21 amp heat proof flex is run to the heater, it is designed in this way to prevent the overheating of the wires. See also storage water heater.

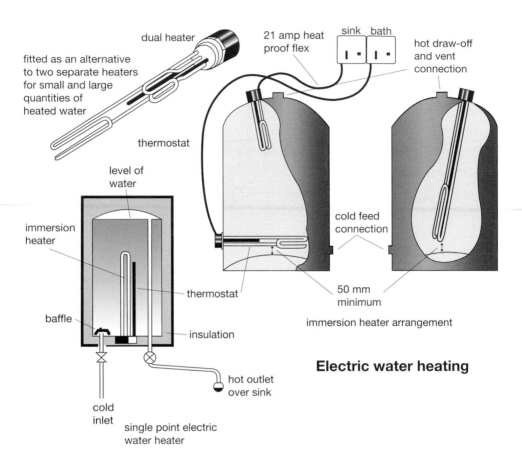

Electric water heating

ELECTRICITY
The flow of infinitely small particles called free electrons that flow in a circuit. This flow of electrons is called a current and is measured in amperes (amps). The initial current flow is created by a reaction such as might be caused by the movement of a magnet past a coil of wire or due to electrolysis (i.e. the reaction between two dissimilar metals), and the potential to cause the reaction is measured in volts. The current developed is possibly passed through specific materials causing a great deal of friction causing them to heat up, being referred to as an element, or the current may be passed through a coil of wire to initiate the rotation of a magnet, etc. As the current flow passes through the workings of a coil or element it uses up the potential voltage and the resultant energy produced is referred to as power, and is measured in watts.

ELECTRO-CHEMICAL CORROSION   See ELECTROLYTIC CORROSION.

ELECTRODE
An electrical conductor, used as either the positive or negative terminal of an electrically conducting medium; it conducts current into and out of the medium, which may be either a liquid, a solid, a gas or a vacuum.

ELECTROFUSION JOINT
A type of joint made on POLYETHELENE (PE) pipe work. It consists of an electric heater coil that is incorporated inside the fitting. When a 40 V output is supplied to two terminals on the outside of the fitting for a period of time that depends on the fitting size, the plastic melts and fuses it on to the pipe.

ELECTRO/HYDRAULIC or MECHANICAL VALVE   See SAFETY SHUT OFF VALVE.

ELECTROLYSIS   See ELECTROLYTIC CORROSION.

ELECTROLYTE
Any liquid or moisture which carries electrically charged particles from an ANODE to a CATHODE. In a plumbing system water would be the electrolyte and if this water was slightly acidic the ELECTROLYTIC CORROSION would be accelerated.

ELECTROLYTIC CORROSION (ELECTROLYSIS/GALVANIC ACTION)
Corrosion of metals brought about by bringing two dissimilar metals in contact with one another via an ELECTROLYTE. One metal (the anode) would be destroyed by the other metal (the cathode). The further the two metals are apart on the ELECTROMOTIVE SERIES the quicker will be the destruction of the anode. It is a water regulation requirement that no metal pipe or fitting shall connect to any other metal pipe or fitting of a dissimilar metal unless effective measures are taken to prevent deterioration due to electrolytic corrosion. This can best be achieved by ensuring that the sequence of metals in any supply or distribution pipe be connected only in the normal direction of water flow.

water
flow
→

galvanised steel
uncoated iron
lead
copper

ELECTROMAGNET   See SOLENOID VALVE.

ELECTROMAGNETIC VALVE   See SOLENOID VALVE.

ELECTROMOTIVE FORCE (EMF)
The charge (measured in volts) used to move the free electrons in a conductor around a circuit. The EMF can be produced by several methods, including the use of thermocouples, batteries, generators and piezo-electric devices, to name a few.

ELECTROMOTIVE SERIES (ELECTRO-CHEMICAL SERIES)
Metals listed in order of their ability to resist destruction by ELECTROLYTIC CORROSION. Metals higher on the list would be the cathodes and destroy those lower on the

list, being the anodes. Those metals high on the electromotive series are sometimes called noble metals.

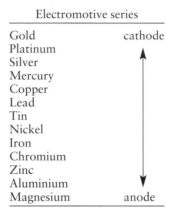

Electromotive series

Gold	cathode
Platinum	
Silver	
Mercury	
Copper	
Lead	
Tin	
Nickel	
Iron	
Chromium	
Zinc	
Aluminium	
Magnesium	anode

ELECTRON   See ATOM.

ELECTRONIC IGNITION
A method employed to ignite gas and oil burning appliances using a spark to jump between two electrodes or between electrode and burner head. Electronic igniters allow for improved appliance efficiency and eliminate the need for pilot flames at times when the appliance is not in use.

ELECTRONIC MANOMETER (DIGITAL DIFFERENTIAL MANOMETER)
A MANOMETER that operates by the use of electricity, sensing the pressure difference between two TRANSDUCERS.

ELEMENTS
In the world there are millions and millions of different substances but all these are made from a selection of only about one hundred basic elements. An element is made up of only one kind of atom, whereas, a substance may be made up of more than one type of atom. For example, WATER in which one MOLECULE is made up from 2 atoms of hydrogen and 1 atom of oxygen (H_2O), hydrogen and oxygen being two different elements.

EMBEDDED PANEL SYSTEM   See RADIANT HEATING.

EMERGENCY CONTROL VALVE
The gas control valve located on the gas pipe at the point of entry to the building.

ENAMEL   See VITREOUS ENAMEL.

END FEED CAPILLARY FITTING   See CAPILLARY JOINTS.

ENERGY
A term meaning the ability to do work or act as a force. Many different forms of energy will be found to include: chemical, thermal, electrical, kinetic, sound and light energy. Energy of one type can be changed to another, for example the chemical

energy reaction of the burning of gas and oxygen produces the heat energy of a pilot flame playing on to a thermocouple to produce electrical energy. The laws of science state that energy cannot be created or destroyed; it can only be converted from one form to another. Thus when we talk of 'losing energy', quite often this energy has been changed into unwanted heat or sound.

ENGINEER'S PLIERS   See PLIERS.

EQUILIBRIUM BALLVALVE

A BALLVALVE of a more sophisticated design than that of the simple lever types. With this design of valve there is a water way passing right through the centre of the valve which exerts a pressure to the other side of the washer, so there is an equal pressure on both sides, thus the washer is in a state of equilibrium. The float on this type of ballvalve has only to lift the weight of the arm, whereas in the case of simple lever types the effort provided by the float not only has to overcome the weight of the arm but also the pressure of the incoming water. The equilibrium ballvalve would be used in areas where the mains water pressure is very high and where persistent WATER HAMMER may be encountered.

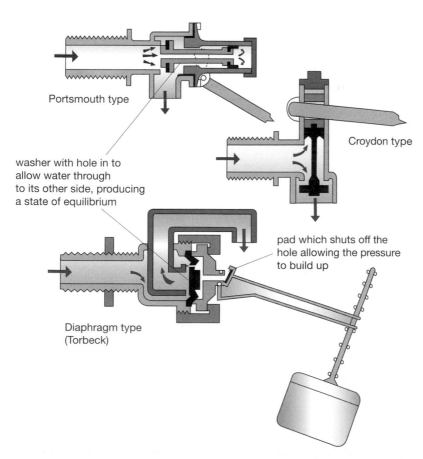

Portsmouth type

washer with hole in to
allow water through
to its other side, producing
a state of equilibrium

Croydon type

pad which shuts off the
hole allowing the pressure
to build up

Diaphragm type
(Torbeck)

Equilibrium ballvalves

EQUIPOTENTIAL BONDING   See CONTINUITY BONDING.

EROSION
The deterioration of a surface brought about by wear, unlike CORROSION which is caused by chemical attack.

ESSEX FIX A TAP   See TAP FIXING SET.

ESSEX FLANGE
The trade name for a type of boss which can be fitted to cylinders and tanks where access is only available from one side. See IMMERSION HEATER BOSS.

ETHENE (ETHYLENE)
A substance produced from crude oil and the building block for a whole range of different plastics.

EUREKA TRUNK VALVE   See ANTI-FLOOD VALVE.

EVAPORATION
The changing of a liquid into a VAPOUR, usually below the liquid's boiling temperature. During evaporation the molecules on the surface of a liquid break away from the COHESION of the other molecules. Water at atmospheric pressure is found in liquid form due to the cohesion of the molecules, water in this form is about 800 times heavier than air, but, surprisingly one molecule of water weighs less than one molecule of air, thus should a molecule of water break free from the liquid it is free from cohesion and pushed upwards to eventually form clouds in the sky.

EXCESS AIR
The additional air required to ensure complete combustion of a fuel, which in practice is slightly more than that predicted in theory.

EXFILTRATION
The unintended leakage of liquid from a drain or sewer pipe, see also INFILTRATION.

EXHAUST PIPE   See VENT PIPE.

EXPANDING TOOL   See SOCKET FORMING TOOL.

EXPANSION
Most materials expand when heated. All substances are made up of molecules and when molecules are heated they move about vigorously and thus move further apart resulting in the material becoming larger. When the material cools the molecules slow down their motion and move closer together thus the material gets smaller or contracts. See also COEFFICIENT OF THERMAL EXPANSION.

(See illustration opposite)

EXPANSION JOINT
A joint designed to allow for expansion and contraction which might take place due to temperature changes. Allowance for movements must be made to prevent damage

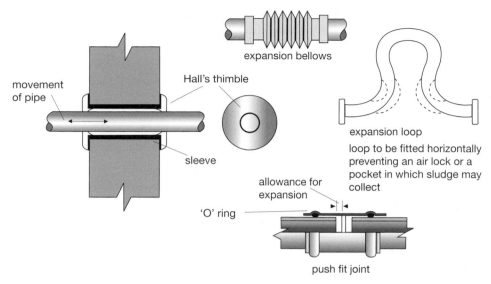

expansion bellows

movement of pipe

Hall's thimble

sleeve

'O' ring

allowance for expansion

expansion loop

loop to be fitted horizontally preventing an air lock or a pocket in which sludge may collect

push fit joint

Allowance for expansion

to the structure, such as a railway line, a bridge or a metallic roof covering, etc. When installing pipework, allowance for expansion due to thermal movement must be considered, otherwise objectionable noises may result. For more examples of expansion joints than those shown see SINGLE and DOUBLE LOCK WELTS; COPPER, LEAD and ZINC SHEET AND ROOF COVERINGS. See also COEFFICIENT OF THERMAL EXPANSION.

EXPANSION PIPE   See VENT PIPE.

EXPANSION VALVE   See PRESSURE RELIEF VALVE.

EXPANSION VESSEL   See SEALED EXPANSION VESSEL. See also FEED AND EXPANSION CISTERN.

EXPLOSION OF BOILERS
A very rare occurrence which can happen should the flow and return pipes to a boiler be blocked, e.g. due to frost or scale. If the pipes are obstructed and the boiler is fired, the water will heat up and expand and would continue to do so until the boiler thermostat shuts off the fuel. As the water expands it creates a pressure within the boiler which eventually causes the boiler to crack or a pipe to be blown from its fitting, this causes boiling hot water or steam to discharge from the fractured point. If it is suspected that the system is frozen, the boiler should never be lit. As a precaution to boiler explosion some plumbers fit a SAFETY VALVE on or near to the boiler which should open due to excessive pressure within the system.

EXTENDED-PLENUM SYSTEM   See WARM AIR CENTRAL HEATING.

EXTENSION BOSS
A fitting used which is sometimes chromium plated, having a female thread at one end and a male thread at the other. It is used for extending the distance between a

BIB TAP and the pipe fitting into which it would screw, thus in most cases extending the tap further away from the wall and discharging the water nearer to the centre of the appliance.

EXTERNAL AND INTERNAL CORNERS

To define whether a corner is internal or external is often open to a lot of debate. With pipe and eaves guttering it depends on whether the section goes around the structure or not, see sketch. With sheet material it depends upon which way up the material is. The term break corner applies to a special corner consisting of both external and internal angles.

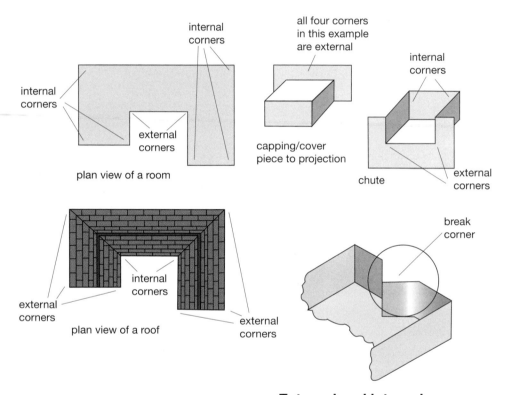

External and internal corners

EXTRACTOR FAN   See MECHANICAL VENTILATION.

EYE WASH FOUNTAIN

A sanitary appliance which consists of a small bowl with an ASCENDING SPRAY, operated by a hand or foot lever. The appliance is installed in work places where there is a risk of eye injury by dangerous liquids and solid particles. The injured person is able to lean over the bowl and bathe the eyes without touching them.

F & E CISTERN
Abbreviation for FEED AND EXPANSION CISTERN.

FAHRENHEIT
The old British Imperial temperature scale. See METRIC SYSTEM.

FALL PIPE
Another name for a rainwater pipe.

FAMILY OF GASES
The range of gases having a WOBBE INDEX within a specified band. The wobbe number indicates the heat released when a gas is burned at a constant gas pressure.

Gas family	Approx. Wobbe value MJ/m^3	Gas type
1	24.4–28.8	Manufactured (town)
2	48.2–53.2	Natural
3	72.6–87.8	LPG

FAN ASSISTED FLUES (FAN DRAUGHT FLUE)
Many burning appliances employ the use of electric fans to create a positive force for the extraction of flue gases. By using a fan, smaller cross-sectional area flue systems can be designed for a given heat input, making for more compact appliance design. Also no draught diverters are necessary as down draughts are eliminated. However, in all cases (i.e. ROOM SEALED or OPEN FLUED appliances) in the event of fan motor failure, the main appliance burners must shut down and not operate as the flue would be ineffective.

FAN DILUTED FLUE
A type of flue sometimes used on boilers installed at ground level in tall buildings. The flue is fan assisted and the products of combustion are discharged at or about the same level as the boiler. To assist this type of flue to function efficiently fresh air is mixed in the flue pipe with the products of combustion. Should this flue arrangement be used, there must be permanent fresh air inlets into the boiler room located at least 300 mm above ground level.

(See illustration over.)

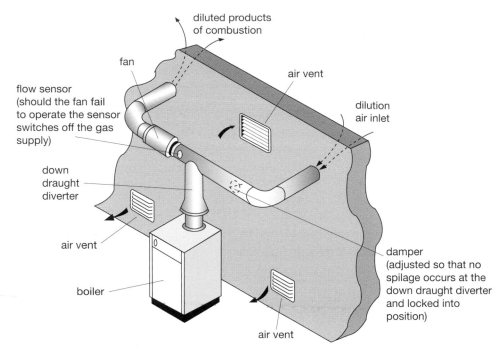

diluted products
of combustion

fan

air vent

flow sensor
(should the fan fail
to operate the sensor
switches off the gas
supply)

dilution
air inlet

down
draught
diverter

air vent

damper
(adjusted so that no
spilage occurs at the
down draught diverter
and locked into
position)

boiler

air vent

Fan diluted flue

FASCIA BOARD
A timber board fixed at the eaves of a building. See BUILDING TERMINOLOGY.

FASCIA BRACKET
A bracket used to hold the EAVES GUTTER in position. A fascia bracket is designed to screw directly onto the fascia board located at the lowest part of a sloping roof (the eaves). See also RAFTER BRACKET.

FATIGUE CRACKING
The fracture or splitting of a material brought about by excessive loads or working of the material. Should a piece of wire be continuously bent back and forth it will eventually break due to the fact that the material has become WORK HARDENED at the point where bending has taken place. Equally with the continuous movement of sheet weatherings, due to temperature changes, the same problems can arise. Therefore when installing any work ensure the correct grades of material are chosen and that expansion of the sheet is allowed for with sufficient fixings.

FAUCET
A term not used in England. In Scotland the term means socket and in America it means tap.

FEED AND EXPANSION CISTERN
A cistern used to feed water and to take up the expansion of water from a vented hot water heating system. It must be remembered that the volume of water in the heating

system would expand by approximately 4%, from cold, therefore the water level in the cistern should be sufficiently low so that when the water heats up and expands it will not rise to a point higher in the cistern than 25 mm below the WARNING PIPE.

FEED CISTERN

A storage cistern which supplies cold water to the hot storage vessel only. It should have a minimum capacity of the contents of the hot storage vessel and in any case should not be less than 100 litres.

FEED PIPE

A supply or distribution pipe conveying water to an appliance, fitting or vessel.

FEMALE THREAD

An internal thread cut inside a hole, tube or fitting into which a male thread can screw.

FERROUS METAL

A metal containing, or made of, IRON, e.g. mild steel, cast iron, etc.

FERRULE

A valve used to connect the SERVICE PIPE to the water main, for example, see SCREW-DOWN FERRULE.

FERRULE

A copper or brass sleeve fitted to the end of some wooden tools such as CHASE WEDGES. The sleeve is fitted in order to avoid the tool splitting when struck with a hammer.

FERRULE KEY

A special key designed to fit and operate a SCREWDOWN FERRULE.

FERRULE SLEEVE (TAILPIECE)

A short piece of brass or copper pipe with a lug on one end. The ferrule is used to make joints between lead and cast iron pipes. A wiped solder joint is made between the straight end of the ferrule and the lead pipe, the other end of the ferrule, with the lug, is CAULKED into a cast iron drainage socket.

FIBRE WASHER

A ring of densely packed fibre material used for various types of washers. Fibre washers are used on tap connectors and flange joints designed so that on contact with water the washer swells up preventing any leakage. Fibre washers are also used on globe valves and some safety valves to replace rubber washers which might otherwise weaken under higher temperatures.

FIELD DRAIN   See LAND DRAIN.

FILAMENT IGNITION   See IGNITION DEVICES.

FILE

A tool used to cut away the surface of the metal to achieve a nice finish to the work in hand. As a point of safety, no file should ever be used without a correctly fitting handle, as the bare tang (the point which goes on the handle) may cause a severe cut to the hand. A file only cuts on the forward stroke therefore pressure should be released on the backward movement. See also RASP.

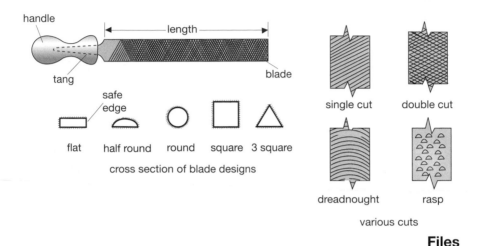

Files

FILLER ROD (WELDING ROD)

The rod or strip of metal used to fill and assist making a welded joint when welding metals together.

FILLET WELDING

The welding of two sheets of material lying at different angles. When welding lead sheet it is often referred to as an ANGLE WELDED JOINT.

FILLING LOOP

A temporary hose filling connection designed to supply water to a sealed system of heating. In addition to the hose, which should be disconnected from the system once it is full, there should be a stop valve and double checkvalve incorporated within the connection.

FILTER (STRAINER)

A device fitted over an inlet or outlet to an appliance or fitted into a pipeline, designed to prevent the passage of unwanted particles or insects, etc. A basic filter could consist of a fine mesh or gauze. See STORAGE CISTERN and TRAPPING SET for examples of different types of filters being put to use.

(See illustration opposite)

FINE SOLDER

Solder which is used for CAPILLARY JOINTS. The term is also applied when solder is used in conjunction with a soldering iron.

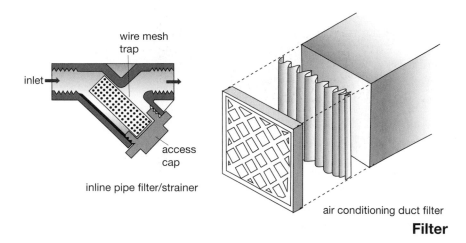

inline pipe filter/strainer

air conditioning duct filter

Filter

FIRE BED

The area within the combustion chamber of a heat-producing appliance where the fuel is burned and consumed.

FIRE BOX

The area in which the combustion of a fuel takes place.

FIRE BRICK

This is a refractory brick that is designed to store heat and prevent it being wasted. Refractory bricks are typically found in appliances such as radiant gas fires where they maintain the radiants at the highest possible temperature and prevent the firebox cracking.

FIRE CEMENT

A cement mixture designed to resist cracking when exposed to heat. Fire cement is used to assist jointing spigot and socket type flue pipes, and other areas where heat may cause problems to the jointing medium.

FIRE COMPARTMENT

A room that has been constructed to prevent the spread of fire.

FIRE HYDRANT   See HYDRANT.

FIRE STOP

A precautionary measure taken where a pipe passes through a wall or floor to prevent the supply of air, or spread of smoke, from one room to another in the event of a fire.

(See illustration over.)

FIRE SURROUND

The decorative finish that surrounds the outer edge and is fitted against the wall of a builder's opening. The surround usually incorporates a hearth.

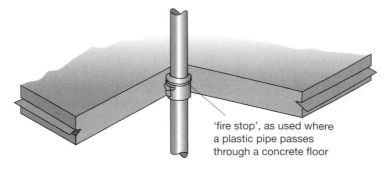

'fire stop', as used where
a plastic pipe passes
through a concrete floor

Fire stop

FIRE VALVE
A special valve designed to automatically shut off the flow of oil to an oil burning appliance in the case of a fire. There are three basic types; the fusible link, the leaded handle and the pressure type. With the fusible link type the valve is held open by a tensioned wire and the arrangement is set up as shown in the sketch. The fusible link has a low melting temperature and in the event of a fire will break, thus the valve will close assisted by a weight or attached spring. The fusible link type would normally only be used in a purpose made boiler house. For domestic oil fired installations the leaded handle or pressure type is generally used. The leaded handle type is probably the most common and consists of a small spring loaded stopvalve in which the thread in the wheelhead is made of lead, which has a low melting point. In the event of a fire the lead melts and the valve is made to close. The pressure type consists of a bellows type valve which is connected to a heat sensitive bulb. In the event of a fire, the bulb overheats and causes an increased vapour pressure in the bellows resulting in its expansion, thus closing the valve.

(See illustration opposite)

FIRECLAY   See CERAMIC WARE.

FIRRING PIECE
The name given to the timber blocks or wedges used to fit under the decking of a flat roof in order to give it the correct fall.

FIRST FIX
A term used to indicate that the pipework, both waste and water supplies have been run to a specific location, such as for the installation of a basin, etc. and tested before the surface finishes, such as the plastering or floor boarding, have been applied. See also SECOND FIX.

FIRST STAGE REGULATOR
The gas regulator located on the gas supply line between a bulk tank and the second stage regulator. It is designed to reduce the very high varying pressure within the tank to a constant 0.75 mbar pressure. See also BULK TANK and REGULATOR.

FISH TAILED CLEAT   See SHEET FIXINGS.

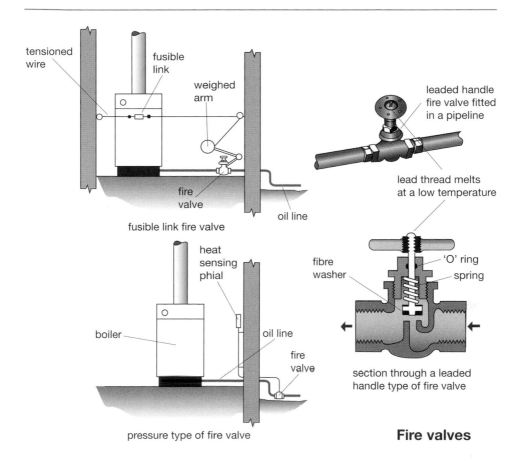

leaded handle
fire valve fitted
in a pipeline

lead thread melts
at a low temperature

fusible link fire valve

pressure type of fire valve

section through a leaded
handle type of fire valve

Fire valves

F

FIXED JUMPER

The JUMPER which holds the washer inside a screw-down stopcock, etc. has been fixed to the spindle. See STOPCOCK.

FIXING CLIP (FIXING CLEAT)

A clip designed to hold sheet weathering materials down onto the roof structure. See SHEET FIXINGS.

FIXING POINTS (FIXING CLAMPS)

A specially designed set of clamps which are designed to hold pipes securely in position while making solder joints to them.

(See illustration over.)

FIXING STRIP   See SHEET FIXINGS.

FLAME ARRESTOR

A device fitted in a gas line to prevent the passing of a flame through into the pipe supply. It generally consists of a meshed steel surface by which the flame is either contained or extinguished.

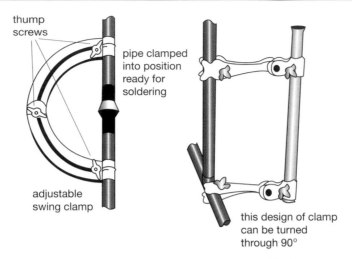

thump screws

pipe clamped into position ready for soldering

adjustable swing clamp

this design of clamp can be turned through 90°

Fixing points

FLAME FAILURE DEVICE (FLAME SUPERVISION DEVICE)

A device fitted to gas boilers to prevent gas from reaching the main burner until the pilot light has become established. There are various types of flame failure devices the most common being the THERMOELECTRIC VALVE, FLAME RECTIFICATION DEVICE and the MERCURY VAPOUR FLAME FAILURE DEVICE. See also BIMETALLIC STRIP FLAME FAILURE DEVICE.

FLAME LIFT   See LIFT OFF.

FLAME RECTIFICATION DEVICE

A FLAME FAILURE DEVICE typically found in gas appliances using ELECTRONIC IGNITION. The device works by allowing the current to flow through a flame making a circuit. If no flame were present, no electron flow could occur and as a result the appliance would shut down the gas supply, thereby preventing large volumes of gas accumulating within the combustion chamber. These devices are called rectification devices because they allow the flow of electricity only in one direction, i.e. AC is passed into the flame (allowing electrons to flow from the electrode direction and the burner direction), but only DC is detected. This is because the 'small' electrode does not sense the electron flow, whereas the burner itself is quite large. (The term 'rectification' means the change of alternating current to direct current.)

(See illustration opposite)

FLAME RETENTION

A method used to avoid gas burners suffering the effects of LIFT OFF. One method uses very small holes in a burner head positioned next to the main burner ports. Thus gas is restricted from passing through these very small holes and as a result a stable flame is produced. The stable flame subsequently secures the main burner flame from lifting off and potentially going out. During any service work on the gas burner these flame retention ports need to be inspected to ensure that they remain effective and are not blocked.

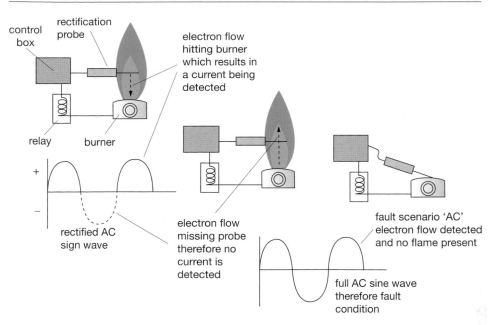

Flame rectification

FLAME SPEED

When a flame is produced by burning fuel it travels through the fuel at a particular speed. Igniting a piece of paper would allow the flame to travel throughout its length and the flame speed could fairly easily be monitored. Fuels such as natural gas, propane, etc. have flame speeds of less than half a metre per second. Petroleum vapour and hydrogen, however, have very rapid flame speeds, which travel faster than the speed of sound.

FLAME SUPERVISION DEVICE   See FLAME FAILURE DEVICE.

FLAMELESS COMBUSTION   See CATALYTIC COMBUSTION.

FLAMMABILITY LIMITS

The band within which a fuel will burn. All fuels have higher and lower flammability limits, for example natural gas will only burn with a gas/air mixture of 5–15% whereas butane will only burn with a gas/air mixture of 1.9–8.5%.

FLAMMABLE

Term used to mean something is combustible and will burn with a flame.

FLANGE JOINT

A round projecting disc usually with bolt holes which may be cast, welded or screwed onto the end of a pipe or vessel, enabling the joint to another pipe or vessel to be made without turning either object. The two sections are bolted together with a gasket, FIBRE WASHER or TAYLOR'S RING between them.

(See illustration over.)

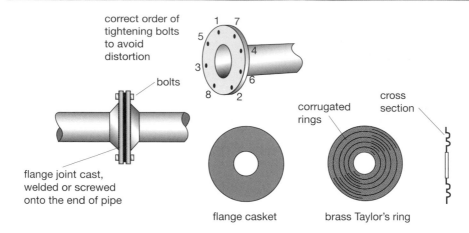

Flange joints

FLAP VALVE

A type of non-return valve in which a disc hinges at the top and rests upon the face of its inlet therefore permitting the flow of liquid in one direction only. For example, see NON-RETURN VALVE.

FLARING TOOL (DRIFT)

A tool made of mild steel supplied in a range of sizes. It is used to form a bell mouth when preparing to make a BRONZE WELDED joint to copper pipe. The tool is also used to flare out the end of copper pipe when making some types of MANIPULATIVE COMPRESSION JOINTS.

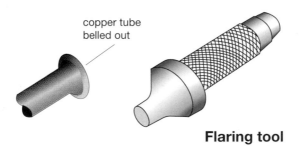

Flaring tool

FLASHBACK

The pre-ignition of gases in the welding tip or nozzle of a blowpipe when using oxy-acetylene welding equipment. It is usually identified by a series of bangs in quick succession. In most cases just turning off the gases and re-lighting the blowpipe cures the problem. But in severe cases the nozzle should be plunged into a pail of water to cool it down, whilst this is done the oxygen valve is left open to prevent the entry of water into the blowpipe. Another type of flashback is known as backfire. This is caused by the pre-ignition of gases in the mixing chamber of the blowpipe and is identified by a loud squealing noise and black smoke issuing from the tip. The most common causes of flashback and back fire are:

(1) The welding tip is not fully tightened into the blowpipe thus air is allowed to enter the mixing chamber.
(2) The welding tip is held too close to the work.
(3) The welding tip has become overheated, due to slag build up on the tip or by working in a confined area.
(4) The regulator pressures set too low.
(5) The welding tip is too large for the job in hand.

See also WELDING SAFETY.

FLASHBACK ARRESTOR

Flashback arrestors are CHECK VALVES which have been fitted onto oxygen and acetylene hoses to prevent the flame, in the event of a FLASHBACK or BACKFIRE, passing into the hoses, this type of arrestor is commonly called a hose protector. A second flashback arrestor should be fitted at the regulator outlet, thus, should the hoses accidently catch fire, the gas supply is automatically shut off protecting the regulator and cylinder. This second type of arrestor differs from a simple check valve in that it is constructed with a non-return valve, a flashback extinguisher and a device to shut down the supply of fuel, which will not allow gas through until manually opened. See WELDING EQUIPMENT.

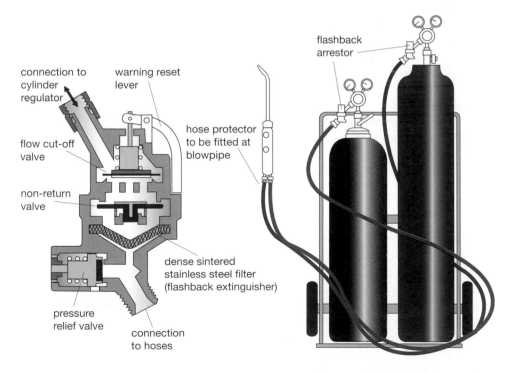

Flashback arrestor

FLASHINGS   See CHIMNEY FLASHINGS and COVER FLASHING.

FLASHPOINT
The lowest temperature at which any petroleum product will ignite and burn when a flame is put to it. Below this temperature insufficient petroleum vapour will be given off to support combustion. See also IGNITION TEMPERATURE.

FLAT DRESSER
A tool made of boxwood or high density plastic used for dressing and assisting in the laying of hard and soft sheet roofing materials.

Flat dresser

FLAT ROOF OUTLET   See BALCONY OUTLET.

FLEXIBLE JOINT
A joint such as a SEMI-RIGID COUPLING that allows for a small degree of angular deflection without any fluid loss.

FLEXIBLE MEMBRANE VESSEL   See SEALED EXPANSION VESSEL.

FLINT GUN (SPARK GUN)
A tool designed with a small sparking device used to light blowlamps and blowpipes therefore avoiding the use of matches, which often become damp in the tool box.

FLOAT
A device which is very buoyant and will always try to rise to the surface if submerged in water. Floats are made from various materials including polythene, expanded ebonite and expanded vinyl-chloride. Floats are used in conjunction with all kinds of fittings. For examples, see BALLVALVE and FLOAT SWITCH.

FLOAT (RANGE)
A horizontal DISCHARGE PIPE which has a series of branches fitted in order to receive a range of, or several, SANITARY APPLIANCES in a row. Should the float be unventilated, its diameter should be at least that shown in the following table. See also COMMON DISCHARGE PIPE.

Appliance	Maximum number to be connected	Minimum pipe size (mm)	Slope
WCs	8	100	between 9–90 mm/m
Washbasins	4	50	between 18–45 mm/m
Bowl urinals	5	50	between 18–90 mm/m
Stall urinals	6	65	between 18–90 mm/m

Unvented discharge pipes serving more than one appliance

F

FLOAT OPERATED VALVE

A valve which is operated by the raising and lowering of a float, for example, see BALLVALVE.

FLOAT SWITCH

A device consisting of a float which is allowed to rise and fall with the water level in a vessel and in so doing operate an electric switch. This in turn may start or stop a motor or pump which may empty or refill the pipe or vessel in which the float is fitted. For examples, see AUTOMATIC PNEUMATIC CYLINDER, PNEUMATIC EJECTOR and PIPELINE SWITCH.

FLOW PIPE

A pipe through which water flows. Examples could be found in a central heating circuit, a PRIMARY CIRCUIT or a SECONDARY CIRCUIT.

FLOW RATE

The volume of fluid (e.g. water or gas) flowing through a pipe during a given period, usually expressed in litres per second or m^3 per hour. Flow rate must not be confused with pressure. For example, a 1 mm diameter pipe supplying a bath would take a very long time to fill the appliance regardless of the water pressure; whereas a 22 mm diameter pipe would allow for a much greater fill-up rate.

FLOW SWITCH

A device fitted in a pipeline to activate an electrical control, such as a pump. Should water be static within the pipeline the flow switch will be at rest, however where movement is detected, such as by opening a DRAW OFF POINT, the flow switch is made to move and in turn makes an electrical contact (in the example suggested above, causing the pump to cut in). The continued water flow holds the switch open.

FLUE

A passage through which the products of combustion are transported to the outside. The requirements for oil and solid fuel are different from those of gas, in which the products of combustion cause less problems.

In general with oil and solid fuel the only openings into the flue should be an inspection or cleaning door and a DRAUGHT STABILISER. Flue pipes should be used to connect the appliance to a lined or factory made insulated chimney.

With gas appliances the only openings into the flue should be an inspection or cleaning door and a DOWN DRAUGHT DIVERTER. Flue pipes are permitted to convey the products of combustion from gas appliances or an unlined chimney may be used in many circumstances, alternatively a chimney as for oil and solid fuel appliances can be used. If the chimney is unlined, a debris collection space should be provided at the base of the chimney, which should extend to a depth of at least 250 mm below the point of entry of the flue pipe. Access should be readily available to this space for the removal of debris. Flues from gas appliances should discharge into the atmosphere away from any opening into the building and be situated so that air may pass freely across the outlet at all times.

In all cases with oil, solid fuel and gas appliances no flue should have an opening into more than one room except for cleaning and inspection purposes. Part J of the Building Regulations now requires a notice plate to be located in a prominent position, designed to specify the standard and category with which the flue system complies. For terminal positions, see FLUE TERMINAL. See also BALANCED FLUE, OPEN FLUE, SHARED FLUE, VERTEX FLUE and SPILLAGE.

FLUE BLOCKS   See PRE-CAST FLUE BLOCKS.

FLUE BOX
A prefabricated box, usually metal, designed to act as the space required for a fireplace recess or builder's opening and used in conjunction with a flue pipe to act as a chimney arrangement. It does not require brickwork.

FLUE BREAK
An opening within the SECONDARY FLUE such as a DOWN DRAUGHT DIVERTER.

FLUE FLOW TEST (PULL TEST)
A test carried out to confirm the up draught and soundness of an open flued appliance. To undertake this test, a visual inspection of the flue route is first made to see if the test is likely to succeed, e.g. check the terminal location, excessive bends, horizontal runs, etc. If necessary, sweep the chimney. Now warm the flue with a blow lamp to get the flow by CONVECTION CURRENTS moving, then light a smoke pellet and place it in the base of the flue pipe or chimney fire back (if applicable, temporarily position a closure plate in front of the builder's opening). The smoke should be clearly seen to be drawn up the flue way and by external examination seen discharging from the terminal. As a precautionary measure it is advisable to have a bucket of water in readiness in which to place the pellet, should the test fail, thus preventing the room filling with smoke. See also SPILLAGE.

FLUE GAS ANALYSER
A device used to determine various qualities of the products of combustion of fuel. Several designs of analyser will be found to include the 'Draeger'; 'Fyrite' and electronic types. The first two listed have the limitation of sampling only one particular flue gas product, whereas electronic analysers sample many types of flue gases, compare these with each other and against different temperatures, etc. and compute the date automatically to give efficiencies and appliance safety information.

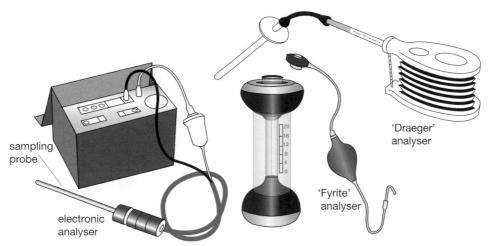

sampling
probe

electronic
analyser

'Fyrite'
analyser

'Draeger'
analyser

F

Flue gas analysers

FLUE LINING (FLUE LINER)

An internal flue passage found inside a chimney. A flue lining is required to prevent water, caused by condensation in the flue, from passing through the brickwork and staining the plaster, the lining also protects the brickwork from acidic attack and prevents any falling masonry within the chimney from blocking the flue passage. Any linings used which have socketed joints should have the sockets looking upwards, this prevents condensate running out of the joint and weakening any jointing material. In many cases the flue lining for gas appliances may be of a flexible steel material.

FLUE PIPE

A pipe which conveys the products of combustion from a heat producing appliance. The materials used for pipes are generally either cast iron, stainless steel or vitreous enamelled mild steel. Any flue pipe with spigot and socket joints should be fitted so that the socket is uppermost, this prevents water, caused by condensation, from running out and weakening the joint. The flue pipe from oil and solid fuel appliances should be as short as possible and connect directly into a chimney. Where a flue pipe is to pass near or through combustible material it should be kept separate from the material as shown in the diagrams; notice the provisions for oil and solid fuel are different from that of gas. See also FLUE.

(See illustration over.)

FLUE SAFETY DEVICE   See ATMOSPHERIC SENSING DEVICE.

FLUE SPIGOT   See SPIGOT.

FLUE SPIGOT RESTRICTOR

A plate sometimes located on the flue outlet from a gas fire to limit the effects of a strong flue draught created within the chimney.

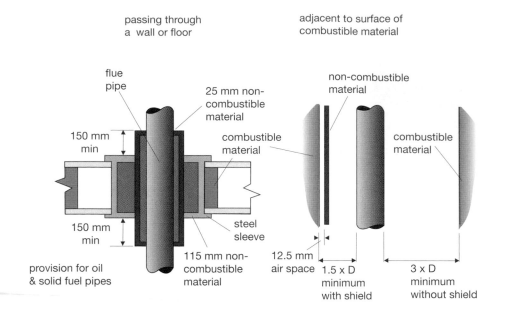

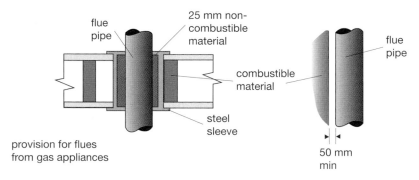

Separating flue pipes from combustible materials

FLUE TERMINAL

The discharge point from the flue of an oil, solid fuel or gas appliance. There are many designs of flue terminal, each designed to suit the flue location and the appliance type, e.g. ROOM SEALED appliance or OPEN FLUE. A few examples of flue terminals are illustrated and the following table indicates a range of location dimensions to be observed, to which the building regulations need to be sought for variables and other positions.

F

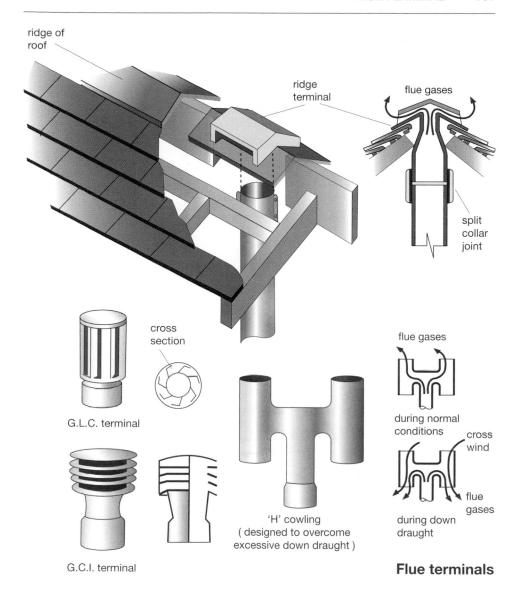

ridge of roof

ridge terminal

flue gases

split collar joint

cross section

G.L.C. terminal

flue gases

during normal conditions

cross wind

flue gases

during down draught

G.C.I. terminal

'H' cowling
(designed to overcome excessive down draught)

Flue terminals

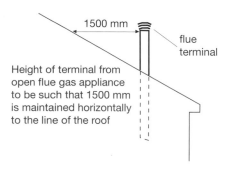

1500 mm

flue terminal

Height of terminal from open flue gas appliance to be such that 1500 mm is maintained horizontally to the line of the roof

Specific terminal height from gas appliance

Location of flue terminals (mm)

Location	Balanced flue 'gas'		Open flue 'gas'		Oil pressure jet burner	Solid fuel
	Forced draught	Natural draught	Forced draught	Natural draught		
1 Above the roof line	As manufacturer's instructions	N/A	150	1500 # Vertical	600	1000
2 From a structure on the roof	N/A	N/A	N/A	1500	1500	N/A
3 Below an opening into the building	300	300 *	300	N/A	600	N/A
4 Above an opening into the building	300	300 *	300	N/A	As manufacturer's instructions	N/A
5 Beside an opening into a building	300	300 *	300	N/A	300	N/A
6 From an internal or external corner	300	600	200	N/A	300	N/A
7 From a vertical discharge stack	150	300	150	N/A	300	N/A
8 Below the eaves	200	300	–	N/A	600	N/A
9 Above ground level	300	300	300	N/A	300	N/A

* For gas appliances above 7 kW this distance is greater than the distance indicated. (See BS.5440.)
# See illustration on previous page.

FLUELESS APPLIANCE
A gas appliance which discharges the products of combustion within the room in which it is installed, for example, see INSTANTANEOUS WATER HEATERS.

FLUID
A substance, such as a gas or liquid that continually deforms or flows to fill the void of the vessel in which it is contained.

FLUID CATEGORY
A term used to indicate the level of quality (i.e. how drinkable) of the water supply. Five fluid categories are identified within the current Water Supply Regulations. Note that the higher the number, the 'greater' will be the health risk.

Fluid Category 1   Wholesome drinking water.
Fluid Category 2   Drinking water that is safe, but which may have been subjected to temperature change or contain a strange taste or odour. Water from a hot distribution pipe may fall within this category.

Fluid Category 3   Water that represents a slight health risk. For example, water from domestic washbasins, baths or dishwashers.

Fluid Category 4   Water representing a significant health risk to include that which contains chemicals, insecticides, etc.

Fluid Category 5   Water representing a serious health risk and includes that which contains faecal or other human or animal waste, for example, that from urinals, WCs and bidets.

FLUORESCEIN   See COLOUR TEST.

FLUSH GRATED WASTE FITTING

The integral grating designed to prevent the entry of large objects. This type of waste fitting is finished flush with the top of the inlet and therefore it is incapable of receiving a waste plug. These waste fittings are commonly fitted to shower trays.

FLUSHING CISTERN

Often known as a water waste preventer (WWP) as it is an appliance designed to prevent the wastage of water. It is a cistern which has a device for rapidly discharging the water it contains. It is used in conjunction with sanitary appliances such as WCs, slop sinks, etc., for the purpose of cleansing the appliance and carrying away its contents to the drain. Two designs of cistern will be found: those that use the principle of SIPHONIC ACTION and those that use a valve. This entry describes the cistern utilising a siphonic action for valved cisterns. See also VALVED FLUSHING CISTERN. The capacity of a modern flushing cistern is 6 litres, however prior to 1993, a 9-litre flush was employed. It must be borne in mind that when replacing 9-litre flushing cisterns it may be necessary to install a new cistern giving an equal 9-litre flush otherwise the volume of water would not be sufficient to remove the contents from the pan which has not been designed to cope with the reduced flush.

Two siphonic cisterns will be found, namely the plunger type and the older cast iron bell type. With the plunger type flushing cistern, when the lever handle or chain is pulled the diaphragm plunger is lifted. As it rises, it pushes some water over into the long leg of the siphon down into the flushpipe taking with it some of the air. This causes a reduction of air pressure in the siphon (a partial vacuum). The greater air pressure which is acting upon the surface of the water in the cistern, created by the atmosphere, forces the water up past the diaphragm washer and through the siphon, this continues until the water level is low enough in the cistern to allow air to enter the siphon, thus breaking the siphonic action. With the bell flushing cistern, when the chain is pulled the bell-shaped dome inside the cistern is raised. When the chain is released, the bell falls and displaces the water which falls down the flushpipe taking with it some air from inside the bell thus creating a partial vacuum, causing siphonic action to occur. For fixing positions of flushing cisterns, see WC SUITE. See also DUAL FLUSH.

(See illustration over.)

FLUSHING TANK

A vessel used to flush a system of drains. The tank could be designed to discharge its contents of water automatically. See AUTOMATIC FLUSHING TANK.

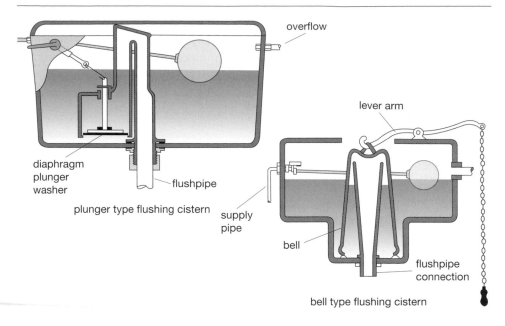

Flushing cisterns

FLUSHING TROUGH (TROUGH CISTERN)

A flushing apparatus which serves several WC pans from one long cistern body. It is designed in this way to allow for more frequent flushing. Flushing troughs are used in places such as schools and factories where the quick refill of the flushing cistern is required. SIPHONIC ACTION is started in the same way as in an ordinary flushing cistern and as the water is siphoned the water level falls only inside a 'timing' box designed to permit the required flush only. Thus when the timing box has been siphoned of water, air gets in through an air pipe and breaks the main bulk of siphonic action. After a short period the timing box refills with water through a hole in its side, the cistern is then ready for re-flushing.

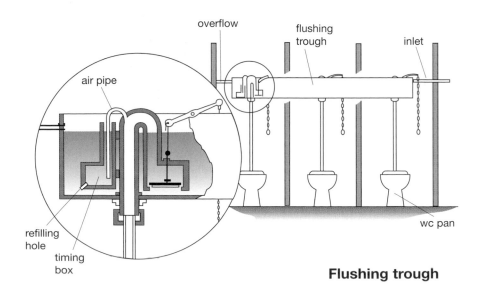

Flushing trough

FLUSHING VALVE

A valve designed to control the flushing water to a SOIL APPLIANCE, it would be operated by either hand or foot control.

The flushing valve works as follows:

(1) Sketch A. Valve shown in the closed position.
(2) Sketch B. When the handle is operated the release spindle is tilted causing water to be discharged from the upper chamber.
(3) Sketch C. Water goes down the flushpipe quicker than through the by-pass therefore the pressure in the upper chamber becomes reduced; this allows the water, at a greater pressure in the inlet pipe to cause the piston to lift allowing the water to discharge at full bore down the flushpipe.
(4) As water travels up the by-pass to the upper chamber, the pressure becomes equalised and the valve falls to close under its own weight.
(5) The amount of water discharging down the flushpipe is determined by the adjustment of the regulating screw, opening or closing the by-pass.

See also VALVED FLUSHING CISTERN.

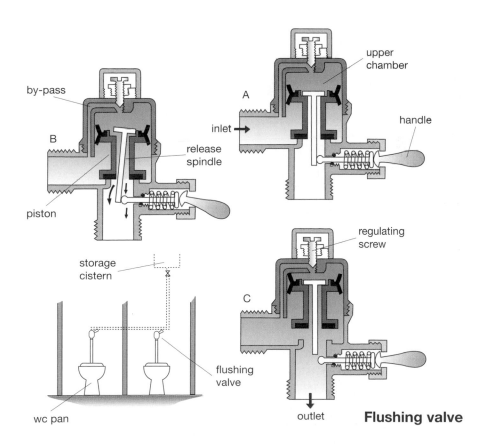

Flushing valve

FLUSHPIPE

A pipe used to convey water from a flushing cistern or flushing valve to a WC pan, slophopper or urinal.

FLUSHPIPE CONNECTOR

A rubber fitting used to join the flushpipe onto a WC pan or similar SOIL APPLIANCE. When joining a flushpipe to its connection it is essential that it enters centrally to ensure an equal flush all around the bowl of the appliance. Before the introduction of rubber connectors the joint was made with RED LEAD and putty which often resulted in leakage due to the movement of the pan; methods such as this are no longer carried out.

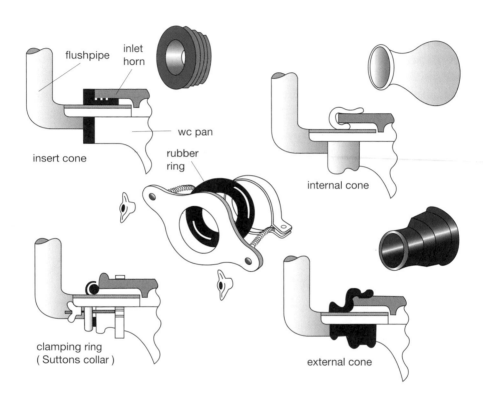

Flushpipe connectors

FLUX

A substance available in many forms from solid, semi-solid, or liquid. It is applied to metals prior to HARD or SOFT SOLDERING and some welding processes. Applying a flux has several important purposes; firstly, it prevents the oxygen in the atmosphere coming into contact with the metal surface and thus prevents OXIDATION. The flux also helps the solder to flow over the material, so that it tins or wets the surface easily. Basically fluxes are classified as either active or non-active. Active fluxes tend to be more corrosive even at room temperature and some are designed to clean the metal as well as prevent oxidation. Active fluxes are unsuitable for electrical work and when using such fluxes it is essential to remove excess flux on completion of any soldering process to prevent further corrosion.

Fluxes in general use		
Flux	Used on	Origin
Tallow	lead	Fat obtained from sheep and cattle.
Rosin	brass and copper	Available in paste and powdered form. Obtained from the gum in the bark of pine trees.
Zinc chloride (killed spirits)	copper bits and dirty zinc	Produced by dissolving pieces of zinc in hydrochloric acid.
Zinc-ammonium chloride	zinc	Produced by adding about 20% salammoniac to zinc chloride.
Borax	copper, brass and iron	A soft and crystalline mineral which occurs in deposits formed by evaporation of hot springs.
Borax and Silicon	aluminium, copper, brass and iron	Borax plus silicon which is a hard dark grey solid found in practically all rocks.

Note tallow and rosin are inactive fluxes.

FOLLOWERS

BOXWOOD beads, slightly smaller in diameter than bobbins. Followers are used to push bobbins around a bend on lead pipe, see BOBBINS.

FOOT BATH

A shallow bath fixed at low level purposely designed for washing the feet. The term foot bath is also sometimes used to mean a walk through foot bath, this is a shallow bath in which a constant water level is maintained, it is positioned so that users to a public swimming pool must walk through it in order to cleanse their feet.

FOOT VALVE

A non-return valve which is fitted at the base of the suction pipe, fitted to a pump. The foot valve is designed to retain the water in the pipe during the cycles of lifting the water. See LIFT PUMP.

FOOTPRINT PIPE WRENCH

A tool used and referred to as 'Footprint' after the name of the company who designed it. See WRENCH.

FORCE CUP   See SUCTION PLUNGER.

FORCE PUMP

A pump in which water is forced into the delivery pipe, where it compresses a section of trapped air, which acts as an air vessel or cushion for the shock waves in the water due to the action of pumping. The air vessel also acts like a PNEUMATIC CYLINDER which ensures a steady continuous flow of water up the pipe. The pump works as follows:

(1) On the upward stroke, the non-return valve A opens and water is sucked into the body of the pump.
(2) On the down stroke, valve A closes and B opens where the water is pushed into the delivery pipe and compresses the air.

(3) As the process is being repeated, the compressed air expands and so forces more water up the delivery pipe.

See also LIFT PUMP.

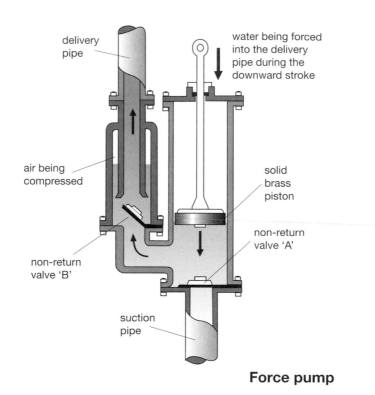

Force pump

FORCED DRAUGHT BURNER (AIR BLAST BURNER, PRE-MIX BURNER)
A gas or oil burner in which the fuel is blown into the heat exchanger by way of a fan and/or pump. Air is also blown into the firebox to give a pre-mix, equal to that required, producing the maximum combustion of the fuel. For one example see PRESSURE JET BURNER.

FORMER
The section of a bending machine which supports the throat of the bend while it is being formed, see BENDING MACHINE.

FOUL WATER DRAIN
A drain used to convey any water contaminated by soil, waste or trade effluent, it is not intended for SURFACE WATER.

FREE AREA   See VENTILATION GRILLE.

FREE STANDING BOILER
An independent boiler which stands on the floor, i.e. not hung on the wall.

FREEZING EQUIPMENT

Pipe freezing equipment consists of a jacket which is fixed around the pipe to be frozen, a cylinder which contains the refrigerant and a hose to connect the cylinder to the jacket. The equipment proves useful when a temporary plug or shut down of water is required to save time draining down a large system, or where a valve is faulty and will not shut off the supply. The refrigerant used could be either freon gas which is supplied in aerosol cans, carbon dioxide gas, or liquid nitrogen. The aerosol cans which are available from most plumbers merchants are generally limited to small pipes, no bigger than 25 mm. Carbon dioxide (CO_2) freezing equipment proves to be more successful and economical than the throw-away canisters although the initial outlay is more. When liquid CO_2 at high pressure is injected into the space between the jacket and the pipe it immediately expands, and forms a solid known as 'dry ice' or 'card ice' which has a temperature of −78°C. This extremely low temperature quickly freezes the water in the pipe and forms an 'ice' plug sealing off the flow of liquid. CO_2 can be successfully used on pipes up to 100 mm. As a point of safety, when using CO_2, freezing equipment protective gloves should be worn as protection from frost bite. CO_2 which is heavier than air often accumulates in low lying and confined spaces therefore care should be taken to ensure good ventilation to disperse the CO_2. For very large diameter pipes, over 75 mm, the refrigerant used is very often liquid nitrogen and this work would be carried out by a team of trained operators. Pipe freezing can also be achieved by the use of an electric freezer, which basically works just like a refrigerator.

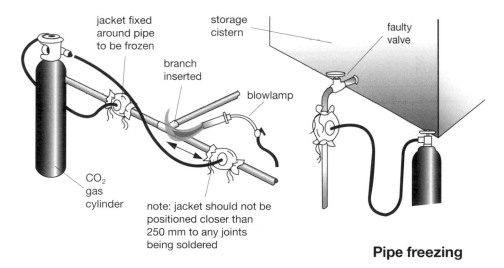

Pipe freezing

FREEZING POINT

The temperature at which a liquid freezes and changes to a solid. Water freezes at 0°C and expands by approximately $1/10$ in volume changing to ice. This is known as A PHYSICAL CHANGE.

FRESH AIR INLET

A terminal designed to allow air to enter a system of below ground drainage. The ventilation of drainage systems is essential to ensure that the internal pressure inside the drain is equal to that of atmospheric pressure, but it also helps to prevent the build-up of possibly dangerous gases. Fresh air inlets are normally only used in conjunction with an INTERCEPTING TRAP, but in most modern developments the intercepting trap is omitted and air enters the drain via the public sewer through holes in inspection covers in the road or sewer ventilating pipes.

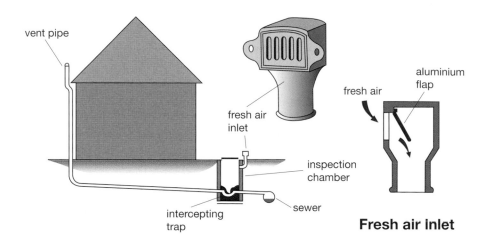

Fresh air inlet

FRICTION
The resistance which a substance encounters in moving across, along or against another substance. For example the brakes applied to a bicycle soon slow down and make pedalling hard to continue. See also FRICTIONAL LOSS.

FRICTION HEAD
The loss of HEAD due to FRICTIONAL LOSS.

FRICTIONAL LOSS
The loss of gas or water pressure caused by the friction created as the gas or water passes along a pipe or around a fitting. See also PIPE SIZING (hot and cold supplies).

FRICTIONAL RESISTANCE
The force due to FRICTION which leads to a FRICTIONAL LOSS.

FRONT APRON (BARGE)
The apron flashing which is fitted to the front of a chimney stack where it passes through the roof. It is designed to prevent the entry of rainwater into the building. See CHIMNEY FLASHING for an example of its application.

(See illustration opposite)

FROST
Particles of frozen moisture formed when the air temperature drops below the freezing point of water, 0°C.

FROST PROTECTION
As a good frost precaution all pipes fitted externally should be laid in the ground at a minimum depth of 750 mm and should not be positioned on external walls or

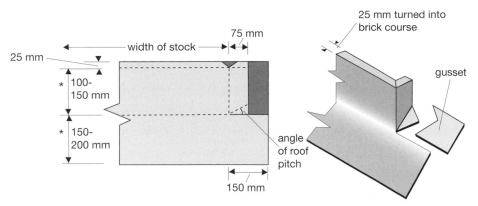

Setting out detail for lead welded front apron

* note: the variation in figures depends
upon the roof pitch and position
of brick course

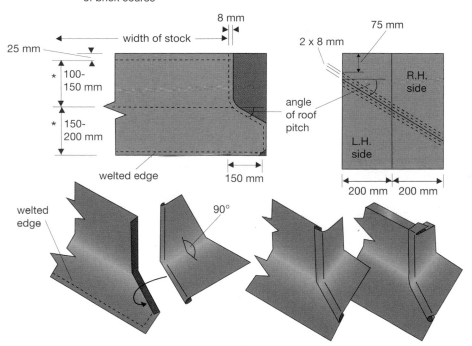

Setting out detail for copper front apron

where they are likely to be subjected to cold draughts, such as near the EAVES. All exposed pipework should be fitted with a good thermal insulation material which must be kept dry. Within a building, the roofspace, and below the ground floor are regarded as exposed positions as are any rooms which are not fitted with some means of central heating. It is possible to buy a special heat tracing wire element which is fixed to or wound around the pipe work, should the temperature fall below 0°C a thermostat cuts in and a small electrical current flows through the element keeping the pipe warm.

FROST THERMOSTAT (FROST STAT)

A thermostat designed to switch on and automatically light the boiler or switch on the power to an electrical element, etc. if the temperature falls to 0°C. If an outhouse is designed specifically to house a boiler a frost thermostat should always be fitted. The frost thermostat should be located in a position most likely to remain the coldest for the longest. For example, should the thermostat be located in a position where the sun can play on it, such as a wall facing east the thermostat may warm up but freezing conditions may still prevail elsewhere in a shaded position. See also PIPE THERMOSTAT.

FULLWAY GATEVALVE

A valve through which, when it is fully open, there is no restriction to the flow of water, for this reason it is recommended to be used where there is a low pressure in the pipeline such as if the supply is fed from a cistern. Gatevalves do not have a washer but a circular wedge-shaped gate which closes into a matching wedge-shaped seating when the valve is closed. It is recommended that if possible these valves should not be fitted in horizontal pipelines because any sludge or grit in the water could settle in the base of the seating thus preventing the gate from fully closing. By installing in vertical pipes, grit is simply washed down. See also SLUICE VALVE.

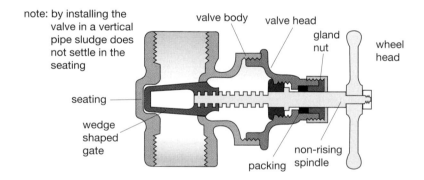

Fullway gatevalve

FULLY PUMPED SYSTEM

A hot water heating system, which also heats up the domestic hot water supply, in which the circulation of water operates fully under the influence of a pump. A system designed in this way does not rely at all upon CONVECTION CURRENTS being set up to circulate hot water to the hot storage vessel, thus a system of this nature could be designed with its boiler in the roof space or anywhere, above or below the hot storage vessel. The motorised valves shown used on the flow pipe are generally wired up to the cylinder or room thermostats and are made to automatically close when the required temperature within the cylinder or room has been reached, see MOTORISED VALVE. See also REVERSED CIRCULATION.

(See illustration opposite)

FULLY VENTILATED ONE PIPE SYSTEM

A ventilated system of above ground drainage. See SANITARY PIPEWORK.

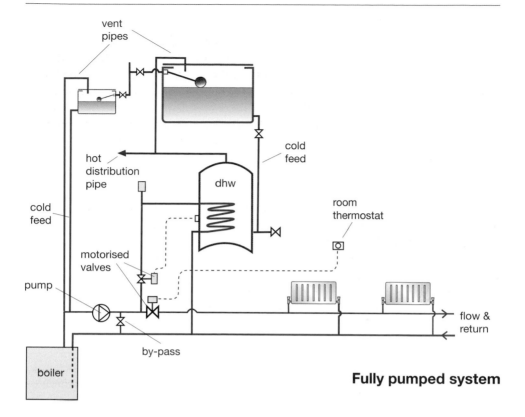

vent pipes

hot distribution pipe

cold feed

cold feed

pump

motorised valves

dhw

cold feed

room thermostat

flow & return

by-pass

boiler

Fully pumped system

FUR   See SCALE.

FUSIBILITY

The melting point of a material; it denotes the temperature at which a solid changes to a liquid. Metals being welded have to be heated to the melting point of the material in order to 'fuse' together with the filler rod.

FUSION WELDED JOINT (PLASTIC)

A 'welded' joint used on certain types of plastics. A tool is heated then inserted into the socket and onto the spigot of the plastic fitting and pipe, this heats up the plastic material melting its surfaces, the tool is removed and the pipe spigot is quickly inserted into the fitting socket whereupon the two become fused together. Plastic pipe can also be fusion welded using a flow of hot air.

FUSION WELDING

A jointing process used to join metal tube and plate. Fusion welding is a form of AUTOGENOUS WELDING in which the filler rod used has roughly the same composition as the parent metal. See LEAD WELDING and LOW CARBON STEEL WELDING.

GABLE END
The triangular section of the end wall of a building going up with the roof slope to the ridge. See BUILDING TERMINOLOGY.

GALVANIC ACTION   See ELECTROLYTIC CORROSION.

GALVANIC CELL SCALE REDUCER   See WATER CONDITIONERS.

GALVANIC PROTECTION   See CATHODIC PROTECTION.

GALVANISED
A zinc coating on FERROUS METALS (iron and steel) to help prevent corrosion problems. The iron is usually dipped into a molten zinc container or the zinc is sprayed onto the object requiring protection. See also SHERADISING.

GARAGE GULLY (YARD GULLY)
A gully designed with a large removable galvanised bucket to collect any grit or silt which might otherwise block the outlet pipe.

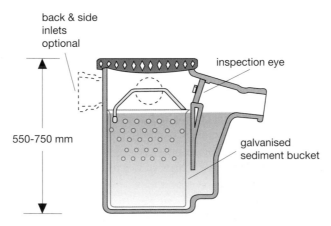

Garage gully

GARSTON BALLVALVE   See BRS BALLVALVE.

GAS BOILER
A gas appliance designed to heat water for domestic hot water purposes and/or for central heating. Gas boilers can be either free standing or wall mounted with a flue

which is either OPEN or of the BALANCED FLUE type. For a gas boiler to function safely and efficiently several devices must be included in its installation, most of which are supplied already fitted to the boiler. These devices consist of a GOVERNOR to reduce the gas pressure to that required, a FLAME FAILURE DEVICE which prevents any gas entering the combustion chamber without the pilot light being established, a thermostat which opens and closes a SOLENOID VALVE allowing the gas to the main burner itself as and when required. There are variations to the arrangements described but for the average gas boiler found in domestic dwellings the governor, flame failure device and solenoid valve are often located within one large control valve called a MULTI-FUNCTION GAS VALVE. Other devices are fitted to assist in the automatic running of the boiler such as programmers and time clocks. See also GAS BURNERS and FLUE.

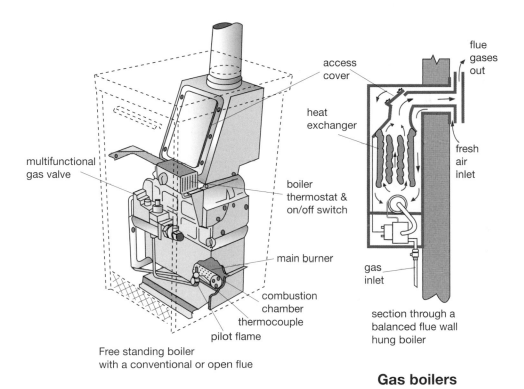

Free standing boiler
with a conventional or open flue

section through a
balanced flue wall
hung boiler

Gas boilers

GAS BOOSTER or COMPRESSOR

Devices designed to increase the gas pressure within a pipeline to that required for the operation of a particular appliance or to overcome problems associated with insufficient pipe sizes. When these devices are used it is essential that the suction effect does not interfere with the correct operation of the supply pipe work 'upstream' of the unit installed, causing a negative pressure within the system, in which ANTI-SUCTION VALVES are occasionally employed.

GAS BURNERS

Gas burners differ depending upon the type of fuel used. The old town gas once commonly used in Britain had the air, required to support combustion, mixed with the gas at a point around the flame. Natural gas on the other hand has had the air mixed with the gas prior to combustion and the type of burner used is called an AERATED BURNER. A second point which influences the design of the burner is the BURNING VELOCITY of the gas used. If the speed at which the gas is travelling through the pipe and from the burner is greater than the burning velocity (the speed the flame flashes through the gas) the flame will lift-off from the burner and burn in the atmosphere some distance away from the burner head. To overcome this problem special enlarged burner heads are designed to cause a turbulence thus slowing down the gas and stabilising the flame holding it onto the burner. Alternatively the gas pressure should be adjusted to ensure the velocity of the gas/air mixture is not more than the burning velocity of the fuel. Should the pressure be too low the flame will be rather smokey and incomplete combustion will result due to insufficient air being drawn in through the primary air port. To maintain the correct gas/air pressure and prevent flame lift-off a small retention flame is sometimes incorporated with some designs of burners.

(See illustration opposite)

GAS CIRCULATOR
A gas circulator is simply a small gas boiler capable of heating water for a domestic hot water supply. Gas circulators are often used in conjunction with a warm air unit to provide hot water.

GAS COCK   See PLUG COCK.

GAS CONSUMPTION   See GAS RATE.

GAS FAMILY   See FAMILY OF GASES.

GAS FIRES
Gas fires are designed to provide both heated CONVECTION CURRENTS and RADIANT HEAT into a room. The gas fire could be either free standing or wall mounted. When a gas fire is connected to a chimney it is essential to ensure that there is a good updraft. The connections to the chimney must be correctly sealed from the room to prevent CARBON MONOXIDE fumes escaping back into the room. If the chimney is unlined a debris collection space must be provided at the bottom of the chimney to a depth of at least 250 mm below the point of entry of the appliance. This space must also be accessible for periodical clearance of debris, this is normally achieved by the removal of the fire itself. See also DECORATIVE FUEL EFFECT FIRE.

GAS GOVERNOR (REGULATOR)
A device for controlling the flow or pressure of gas in a pipeline. The gas 'mains' pressure varies from area to area and in any case is higher than that required by the consumer, therefore a pressure reducing governor is situated on the inlet side of the gas meter and should not be tampered with or altered by anyone except those authorised by the gas supplier. To get the best performance from any gas appliance equipment is often provided with its own governor. This prevents pressure

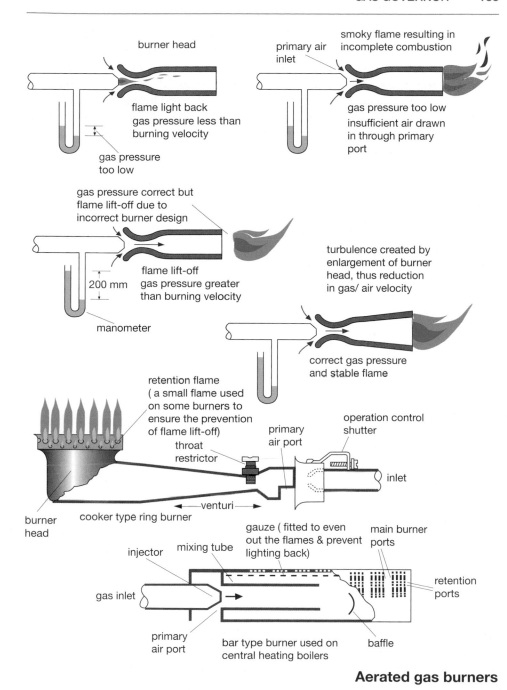

burner head

primary air inlet

smoky flame resulting in incomplete combustion

flame light back
gas pressure less than burning velocity

gas pressure too low

gas pressure too low
insufficient air drawn in through primary port

gas pressure correct but flame lift-off due to incorrect burner design

flame lift-off
gas pressure greater than burning velocity

200 mm

manometer

turbulence created by enlargement of burner head, thus reduction in gas/ air velocity

correct gas pressure and stable flame

retention flame
(a small flame used on some burners to ensure the prevention of flame lift-off)

operation control shutter

primary air port

throat restrictor

inlet

burner head

cooker type ring burner

venturi

gauze (fitted to even out the flames & prevent lighting back)

main burner ports

injector

mixing tube

gas inlet

retention ports

primary air port

bar type burner used on central heating boilers

baffle

Aerated gas burners

fluctuations caused by several installations being used simultaneously affecting the efficiency of the appliance.

There are two basic types of governor, those which are fitted with a flexible spring loaded diaphragm and those which lift a metal disc. With the flexible diaphragm type when the gas exerts a force on the underside of the diaphragm it causes it to lift carrying with it the valve, thus reducing the gap through which the gas can pass to the appliance. Should the gas pressure drop the valve drops increasing the gap allowing more gas through, conversely should the pressure increase the diaphragm is lifted higher closing the valve gap even more. This continuous up and down movement of the valve ensures that a balance is maintained between the inlet and outlet to the governor. The outlet pressure from this type of governor should be checked using a MANOMETER and its pressure increased or decreased by adjusting the spring pressure screw as necessary, turning it clockwise to increase the pressure. When adjusting the outlet pressure from this governor the burner should be alight for 8–10 minutes to allow the expansion of the injector and for the pressure to stabilise.

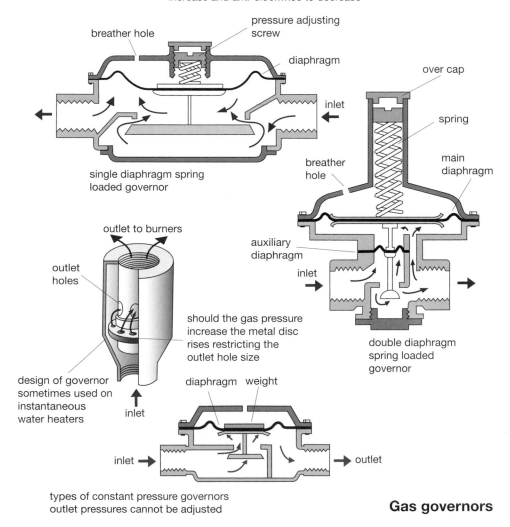

note: the outlet pressure from the spring loaded governor
is adjusted by turning the the adjusting screw clockwise to
increase and anti-clockwise to decrease

breather hole

pressure adjusting
screw

diaphragm

over cap

inlet

spring

single diaphragm spring
loaded governor

breather
hole

main
diaphragm

outlet to burners

auxiliary
diaphragm

outlet
holes

inlet

should the gas pressure
increase the metal disc
rises restricting the
outlet hole size

double diaphragm
spring loaded
governor

design of governor
sometimes used on
instantaneous
water heaters

inlet

diaphragm   weight

inlet

outlet

types of constant pressure governors
outlet pressures cannot be adjusted

Gas governors

The second type of governor is often called a constant pressure governor because the outlet pressure cannot be adjusted. This type of governor has a metal disc which can rise up and down its outlet pipe. This governor works by restricting the flow of gas through a series of small holes. Should the gas pressure increase the metal disc is lifted upwards which in turn restricts the outlet hole size (see sketch). This type of governor is used on some types of INSTANTANEOUS WATER HEATERS. See also REGULATOR.

GAS METER

A device which records on a counter or dial the flow of gas which passes through it. Gas meters are generally referred to as either primary or secondary gas meters. Primary gas meters are the property and responsibility of the gas supplier. Secondary meters are the responsibility of the owner or occupier and are used in separate parts of large buildings or property belonging to the same owner. When secondary meters are installed, their location must be indicated on a permanent notice located on or near the primary meter. The siting of any gas meters is subject to strict control with regard to fire precautions as laid down in the GAS SAFETY REGULATIONS but, in general, meters should be installed in dry, accessible positions which are well ventilated. Where a meter box is attached to or built into the external surface of an outside wall, the box should be constructed so as to ensure that any gas escaping within the box cannot enter the building or any cavity wall but must disperse to the external air. Meters installed in hospitals and larger buildings are usually fitted with a by-pass to ensure that at times when the meter may have to be replaced the gas supply is not shut down.

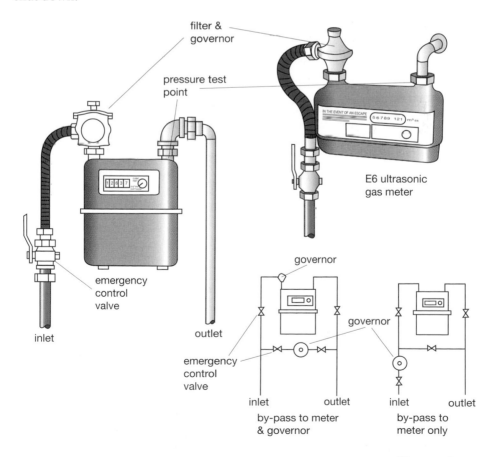

Gas meters

GAS MODULUS

A numerical figure that can be used to determine the gas pressure adjustment where one gas type is changed to another. The gas modulus is expressed as $\sqrt{\text{pressure}} \div$ WOBBE INDEX. Thus, for example, where an appliance burning natural gas with a Wobbe index of 53, supplied at a pressure of 20 mbar, is converted to burn LPG with a Wobbe index of 76 the following pressure adjustment would need to be made:

$$\frac{\sqrt{20}}{53} = 0.08 \text{ gas modulus}$$

Therefore, by transposition of the formula (gas modulus × Wobbe index)^2 = pressure:

$$(0.08 \times 76)^2 = 37 \text{ mbar revised gas pressure.}$$

See also FAMILY OF GASES.

GAS PLIERS   See PLIERS.

GAS PRESSURE

The pressure generated within a gas installation. This is generally registered in milli-bars and would typically be 21 mbar ± 2 mbar for a Natural gas installation and 37 mbar ± 5 mbar for a Propane (LPG) installation. However, there are several definitions in relation to gas pressures that further identify a specific pressure, these include among others the following:

- **Maximum Incidental Pressure (MIP)** The maximum pressure that could be experienced within a gas installation under fault conditions.
- **Maximum Operating Pressure (MOP)** The maximum pressure within a gas installation under normal operating conditions.
- **Operating Pressure** The general working pressure to be found within the gas installation, such as identified above – or as recommended by the manufacturer.
- **Burner Pressure** The pressure as recommended by the manufacturer.
- **Absorption Pressure** Term used to indicate the pressure loss across a gas installation.

GAS RATE (GAS CONSUMPTION; HEAT INPUT)

The amount of gas an appliance consumes over a period of time. All appliances must be adjusted to ensure that they do not consume more gas than is required, failure to do so could result in incomplete combustion of the fuel and the production of CARBON MONOXIDE gas. The method used to determine the gas rate depends upon the type of meter fitted. With older Imperial measurement meters, the method employed is as follows. Ignite the appliance and bring up to maximum heat input, then observe the time in seconds it takes the test dial on the meter to burn 1 ft^3 of gas. This period is then divided into the number of seconds in one hour to find the amount of gas used over this period. The heat input is then simply found by multiplying the answer by the CALORIFIC VALUE of the gas (1035 BTU/ft^3).

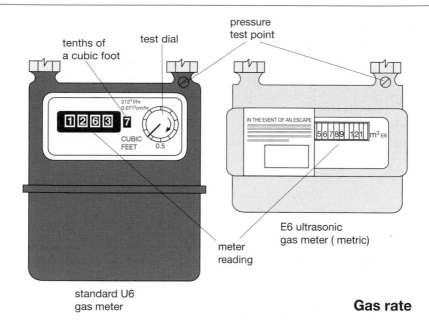

standard U6
gas meter

Gas rate

Example

Assume it takes 54 seconds for the test dial to complete 1 ft³ with the gas burning full rate.

Therefore $\dfrac{3600}{54} = 66.667 \text{ ft}^3 \times 1035 = \underline{69\ 000 \text{ BTU/h}}$

Note: to convert BTU to kW divide by 3412.

With the new E6 ultrasonic gas meters which measure in m³, the gas rate is found by observing the meter reading over a two-minute period. The initial reading is then deducted from the second reading to give the gas flow in m³ used over the two-minute test period. *A two-minute period is used to allow for some degree of accuracy.*

Example

Initial reading 00894.176; second reading 00894.237.

Therefore $894.237 - 894.176 = \underline{0.061 \text{ m}^3}$

Having found the gas flow over two minutes, the gas rate (heat input) is found by multiplying by 323.308 (*this figure is constant and based on the calorific value of gas being 38.8 MJ/m³*).

Therefore $0.061 \times 323.308 = \underline{19.72 \text{ kW}}$

The gas rate/heat input found is referred to as the 'gross input'. Should the net input be required, such as when identified by a manufacturer, the gross input is simply divided by 1.1.

GAS REFRIGERATOR

The gas refrigerator works on the principle that when liquid evaporates it extracts heat from the surrounding or ambient air within the refrigerator. The gas refrigerator works as follows:

(1) A solution of ammonia and water is heated in the small boiler; this causes an ammonia gas to be given off which rises up to the air-cooled condenser.
(2) Here the ammonia vapour cools and changes back into a liquid and it falls by gravity to the evaporator which is inside the refrigerator.
(3) Hydrogen gas present in the evaporator tends to lower the pressure sufficiently to allow the liquid ammonia to evaporate (when a substance evaporates, heat is drawn from its surroundings).
(4) Finally, the ammonia/hydrogen vapours pass to the absorber where they meet cooled water from the boiler, the ammonia changes back to a liquid and is reabsorbed into the water and passes back to the boiler for re-heating. The released and liberated hydrogen passes up through the absorber to return to the evaporator and so the cycle continues.

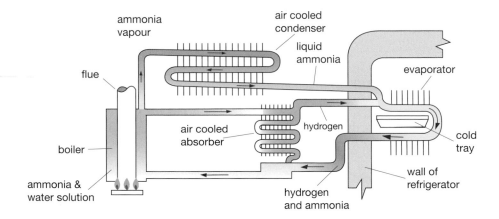

Gas refrigerator

GAS RISER
A gas pipe which rises up through a high rise building to the various floors. The gas pipe should be run up a specially designed duct which is ventilated to the outside of the building at both high and low levels. Where the pipe passes horizontally through the wall of the shaft the space around the pipe must be plugged to provide a fire stop. To gain access to the shaft, a half-hour fire-resistant panel should be fitted.

GAS SAFETY REGULATIONS
The statutory requirements which must be observed, by law, when carrying out any gas installation or repair work. See also BUILDING REGULATIONS and CORGI.

GAS SHIELDED ARC WELDING   See TUNGSTEN INERT GAS WELDING and METAL ARC GAS SHIELDED WELDING.

GAS STORAGE HEATERS
Water heated by gas for a stored supply of domestic hot water. It generally consists of either a GAS CIRCULATOR connected up to a hot storage vessel by primary flow and return pipes or a purpose made vessel with a gas burner installed directly underneath as shown in the sketch opposite. See also STORAGE WATER HEATER.

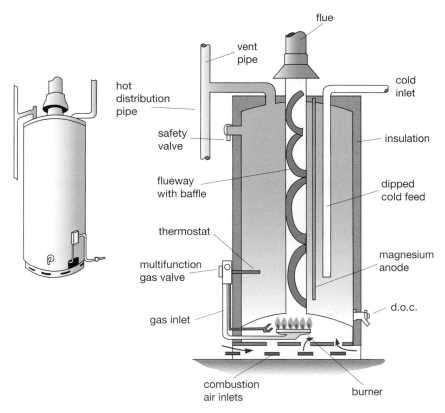

Gas storage heater

GASCOSEEKER
The trade name for a multi-purpose gas detector designed to undertake an assortment of gas checks to include tightness testing, gas in air sampling and purging operations.

GASKET
A piece of compressible material used to make the joint between two surfaces, for example, see FLANGE JOINT.

GASKIN   See YARN.

GATEVALVE   See FULLWAY GATEVALVE.

GAUGE
(1) An instrument which indicates pressure or temperature, etc., see BOURDON PRESSURE GAUGE for example.
(2) The thickness of a material. Since the introduction of the metric system in this country sheet material has been expressed in millimetres thickness, but occasionally the term standard wire gauge (swg) is used, this being an old British Imperial BS number

	Some standard wire gauge/metric equivalents				
swg	mm	swg	mm	swg	mm
16	1.6	20	0.9	24	0.56
17	1.4	21	0.8	25	0.5
18	1.2	22	0.7	26	0.45
19	1.0	23	0.6	27	0.4

Note the bigger the number, the smaller the sheet thickness.

(3) The exposed depth of a slate or tile which can be seen on a roof. This distance is the same as the distance between the centres of the roof slate battens. See also SOAKER.

GAUGE PRESSURE
The pressure in a system created by the HEAD of water. See ABSOLUTE PRESSURE. See also BOURDON PRESSURE GAUGE.

GCI TERMINAL   See FLUE TERMINAL.

GEAR PUMP   See PUMP.

GEYSER
The first design of INSTANTANEOUS GAS WATER HEATER invented by Benjamin Maughan in 1868. This invention proved very popular although in the early days geysers had no flue therefore were extremely dangerous.

GIMLET
A tool which looks like a corkscrew and is used for making holes in soft metal or used to form a starting hole in timber to assist the entry of a wood screw.

GLAND COCK
A plug cock in which the tapered plug is held in the body by means of a packing gland. For example, see PLUG COCK.

GLAND NUT
The nut which holds the packing into the gland of a valve and is designed to prevent the leakage of water or gas where a spindle may turn or ride up and down, for example, see BIB COCK.

GLAND NUT PLIERS   See PLIERS.

GLC TERMINAL   See FLUE TERMINAL.

GLOBE TAP
A screwdown valve with a horizontal female thread as its inlet. Globe taps were designed to be connected through the side wall of baths. Due to the fact that the spout outlet into the bath discharged water below the spillover level of the appliance they are prone to BACK-SIPHONAGE and therefore are no longer made or installed.

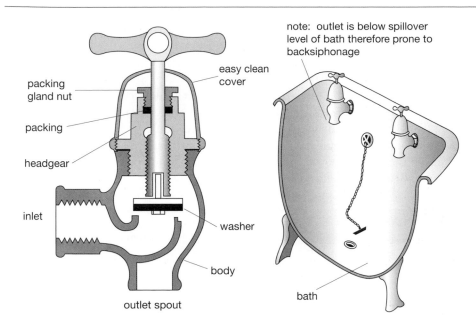

note: outlet is below spillover
level of bath therefore prone to
backsiphonage

easy clean
cover

packing
gland nut

packing

headgear

inlet

washer

body

outlet spout

bath

Globe tap

G

GLOBE VALVE   See DISC GLOBE VALVE.

GOOSE NECK (SWAN NECK)
A curve or bend which is purposely put into the COMMUNICATION PIPE to allow for
ground movement. Therefore should subsidence or shrinkage of the soil which surrounds the pipe occur the pipe would be allowed to straighten out and thus prevent
it being pulled out of its fitting. For example, see COLD WATER SUPPLY.

GOVERNOR
A device fitted in a pipeline to automatically control the pressure and flow of gas or
water. See GAS GOVERNOR, PRESSURE REDUCING VALVE and REGULATOR.

GRADIENT
An incline rising or falling by regular degrees. A gradient is usually expressed either
in degrees or one in x amount (for example 1 in 40). If the gradient is expressed in
degrees a protractor should be used but for small degrees a rule of thumb method
can be adopted which allows 18 mm for each metre run.

Example
Should a gradient of $2^{1}/_{2}°$ be required you would allow $2.5 \times 18 = 45$ therefore $2^{1}/_{2}°$
equals 45 mm per metre run (45 mm/m). Should the gradient be expressed as, say, 1
in 40, then you would divide 1 metre by 40

Example

$$1 \text{ in } 40 = \frac{1000}{40}$$
$$= 25 \text{ mm per metre } (25 \text{ mm/m})$$

To find the required gradient for drainage and discharge pipes see LOADING UNITS, CHEZY FORMULA or MAGUIRE'S RULE.

GRADIENT BOARD (INCIDENCE BOARD)

A plank of wood cut to the shape of the required GRADIENT of a drain, and used in conjunction with a spirit level when levelling in short branch drains. If a drain run is to have a fall of 1 in 40, the drop in depth over 1 metre would be (1 m ÷ 40 = 0.025 m) 25 mm. So to make a simple gradient board, take a straight plank 1 metre long and cut it at an angle as shown in the sketch. For long drain runs, to ensure accuracy SIGHT RAILS AND BONING RODS should be used.

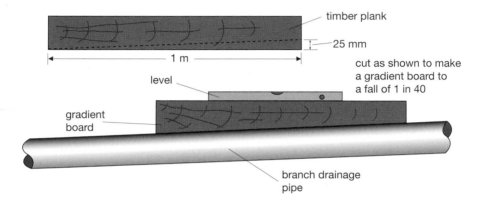

Gradient board

GRAPHITISATION

A peculiar form of corrosion to cast iron and steel pipes which sometimes happens as the result of anaerobic bacteria found in waterlogged sulphate bearing soils. The corrosion is identified by the pipe breaking down into a black graphitic form which is very soft. Therefore when laying FERROUS METAL pipes in this type of soil, precautions must be taken to protect them.

GRATED WASTE   See FLUSH GRATED WASTE FITTING.

GRAVITY CIRCULATION

The circulation of hot water caused by CONVECTION CURRENTS.

GRAVITY TANK

A cold water storage vessel sited at a high position to provide the required flow rate and pressure at the point of draw-off. Gravity tanks are sometimes used to supply water to SPRINKLER SYSTEMS.

GREASE TRAPS AND GULLIES

A grease trap is a device which houses a quantity of water and is designed so that should grease from canteen kitchens be discharged into the drains it would, when reaching the water, cool and solidify. Lighter solids would rise to the surface and collect in a solid cake, heavier solids would sink and collect in the GALVANISED tray.

Periodically the tray should be lifted out, removing all the grease which has accumulated. Some grease traps are designed to be fitted under sinks but in most cases the trap is fitted externally to the building.

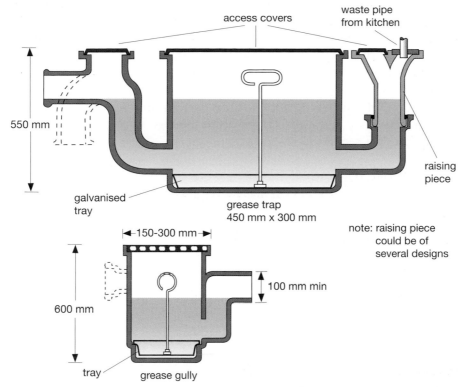

Grease traps and gullies

GREVAK RESEALING TRAP   See RESEALING TRAP.

GREY WATER
Water from baths, sinks and basins that has not been contaminated by faecal matter or urine. It gets its name from the fact it generally has a cloudy appearance, unlike that of fresh POTABLE WATER or water that is heavily polluted such as is found with BLACK WATER.

GRIDDLE
A gas or electric hot plate, that is often designed with a channel to drain off fat from cooking. Griddles are used to dry-fry food such as eggs, bacon and pancakes, etc. and are often found in fast-food restaurants.

GRILL   See VENTILATION GRILLE.

GRIT GULLY (DETRITUS GULLY)
A GULLY which has a deep (SUMP) section for the collection of any grit or silt, etc. See also SILT TRAP.

GROMMET (GRUMMET)

A small ring or washer of HEMP made by the plumber to assist in making a water tight joint on tank connectors. Nowadays, in most cases rubber or polythene washers can be used but the occasional need for a grommet exists such as when making connections with a LONGSCREW. It is made by rolling a few strands of hemp between the hands with TALLOW or a JOINTING COMPOUND to produce a string, it is then wound onto the fitting or is held in a circle to a size just bigger than required and interwoven to produce a hemp washer. By intertwining the hemp the washer, when finished, should be the correct size.

GROOVING TOOL (SEAMING TOOL/TINMANS GROOVE PUNCH)

A tool sometimes used to consolidate and define the edges of single and double lock welts. For example, see SINGLE LOCK WELT.

GROUP CONNECTOR

A branch fitting with two or more branches. See also MULTI-BRANCH FITTING.

GRP

Glass reinforced plastic which is a type of THERMO-SETTING PLASTIC.

GUIDE BRACKET

A loosely fitting pipe support which does not restrict any thermal movement of the pipe.

GULLY

A drainage fitting designed to receive surface and/or foul water from waste pipes. The purpose of a gully is to provide a trap preventing odours from the drain entering the atmosphere. All new works must discharge the water from appliances below the cover or grating and in many cases a back or side inlet gully is used to assist in this design. The old method of discharging the water into a chamber above the grating is no longer permissible. Trapless gullies are also available but should only be used for SURFACE WATER drains. See also YARD GULLY and GREASE TRAPS AND GULLIES.

(See illustration opposite)

GULLY GRAB

A GULLY cleaning tool which consists of a long handle with a hinged bowl and lever. The grab is lowered into a gully with the bowl open, the lever is pulled which closes the bowl grabbing hold of some grit from within the SUMP of the trap. For example, see DRAIN CLEANING.

GULLY RISER   See RAISING PIECE.

GUNMETAL (G METAL)

An ALLOY consisting mostly of copper and tin. Gunmetal is a variety of BRONZE, originally used for ordnance, and is used where resistance to wear or corrosion is required, for example steam pipe fittings.

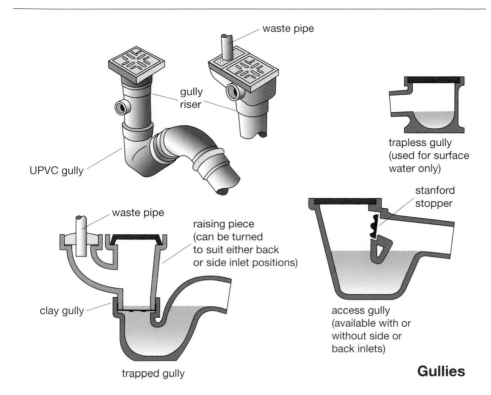

waste pipe

gully riser

UPVC gully

trapless gully
(used for surface
water only)

stanford
stopper

waste pipe

raising piece
(can be turned
to suit either back
or side inlet positions)

clay gully

access gully
(available with or
without side or
back inlets)

trapped gully

Gullies

GUNSTOCK LAP

Name sometimes used to identify a COVER FLASHING on a flat roof which overlaps another cover flashing a brick course lower. It is necessary to step cover-flashings down with the brickwork in order to facilitate the slope on supposedly flat roofs.

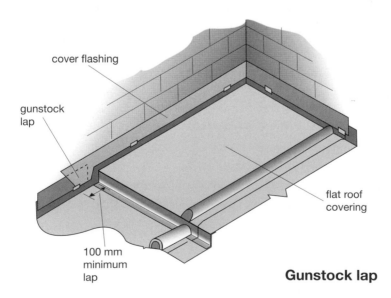

cover flashing

gunstock lap

flat roof covering

100 mm minimum lap

Gunstock lap

GUSSET

A piece of sheet material inserted into a sheet weathering detail to assist in forming an external corner.

GUTTER

A channel for collecting surface water. The design would vary depending upon the material used and its position on the roof. The gutter should be located at the lowest point of any roof. If the gutter is to be made on site using metallic roofing materials then allowances must be made for expansion. See COPPER, LEAD and ZINC SHEET AND ROOF COVERINGS. The gutter could take the form of a valley, or tapering valley positioned between two pitched roofs, or a parapet wall gutter having one side meeting a parapet wall, or it could be in the form of a long box, which would serve two flat roofs. At the lower end of the gutter either a catchpit or a rainwater chute should be sited to redirect the water down into the rainwater pipe. See also EAVES GUTTER.

G

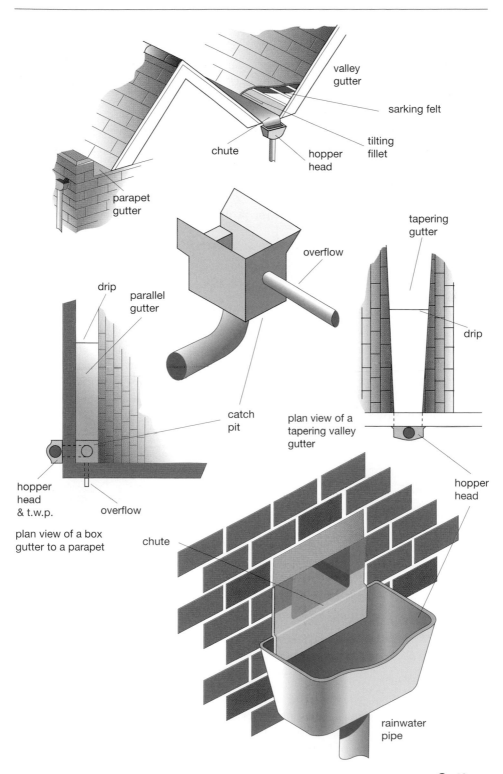

valley gutter

sarking felt

tilting fillet

chute

hopper head

parapet gutter

overflow

tapering gutter

drip

parallel gutter

drip

catch pit

plan view of a tapering valley gutter

hopper head & t.w.p.

overflow

plan view of a box gutter to a parapet

chute

hopper head

rainwater pipe

Gutters

H COWLING
A flue terminal designed to overcome the problems of excessive DOWN DRAUGHTS. For example, see FLUE TERMINAL.

HACKSAW
A tool used to cut metal. The hacksaw comes as a frame which holds blades which can be replaced as and when required. The blades used for hacksaws are supplied with 'x' amount of teeth for every 25 mm of the blade. For example, 18 teeth per 25 mm (a coarse blade) or 32 teeth per 25 mm (a fine blade). At least three teeth should be in contact with the section being cut but in general a fine blade is chosen when cutting thin material such as copper tube and a coarser blade when cutting material such as mild steel pipe. When fixing in a blade it is essential to face the teeth forward and only to apply pressure on the forward stroke. To obtain the maximum life from a blade, do not saw at too fast a speed as the blade overheats.

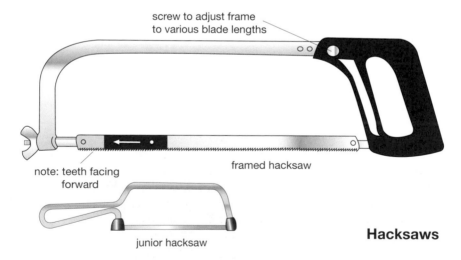

screw to adjust frame to various blade lengths

note: teeth facing forward

framed hacksaw

junior hacksaw

Hacksaws

HAIRFELT
A cheap form of thermal insulation material which is wrapped around pipes. Its main disadvantages are the fact that it is not fireproof and it makes a good bedding material for vermin and rodents.

HALF MOON WELD   See THUMB NAIL WELD.

HALK WHITE   See JOINTING COMPOUND.

HALL'S THIMBLE

A cover fitted to pipe sleeves which pass through walls. The thimble cover gives a better decorative finish and hides the space between the sleeve and the pipe passing through it. For example, see EXPANSION JOINT.

HAMMER

Tool consisting of a hardened and tempered steel head with a wooden handle usually ash or hickory. There are many types of hammers to choose from, the claw hammer proves useful when removing nails or floorboards, the cross and straight pein hammers often prove useful when forming sheet metal, the ball pein is used when securing rivets and the club hammer is invaluable when striking heavy blows.

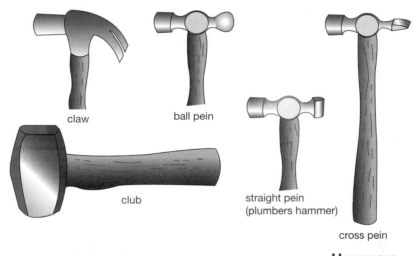

claw

ball pein

club

straight pein
(plumbers hammer)

cross pein

Hammers

HAND DUMMY   See DUMMY.

HARD SOLDERING

A CAPILLARY JOINT in which the solder used has a much higher melting temperature than the solder used for SOFT SOLDERING, therefore a very secure job is made, note the spigot and socket type joint does not need to be the same depth as that required for soft soldered joints. There are several types of hard solders including silver, or silver alloys with varying percentages of copper, or copper/phosphorus alloys (cupro-tected) which are cheaper to purchase. Hard soldered capillary joints are made in much the same way as soft soldered joints the main difference being the temperature required which ranges from 600°–850°C. See also BRAZED JOINT, SILVER SOLDERING and BRONZE WELDING.

HARD WATER

WATER which has fallen on and filtered through chalk and limestone, taking it in suspension. The dissolved solids in the water make the lathering of soap difficult and the more soap used causes a scum to form on the surface of the water and sanitary appliances. The water could be of TEMPORARY or PERMANENT HARDNESS, which if required could be removed by the use of a WATER SOFTENER.

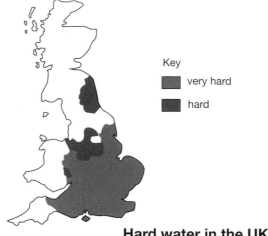

Hard water in the UK

Key
■ very hard
■ hard

HARDCORE
The old brick rubble, etc. used beneath a solid concrete floor.

HARDENING
The tempering of steel tools, see TEMPERING. See also TEMPER and WORK HARDENED.

HATCHET BOLT
A large soldering iron in which the copper head is at right angles to the handle, see SOLDERING IRON for example.

HAUNCHING
The concrete which slopes from the top of a drainage pipe down to the bedding. For example, see BEDDING.

HEAD
The height of water above a specified point. See WATER PRESSURE.

HEADER CISTERN (HEADER TANK)
A feed and expansion cistern, or storage cistern, this being the cistern fitted at the highest point and therefore creating a HEAD of water.

HEADER PIPE
A large pipe which holds a reserve quantity of drinking water in high rise buildings. See BOOSTED SYSTEM OF COLD WATER SUPPLY.

HEADER TANK
The AUXILIARY TANK in a SUPPLEMENTARY STORAGE SYSTEM of domestic hot water supply.

HEALTH AND SAFETY AT WORK ACT
An Act of Parliament which covers the whole spectrum of health and safety. Anyone found not observing the requirements contained in this Act and its subsequent

regulations will be liable on conviction to a fine. Serious breaches can lead to imprisonment. One of the main features of the act is to make employers, employees and anyone having control over premises where work is carried out, jointly responsible for site safety.

HEARTH

A solid non-combustible material a minimum of 125 mm thick upon which an oil or solid fuel free standing boiler stands. A hearth may also be required for gas appliances depending upon the design. Where a hearth has been installed under Current Building Regulations a hearth notice plate must also be installed. See FLUE.

HEAT EMITTER

Heaters used in hot water heating systems, there are two basic types, RADIATORS and CONVECTORS. Hot water is allowed to flow through these heaters and the temperature could then be controlled by radiator valves. See also RADIANT PANELS.

HEAT EXCHANGER

Heat exchangers can be found in a variety of places from localised hot water heaters, boilers and hot storage vessels, they are devices to transmit heat from one source to another. A heat exchanger could consist of a simple pipe coil. For an example see INDIRECT CYLINDER. See also PLATE HEAT EXCHANGER.

HEAT INPUT   See GAS RATE.

HEAT LEAK (HEAT SINK)

When using a solid fuel boiler which cannot be completely shut down by a thermostat a HEAT EMITTER or TOWEL RAIL should be run from the PRIMARY CIRCUIT. Any excess heat produced by the mass of fuel in the boiler can circulate through the heat emitter by CONVECTION CURRENTS, thus the excess heat can escape, hence the term heat leak. Failure to provide a heat leak may result in overheating and possible boiling of water in the system at times when cylinder thermostats or pumps prevent the flow of hot water from the boiler.

HEAT OF COMBUSTION   See CALORIFIC VALUE.

HEAT OUTPUT

A term used to indicate the heat available from an appliance to warm, for example, a central heating system. When fuel is supplied to an appliance it is consumed at a specific rate and is generally referred to as the 'heat input'. Owing to the EFFICIENCY of the appliance, some of this heat is lost, such as that lost up a chimney, therefore the heat output will be lower than the heat input.

HEAT PUMP

A device for transferring heat from one space to another. It is best described as a refrigerator working in reverse. A compressor delivers a vaporised refrigerant such as carbon dioxide, ammonia or freon gas to a condenser positioned in a room to be heated. The vaporised refrigerant is under a high pressure and temperature when it reaches the condenser. The cool air surrounding it condenses the vapour into a liquid and the air in the room is heated in the process. The liquid refrigerant then enters a valve where it expands and comes out as a liquid vapour mixture, at a lower temperature and pressure, it then enters the evaporator where the liquid is evaporated by contact with the comparatively warmer space outside the building. The vapour then returns to the compressor for re-cycling. The heat pump becomes a refrigerator if the evaporator is located in the ice box compartment. The condenser can be used to warm up air or water for central heating purposes.

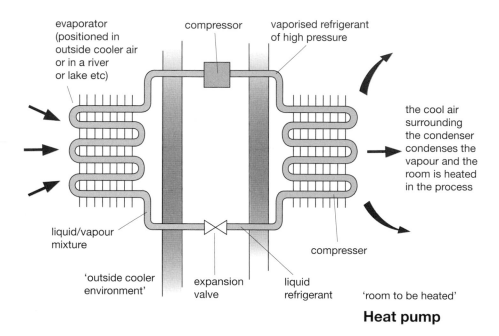

Heat pump

HEAT RECOVERY
Term used to indicate a method employed to obtain heat, from sources, such as through cooker extract hoods and light fittings, that would otherwise be discharged from the building.

HEAT RECOVERY TIME   See RECOVERY TIME.

HEAT SHRUNK JOINT   See SHRUNK RUBBER JOINT.

HEAT TRACING TAPE
A special heating element in the form of a wide wire which is fixed to pipes by special bands or adhesive tape. Should the temperature fall below freezing point the power supply can be switched on to the tape either manually or automatically via a FROST THERMOSTAT. The electric current flows through the wire which warms it up and consequently prevents the pipe from freezing. See also FROST PROTECTION.

HEATING CIRCUIT
The flow and return pipes to water-filled HEAT EMITTERS.

HEATING PANEL   See RADIANT PANEL.

HEMP

Strong fibres used to make GROMMETS and assist making joints to screwed pipework. Hemp is yellow/brown in colour and looks very much like horse hair but in fact it is obtained from dried crushed stalks of the hemp plant, which is also grown for its seed, used to make the drug cannabis. Hemp should not be used on steam pipework because the high temperatures tend to render it brittle and burn it out. See also JOINTING COMPOUND.

HEPSEAL

A type of flexible joint used on clay pipes, for example, see CLAYWARE PIPES.

HEPv0 VALVE   See SELF-SEALING VALVE.

HERRINGBONE FLASHING   See CONTINUOUS STEP FLASHING.

HERRINGBONE WELD

The style of a finished weld pattern. See LEAD WELDING.

HERTZ (Hz)

A term used to denote the frequency of an electrical supply. The frequency relates to the number of times a coil rotates between the magnetic poles of a generator in one second. In the case of a 230 V AC supply distributed by the electricity generating board the generator cycles at 50 times/second (i.e. 50 Hz).

HEXAGON NIPPLE

A nipple pipe fitting used on mild steel pipe in which a spanner can be fitted onto a nut between the two male threads. For example, see LOW CARBON STEEL PIPES AND FITTINGS.

HIDE-AWAY FLUSHING CISTERN

A WC FLUSHING CISTERN which is located behind a partition wall for reasons of appearance and to make it vandal proof. The flushing lever arm is extended through the partition wall and is terminated in the WC compartment by either a hand or foot flushing handle.

HIGH LEVEL FLUSHING CISTERN   See FLUSHING CISTERN and WC SUITE.

HIGH LIMIT THERMOSTAT

A thermostat set to break the electrical supply to an appliance at high temperature (e.g. 90–95°C) that is used as a back up/fail safe should the normal operating thermostat not function correctly. High limit stats, when they operate, shut down the appliance completely and require manual intervention to reinstate the appliance.

HIGH OUTPUT BACK BOILER

A solid fuel BACK BOILER with a much larger surface area than the smaller block type. See BOILER.

H

HIGH PRESSURE CUT-OFF DEVICE

A device to prevent high pressure entering a property that is operating on a lower pressure, i.e. after it has been regulated. See OVER PRESSURE/UNDER PRESSURE SHUT OFF.

HIGH PRESSURE HOT WATER HEATING SYSTEM

A hot water heating system which is not open to the atmosphere, thus the water being heated forms part of a CLOSED or SEALED SYSTEM. Water in such a system is capable of boiling at higher temperatures than 100°C. For example, see SEALED SYSTEM OF HEATING. Sometimes the term medium pressure installation is used to identify systems in which the water does not exceed 121°C, high pressure systems are those in which the water is heated up to 204°C. See also LOW PRESSURE HOT WATER HEATING SYSTEMS.

HIGH PRESSURE STAGE

Term usually relating to LPG in which the gas has not been regulated. It is the part from the cylinder valve to the regulator that in the case of propane would typically be 6–7 bar in pressure. See PIGTAIL.

HIGH PRESSURE STEAM HEATING

Before reading this entry, see STEAM HEATING. In large installations the pressure produced in low pressure steam heating systems is generally insufficient to pass along the long pipelines and remove the condensate from the STEAM MAIN, therefore it is sometimes necessary to supply water to the boiler under pressure. Once the condensate has been removed from the steam mains it is returned to the boiler house either by gravity or by pumping. If pumping is required the water from condensation is collected in a condensate receiver. When sufficient water has been collected it is pumped by either an electric pump or a mechanical pump in which the power is supplied from the steam in the mains itself, thus reducing its operating cost. At the boiler house the water is stored in a BOILER FEED TANK (often called a hot well tank) until required to make up the water in the boiler. The water level in this tank is also maintained by treated cold mains water from the feed cistern. As and when necessary, the water is pumped from the boiler feed tank into the boiler at a force greatly in excess of the boiler pressure to overcome the pressure of the steam generated.

(See illustration opposite)

HIGH TENSION LEAD (HT LEAD)

The wire found in an appliance that transmits the high voltage spark from its generator to the electrode. HT leads are the highly insulated wires from transformers.

HIGHER EXPLOSIVE LIMITS   See FLAMMABILITY LIMITS.

HIP

The intersection of two sloping roofs, for example, see BUILDING TERMINOLOGY.

HOLDERBAT

A pipe bracket designed to support a pipe clear of the structure. See PIPE SUPPORT.

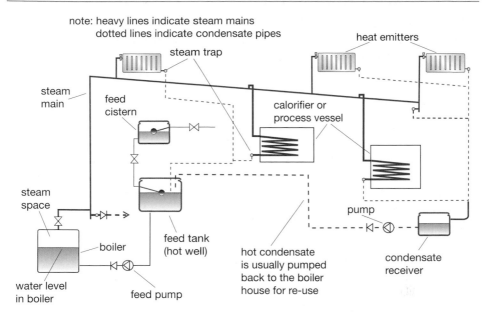

note: heavy lines indicate steam mains
dotted lines indicate condensate pipes

heat emitters

steam trap

steam main

feed cistern

calorifier or process vessel

steam space

pump

boiler

feed tank (hot well)

hot condensate is usually pumped back to the boiler house for re-use

condensate receiver

water level in boiler

feed pump

High pressure steam heating

HOLE CUTTERS AND HOLE SAWS

An assortment of tools used to cut holes of various sizes through a flat surface, e.g. the side walls of cisterns. The 'Cook's' tank cutter shown differs from the others in that, after an initial pilot hole has been drilled, the tank cutter is clamped as shown and is slowly rotated clockwise by hand, forcing the teeth into the metal and cutting the hole. With the others the pilot hole is drilled followed by the cutting blade.

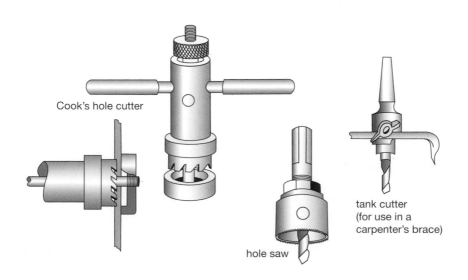

Cook's hole cutter

tank cutter (for use in a carpenter's brace)

hole saw

Hole cutters and saws

HOLLOW ROLL

An EXPANSION JOINT used on lead roofs. It should only be used on roofs with a steep pitch, as foot traffic would flatten the roll. The formation of the roll is made by turning the lead over a BENDING SPRING or wooden core which on completion is removed.

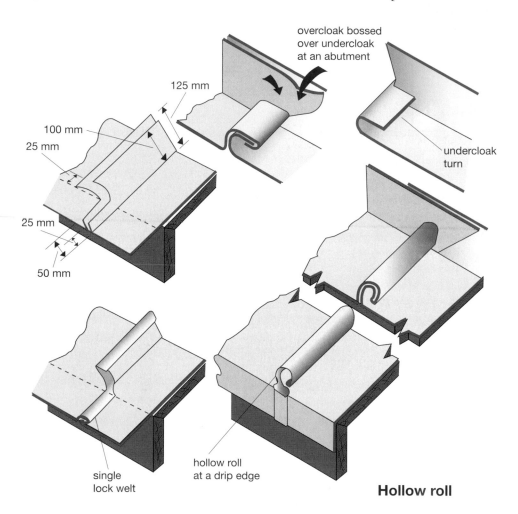

Hollow roll

HOPPER (GULLY INLET)

A funnel shaped drainage fitting used in conjunction with the trap of a GULLY. The hopper may have some branch inlets and is similar in use to a GULLY RAISING PIECE.

HOPPER FED BOILER

A boiler in which SOLID FUEL is used with graded fuel to suit the boiler, such as ANTHRACITE peas. The fuel is put into a large hopper and can slowly fall by gravity to the fire bed. To control the rate of burning an electric blower is used, which is operated by the boiler thermostat. For example, see BOILER.

HOPPER HEAD

A tapered box with an outlet connected to a waste DISCHARGE STACK. Hopper heads were used to receive WASTE WATER from first floor bathrooms but except for repair work this is no longer permitted due to its unsanitary design, see TWO PIPE SYSTEM OF ABOVE GROUND DRAINAGE.

HOPPER HEAD (RAINWATER HEAD)
A tapered box with an outlet connected to a rainwater pipe, designed to receive water from a gutter. For example, see GUTTER.

HORIZONTAL
A line running parallel to the Earth, e.g. going from left to right.

HORIZONTALLY FITTED CYLINDER
Sometimes due to lack of room it becomes necessary to install the domestic hot water cylinder in a horizontal position. This can simply be achieved as shown in the sketch. Note how, as in all HOT STORAGE VESSELS, the cold water enters into the bottom and the hot water is taken from the highest point. HEAT EXCHANGERS fitted to horizontal cylinders must be of the ANNULUS type with connections to either side of the vessel, otherwise an air lock would be caused. For the same reason a coil type heat exchanger cannot be fitted.

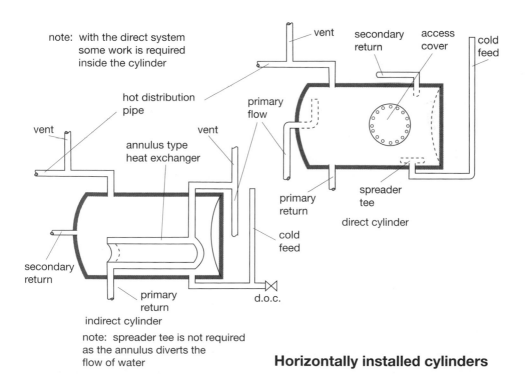

Horizontally installed cylinders

HORSESHOE WC SEAT
An open front WC seat.

HOSE PROTECTOR   See FLASHBACK ARRESTOR.

HOSE REEL INSTALLATION

A 'first aid' means of fire fighting. Hose reels should be located at prominent positions and adjacent to exits. The hose should be capable of reaching every part of the building being protected and in all cases be within 6 m of any fire. Two designs of hose reel will be encountered, either automatic or manually operated. In the case of a manually operated hose a notice must be provided next to the hose reel to indicate the need to turn on the water supply prior to running out the hose. The water supply to the hose is usually via a system of boosted cold water supply and must be capable of delivering a minimum flow of 0.4 litres per second.

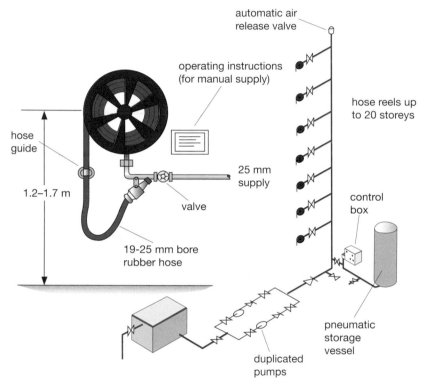

Hose reel installation

HOSE UNION

A fitting with a loose nut at one end for connecting onto a hose union tap and a serrated tail at the other end for insertion into a hose pipe. See BIB TAP.

HOSE UNION TAP

A bib or PILLAR TAP in which the spout or nozzle incorporates a male thread for the attachment of a HOSE UNION. See BIB TAP.

HOSPITAL APPLIANCES

Shown in the diagram is a selection of specialist SANITARY APPLIANCES which are sometimes found in hospitals, etc. for specific purposes. For a description of each appliance shown see under its own specific title.

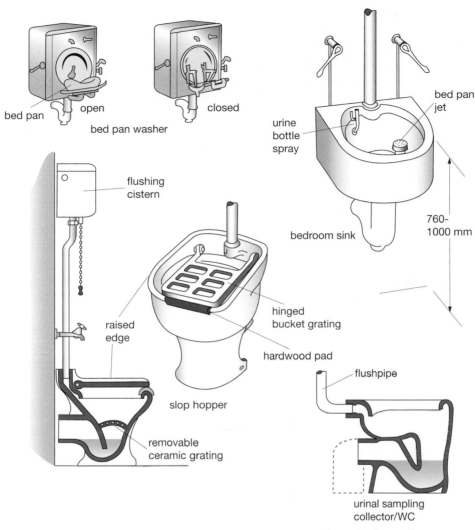

bed pan   open

bed pan washer   closed

urine bottle spray

bed pan jet

bedroom sink

760–1000 mm

flushing cistern

raised edge

hinged bucket grating

hardwood pad

slop hopper

removable ceramic grating

flushpipe

urinal sampling collector/WC

Hospital appliances

H

HOSPITAL RADIATOR   See RADIATOR.

HOT STORAGE VESSEL
A cylindrical or rectangular vessel used to hold a stored supply of hot water. It could be of a direct or indirect type depending on whether or not a HEAT EXCHANGER is fitted. The capacity of a hot storage vessel would depend upon the maximum anticipated demand of hot water required over a period of time, and in any case should not be less than 100 litres.

HOT WATER HEATING SYSTEM
A means of heating up a building by circulating hot water through HEAT EMITTERS. Originally systems were designed to circulate the hot water by CONVECTION CURRENTS but because these systems had a very low CIRCULATING PRESSURE larger pipe sizes were required. More modern systems use a pump to speed up this circulation. By

having a faster velocity, smaller pipe sizes could be used therefore leading to smaller volumes of water in the system, thus giving a quicker heating up time. The water which is heated in the boiler is often also used to circulate, either by pump, or by convection currents, to an indirect hot water storage vessel, giving a domestic hot water supply. It must be noted that the water used in the central heating system should never be allowed to mix with the water used for the domestic hot water supply otherwise the gases (oxygen, carbon dioxide, etc.) which are continually being brought in with the fresh water would soon lead to corrosion in the radiators which are made of steel. To overcome many corrosion problems in the PRIMARY CIRCUIT it is always advisable to add to the system a CORROSION INHIBITOR.

There are basically two main designs of heating systems, the ONE PIPE SYSTEM and the TWO PIPE SYSTEM. It must be stressed that there are many variations to these systems. There are systems which work under the influence of atmospheric pressure using a FEED AND EXPANSION CISTERN, being a vented supply, or systems which work under a higher pressure, known as SEALED SYSTEMS, being unvented supplies. There are systems which use very small pipes, as small as 6 mm in diameter known as MICRO-BORE or systems which are fully pumped, pumping water to the hot storage vessel and heat emitters. The system will incorporate a ROOM THERMOSTAT, MOTORISED VALVES, PROGRAMMERS, etc. but whichever control or system variation is used, it is important to remember that most are either one or two pipe design.

(See illustration opposite)

HOT WATER SUPPLY
A means of heating water for domestic purposes. The method of heating the water could be either a CENTRALISED or LOCALISED HOT WATER SUPPLY. See also TEMPERATURE CONTROL FOR DOMESTIC HOT WATER.

HOT WELL TANK   See HIGH PRESSURE STEAM HEATING.

HOUSEMAID'S CLOSET   See SLOP HOPPER.

HT LEAD   See HIGH TENSION LEAD.

HUMIDIFICATION
An air conditioning unit that is designed to adjust the amount of relative humidity (RH) within a room. Relative humidity is a measure of the percentage of water vapour within a room. If there were no moisture, the RH would be 0% whereas the RH for air that is saturated is 100%. Generally, rooms need to be between 40 and 70% RH to be comfortable. Should the RH be less than 40%, one would experience a dry throat and eyes. Where the RH increases above 70%, discomfort is experienced by way of clamminess and overheating, this is due to the fact that the body's normal method of cooling, via sweating, is ineffective. Also, a high prolonged RH of above 80% tends to lead to mould growth.

HUMIDIFIER
An air-handling unit that increases or decreases the amount of relative humidity within a room. See HUMIDIFICATION.

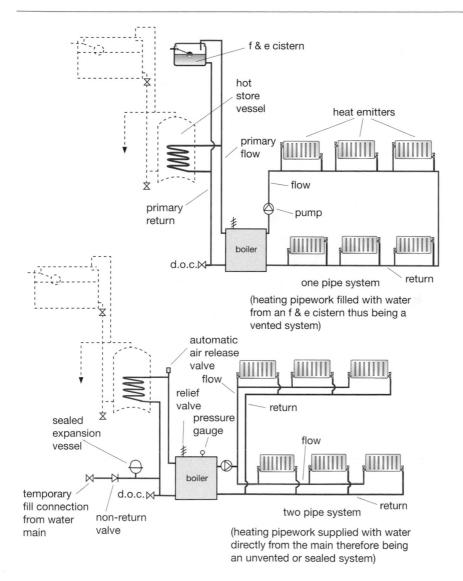

Hot water heating systems

HUMIDITY

A term used to describe the water vapour content present in the air. This vapour is the result of evaporation of water. The higher the air temperature, the more water the air will hold, when it can hold no more, it is said to be saturated. If the temperature of the air drops, some of the vapour will condense back to liquid water and form on cooler surfaces such as windows and pipes.

HYDRANT

A valve from which water can be discharged in large quantities and at high pressures. It is designed with a coupling for the attachment of a STAND PIPE or fire brigade hose. See WET and DRY RISER.

HYDRAULIC FORCE PUMP

A hand pump, similar to a bicycle pump, with a large rubber cone on one end. It is used to force a pressure down a waste pipe in order to remove a blockage. See also SUCTION PLUNGER.

Hydraulic force pump

HYDRAULIC GRADIENT

The loss of water head experienced when water is flowing, as against the STATIC HEAD when water is not flowing. By referral to illustration 'A' it can be seen that when water is at a state of rest the water level is at the same height for each column. However, in picture 'B' where the water is flowing the water level drops steadily in each column.

When installing a system of hot or cold water supply this design concept sometimes needs to be considered, especially where branch connections are made in high sections of pipework. Failure to observe this may lead to an ineffective system or intermittent supplies at the appliance.

(See illustration opposite)

HYDRAULIC JUMP

A situation where water flowing through a DISCHARGE PIPE has formed a solid plug of water, which can lead to problems of self- or induced siphonage. For example, see TRAP SEAL LOSS.

HYDRAULIC MEAN DEPTH

A factor which is used to assist in calculating the rate of flow of liquid along a pipe or channel. The hydraulic mean depth is found by dividing the cross-sectional area of a liquid by the length of the wetted perimeter of the pipe. See CHEZY FORMULA.

HYDRAULIC POWER

Power supplied by a device which pumps oil, which is incompressible, from one chamber to another, thus exerting a pressure against a piston which is forced outwards. Hydraulic power is used to apply force on all kinds of machinery from bulldozers to bending machines used by the plumber. The principle of hydraulic power can be explained as follows (see sketch).

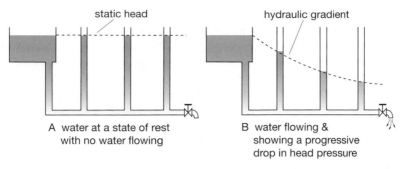

A water at a state of rest
with no water flowing

B water flowing &
showing a progressive
drop in head pressure

Hydraulic gradient

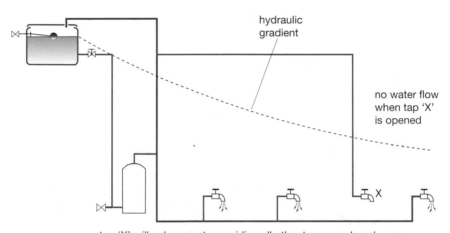

tap 'X' will only operate providing all other taps are closed

System giving rise to no water flow

(1) When the plunger handle is lifted, oil is sucked from chamber A through a NON-RETURN VALVE and held in the plunger chamber.
(2) As the handle is lowered, the oil is forced out and can only go into chamber B through another non-return valve.
(3) The process is continued and as chamber A has its oil removed, chamber B has more thus forcing the ram outwards.
(4) To relieve the pressure, a valve is opened which by-passes the two non-return valves and the spring pushes the piston back, which simultaneously forces the oil from chamber B to A.

(*See illustration over.*)

HYDRAULIC PRESS BENDER

A bending machine which works on the principle of HYDRAULIC POWER. It is used to bend heavy duty pipe such as LOW CARBON STEEL. To use a hydraulic press bender, a FORMER is selected and fitted onto the end of the ram. The stops are positioned into the correct pin holes as marked on the machine, then the pipe is inserted into the former and when the pump is operated the ram moves forward forcing the pipe

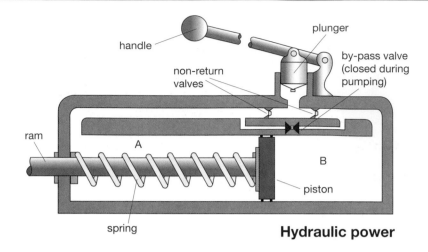

Hydraulic power

against the stops, as the pumping continues the pipe is forced to bend. Unlike machines used on light gauge tube, no BACKGUIDE is required to support the pipe, for these machines see BENDING MACHINE. To bend pipe accurately on the hydraulic press bending machine see PIPE BENDING TECHNIQUE (heavy gauge tube).

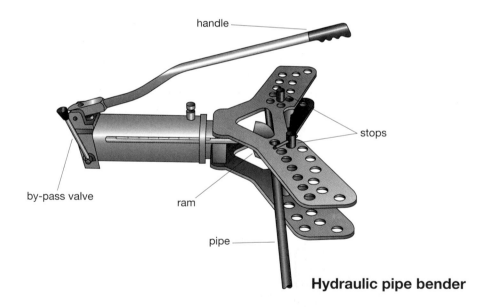

Hydraulic pipe bender

HYDRAULIC PRESSURE TEST

A pressure test carried out on sections of completed hot and cold water pipework to test for leaks which might prove hard to locate and repair. The test is carried out as the work progresses on pipework such as under floors and in pipe ducts. Testing is simply carried out using a special FORCE PUMP as shown; water is sucked from the container and forced into the installation compressing the water and air within the system. The work should be tested to $1\frac{1}{2}$–2 times the WORKING PRESSURE of the completed installation and should hold for at least thirty minutes. A different type of hydraulic test can be carried out on below ground drainage systems, for this, see WATER TEST.

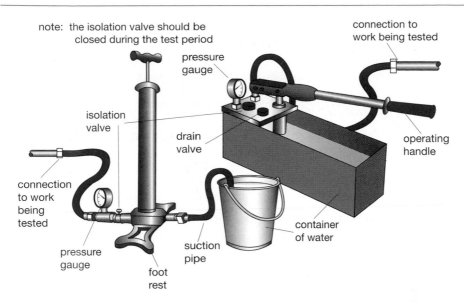

Force pumps used for hydraulic pressure testing

HYDRAULIC RAM

An automatic pumping appliance which utilises the energy of flowing water to pump it to great heights. The flowing water gains enough momentum to lift and close the pulse valve. When the pulse valve is lifted it causes a pressure wave to be set up which lifts the NON-RETURN VALVE, thus some water passes to the delivery pipe and the air vessel. When the pressure is expended the non-return valve closes and the water flows to waste until the process is repeated. See also FORCE PUMP.

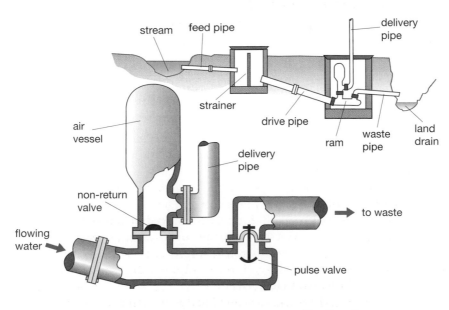

Hydraulic ram

HYDRAULICALLY OPERATED VALVE

An automatic on–off valve fitted in a pipeline to control the flow of water to an appliance. The most common use of such a valve would be to control the flow of water to an AUTOMATIC FLUSHING CISTERN. The type shown works as follows:

(1) When water is supplied to the valve its pressure holds the valve shut down onto its seating, water can flow through the filter and enter the bellows chamber compressing the bellows and exerting an equal pressure to the underside of the valve. The valve is now only held shut by the valve spring.

(2) Should the pressure drop on the inlet side of the valve, such as may be caused by the opening of a wash basin tap, the pressure exerted by the bellows will be greater than that of the spring, thus the valve is pushed open and stays open until the water flows back through the filter to build up a pressure on top of the valve reclosing it, until the process is repeated.

(3) The length of time for which the valve stays open is controlled by the adjustment of a restrictor screw which alters the size of the waterway into the bellows chamber.

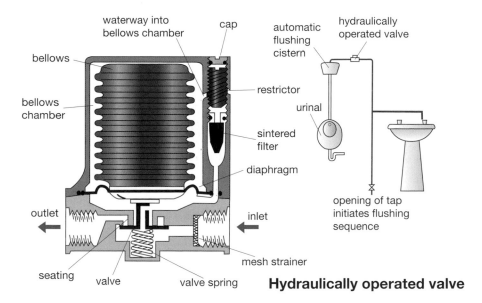

Hydraulically operated valve

The advantage of fitting this valve is that should the building be unoccupied, at weekends, etc., and therefore no taps are opened, the pressure within the pipe remains constant so the valve stays closed, saving water.

HYDROCHLORIC ACID

An acid comprised of hydrogen chloride gas dissolved in water. It is used by the plumber to assist in CHEMICAL PIPE CLEANING or to assist in the making of ZINC CHLORIDE.

HYDROPNEUMATIC ACCUMULATOR

A vessel containing a flexible bag which is filled with nitrogen. The purpose of an accumulator is to take up shock waves in water created by pressure surges in the pipeline. Hydropneumatic accumulators are sometimes installed in large buildings or a block of dwellings with the same main to avoid WATER HAMMER problems. Nitrogen is used in the bag because it is capable of being compressed if subjected to a greater pressure. See also AIR VESSEL.

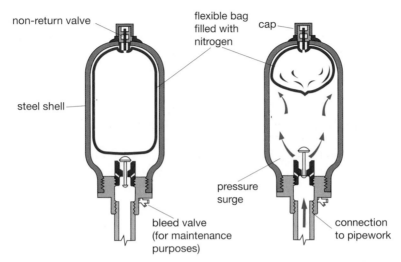

Hydropneumatic accumulator

HYDROSTATIC

Term which relates to the pressures exerted by a liquid when at rest. A liquid exerts a pressure due to its weight 'vertically'; in other words the higher the column of a liquid the greater will be the pressure exerted at its base over a specified area. The pressure caused by a liquid is transmitted equally and undiminished upon all surfaces on which the liquid rests and the free surface of the liquid will always be horizontal as a result. See WATER PRESSURE.

HYDROSTATIC PRESSURE

The pressure exerted by the weight of water in a plumbing system. See WATER PRESSURE.

I

ICE
Water in solid form. Water changes to ice at a temperature of 0°C with an immediate expansion of about $^1/_{10}$ its volume. See also PHYSICAL CHANGE.

IDENTITY BADGE See DATA BADGE.

IGNITION DEVICES
There are several types of ignition devices to light different appliances and the method of ignition depends upon the fuel used. Many gas appliances employ the use of a pilot flame which is kept alight at all times to ignite the gas at the main burner. The pilot flame could be lighted manually with a lighted taper or a special ignition device could be employed. There are two basic types of ignition devices, those which generate a spark and those which heat a filament.

The spark ignition type works by causing a spark to jump between two ELECTRODES, the power to generate the spark being supplied by a mains power supply via a TRANSFORMER or by a battery. Another method sometimes employed to generate a spark is to use a piezoelectric device, which differs from the type just described by generating its own electrical charge. This is achieved by applying stress or pressure on certain non-conducting crystals such as quartz. The filament type of ignition device is also powered by either mains or battery, but this time the electrical charge is allowed to pass through a coil of wire. This causes a resistance to the flow of electricity and as a result the coil will glow red hot, just like an electric fire.

IGNITION TEMPERATURE
The temperature which a combustible material must reach before it will burn, for example wood has to be heated to approximately 200°C before it will ignite, and natural gas to 704°C. See also FLASHPOINT.

IMMERSION HEATER See ELECTRIC WATER HEATING.

IMMERSION HEATER BOSS
The 2$^1/_4$″ BSP connection for an IMMERSION HEATER, usually fitted to HOT STORAGE VESSELS. Should the connection be missing a BOSS can be fitted as shown in the sketch.

(See illustration opposite)

IMMERSION HEATER SPANNER
A large spanner designed specially to fit IMMERSION HEATERS.

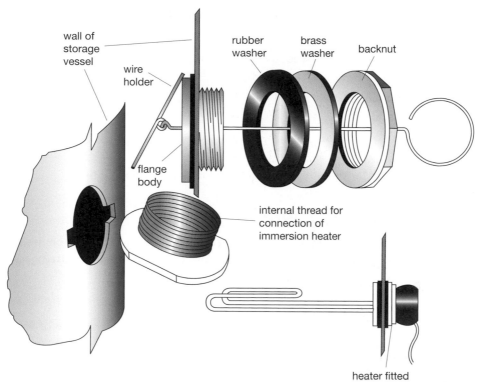

Immersion heater boss

note: these bosses are available in a range
of sizes and are designed to fit either
flat or curved surfaces

Immersion heater spanner

IMPERMEABLE FACTOR

A factor used in designing the volume of flow, in litres per second, of SURFACE WATER through drains. A drain which receives surface water must be large enough to carry away the maximum rainfall, with an allowance for that which soaks into the surface. For calculation purposes a rainfall intensity of 0.05 metres per hour is usually allowed for.

Surface	Impermeable factor
Roofs	0.8–0.95
Asphalt pavements	0.8–0.9
Jointed stone pavements	0.8–0.85
Tarmacadam roads	0.3–0.5
Lawns	0.05–0.25
Wooded areas	0.01–0.2

In areas where water drains readily away into the ground the lower number is chosen.

The volume of flow is expressed as Q and is found using the formula

$$Q = \frac{A(\text{m}^2) \times RI(0.05 \text{ m/h}) \times IF}{\text{seconds in one hour}} \times 1000$$

Q = volume of flow
A = area to be drained
RI = rainfall intensity
IF = impermeable factor

Example
Find the volume of flow in litres per second of a roof measuring 72 m².

$$\frac{72 \text{ m}^2 \times 0.05 \text{ m/h} \times 0.9}{3600} \times 1000 = 0.9 \text{ l/s}$$

Knowing the volume of flow in litres per second, the information can be used to transpose the formula from $Q = VA \times 1000$, found in the section on CHEZY FORMULA, to $Q/V \times 1000 = A$ thus giving the diameter of the pipe.

IMPERVIOUS
Any material or rock strata which will not allow admittance or passage of water through it. See WATER CYCLE.

IMPLOSION
The collapse of a hot water cylinder caused by a reduction of pressure inside the cylinder. Such a situation could occur where the cold feed and vent pipes of an open vented hot water system become blocked due to a plug of ice forming within the pipework. As the water in the cylinder cools or is drawn off, it leaves a partial vacuum, i.e. less than atmospheric pressure and therefore is crushed.

IMPULSIVE NOISE   See WATER HAMMER

INACTIVE FLUX (NON-ACTIVE OR SAFE FLUX) See FLUX.

INCANDESCENT
The release of thermal radiation in the form of a bright light from a body due to its temperature created by being heated.

INCIDENCE BOARD   See GRADIENT BOARD.

INCLINE
A slope going up or down, see GRADIENT.

INCLINED LEAD WELDING
Two different techniques used in lead welding. One is a true incline joint in which the seams to be welded are either butted or lapped together and welding of the sheet is carried out by depositing the lead up the incline in the form of overlapping semi-circular beads. The second type of incline joint is really a vertical welded joint in which the two adjoining sheets overlap one another at an angle; lead welding

vertically in this fashion tends to be easier and stronger than a true vertical lead weld. See also LEAD WELDING.

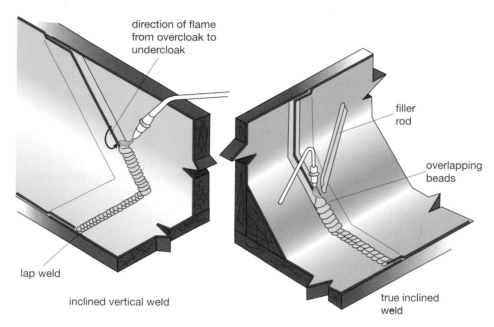

Inclined lead welding

INCLINED MANOMETER (DRAUGHT GAUGE)

A device used to measure the draught within the flue system from a solid fuel or oil burning appliance. For example, see COMBUSTION EFFICIENCY TESTING.

INCOMPLETE COMBUSTION

A situation in which fuel being burnt is not fully converted to the final products of combustion such as carbon dioxide (CO_2) and water vapour (H_2O). Should the process of combustion be interrupted, such as by cooling the flame or starving it of oxygen, products such as smoke and CARBON MONOXIDE will be produced, resulting in a potentially dangerous situation and a waste of fuel.

INDEX CIRCUIT

The distance between the boiler and the furthest HEAT EMITTER, or the heat emitter which offers the most resistance to the flow of water, and back to the boiler again. If the pipework is sized correctly and the pump adjusted to ensure a good circulation of water through the index heat emitter an adequate circulation to the remaining heaters is assured.

INDIRECT CYLINDER

A HOT STORAGE VESSEL used with an INDIRECT SYSTEM OF HOT WATER SUPPLY. It is a cylinder fitted with a HEAT EXCHANGER, the most common types being of either a coil or an annulus design. See also DIRECT CYLINDER and HORIZONTALLY FITTED CYLINDER.

(See illustration over.)

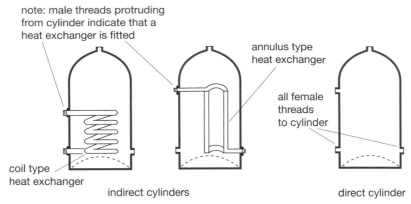

note: male threads protruding from cylinder indicate that a heat exchanger is fitted

annulus type heat exchanger

all female threads to cylinder

coil type heat exchanger

indirect cylinders

direct cylinder

Hot storage cylinders

INDIRECT SYSTEM OF COLD WATER SUPPLY (STORED COLD SUPPLY)

A system of cold water supply where all the cold water in the building is fed from a STORAGE CISTERN, except one draw off point which would be in the kitchen for drinking purposes. Therefore the cold water 'indirectly' reaches the taps via the storage cistern. With this type of system, DRAIN OFF COCKS must be fitted at all low points. See also DIRECT SYSTEM OF COLD WATER SUPPLY.

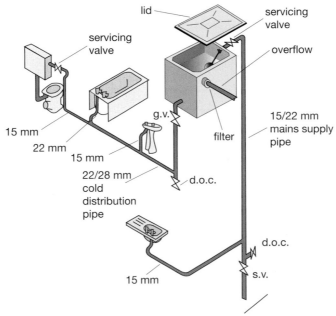

lid

servicing valve

servicing valve

overflow

15 mm

22 mm

15 mm

g.v.

filter

15/22 mm mains supply pipe

22/28 mm cold distribution pipe

d.o.c.

d.o.c.

15 mm

s.v.

Indirect system of cold water supply

INDIRECT SYSTEM OF HOT WATER SUPPLY

A method of CENTRALISED HOT WATER SUPPLY in which the water drawn off at the appliances has been heated in the hot storage vessel by means of a HEAT EXCHANGER. The boiler is supplied from a different supply of water from that in the hot storage vessel and the two systems are separate. The water is heated in the boiler and is then allowed to circulate through a coil or annulus cylinder within the hot storage vessel thus indirectly heating the water. See also SINGLE FEED HOT WATER SUPPLY and DIRECT SYSTEM OF HOT WATER SUPPLY.

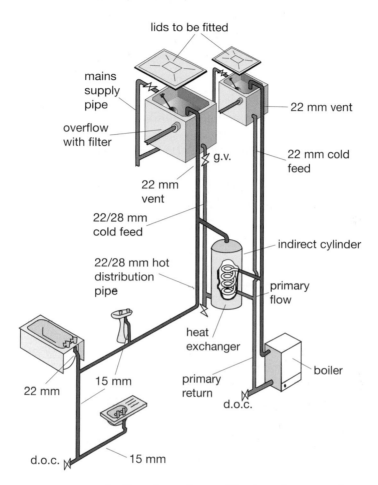

Indirect system of hot water supply

INDUCED SIPHONAGE   See TRAP SEAL LOSS.

INERT GAS   See NOBLE GASES.

INFILTRATION
The unintended leakage of ground SURFACE WATER into a drain or sewer pipe. See also EXFILTRATION.

INFLATABLE STOPPER (TESTING BAG)
A canvas bag which can be inserted into a drain or discharge pipe usually to facilitate testing. Once the bag has been inserted into the pipe it is pumped up sealing off the end. These stoppers prove very handy in awkward positions. For example, see DRAIN PLUG.

INFRARED
Electromagnet radiation beyond the wavelength of visible light, but shorter than radio waves. The name basically means below red. Infrared sensors are often found as motion sensors in public toilets to permit the operation of a flushing cistern. Sometimes they are used to automatically operate a tap as someone holds their hands in close proximity to the sensor. They work by sensing low emissions of heat given off by the human body.

INHIBITOR
A substance that is sometimes added to a wet central heating system to help overcome the problems of corrosion and noise. Inhibitors tend to neutralise the water content and coat the insides of pipes and vessels, etc. to assist in the prevention of corrosion. They also aid in the lubrication of pumps and circulators.

INJECTOR
A device that limits the volume or flow of fluid through a pipe. Examples would include a gas injector and found in the burner assembly or as used in an INJECTOR TEE.

INJECTOR TEE
A tee fitting which is designed to assist the suction of air or water from its branch connection. The fitting is designed to form a vortex, drawing the liquid or gas into the main flow pipe.

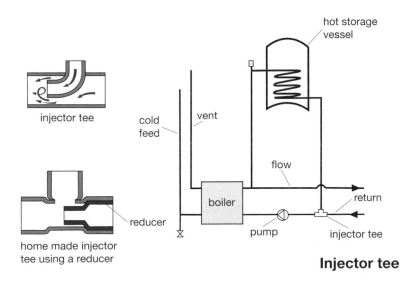

injector tee

home made injector tee using a reducer

reducer

hot storage vessel

cold feed

vent

flow

boiler

pump

return

injector tee

Injector tee

INLET HORN
The socketed projection formed on WC PANS or SLOP HOPPERS for the connection of a flush pipe. See FLUSHPIPE CONNECTOR, see also VENT HORN.

INODOROUS FELT
A brown hairy textured felt about 3 mm thick used as an underlay for sheet weathering materials. Contrary to its name, it smells awful. See UNDERLAY.

INSERT

A piece of copper pipe which is inserted into the end of plastic pipe and is used to give strength to the tube when making a COMPRESSION JOINT on plastic pipe subjected to high pressures.

INSET FIRE

Term relating to a gas fire such as a live fuel effect fire that is fitted into the BUILDER'S OPENING rather than against it. The fire includes a heat exchanger to direct all hot air flow into the room.

INSET LIVE FUEL EFFECT FIRE (ILFE)

A gas appliance designed to simulate the effect of burning SOLID FUEL. It is designed so that the flue products pass to the flue via a restricted opening thereby improving the efficiencies over that of a DECORATIVE EFFECT FIRE (DFE).

INSET SEAT (INSERT OR INSERTA PAD)

Two pads of IMPERVIOUS material fixed onto the top of a WC pan used in schools and factories as an alternative to the normal WC seat. For example, see WC SEAT.

INSPECTION CHAMBER

A chamber constructed over a drain or sewer. Fitted with a cover to provide access for inspecting, testing and clearing pipes should they become blocked.

(See illustration over.)

INSPECTION CHAMBER (CAST IRON)

A cast iron drainage fitting which provides a bolted on access cover to gain access into the drain or sewer pipe. Cast iron inspection chambers prove useful when running cast iron under buildings, or as a suspended drainage system in a basement.

INSPECTION COVER

A cover which is fitted to an INSPECTION CHAMBER. The cover may simply be placed onto a frame and held in position by its weight alone, or it may be bolted down. Some inspection covers are sealed with grease or rubber seals forming an airtight and watertight joint; if an inspection cover is used internally, it should be of this type, preventing odours seeping through the joint into the buildings.

INSPECTION EYE   See ACCESS COVER.

INSPECTION GULLY   See ACCESS GULLY.

INSPECTION OF DRAINS

Occasionally drains need to be inspected to look at and test their condition. To internally look at the condition of a drain can be difficult although using a mirror and torch often proves a success; it is possible to pull a special video camera through the pipe using a cord, this will show any fault and locate its approximate distance from the position where it was inserted. Other tests can be carried out to check the condition of a drain see TESTING OF DRAINS. See also ALIGNMENT TEST.

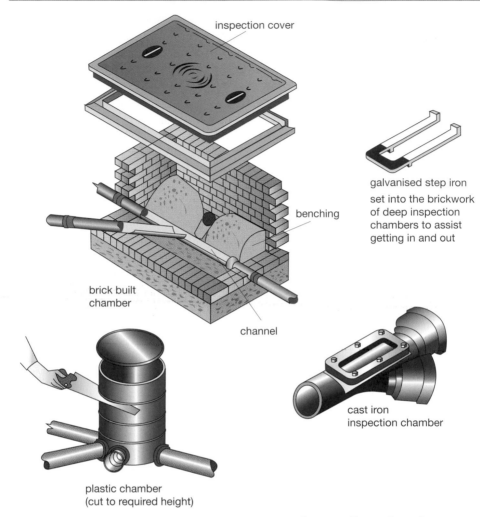

galvanised step iron
set into the brickwork
of deep inspection
chambers to assist
getting in and out

Inspection chambers

INSTANTANEOUS SHOWERS

A shower in which the water is heated locally as and when required. The method of heating the water can be either by gas or electricity. One of the major advantages of instantaneous showers, apart from the fact that fuel is used only to heat water for each shower, is that they can be used on mains pressure; this overcomes the problem of trying to achieve a minimum HEAD of water (see SHOWER). One of their biggest drawbacks is that a localised hot water heater is required and this often proves unsightly. With electrically heated showers the power cable used should be run directly from the consumer unit fuse box. Showers are rated from 4–7 kW and for best results the one with the highest rating should be chosen. When installing an instantaneous shower it is essential that all works comply with the relevant regulations in particular the Institute of Electrical Engineers (IEE) wiring regulations.

INSTANTANEOUS WATER HEATERS

Appliances fuelled either by gas or electricity and used as a LOCALISED HOT WATER SUPPLY and heat water only when it is required. There are two basic types, single-point and multi-point. Single point heaters serve only one draw off point and are usually fitted with a swivel spout, whereas multi-point heaters serve several draw off points. See also INSTANTANEOUS SHOWERS and STORAGE HEATERS.

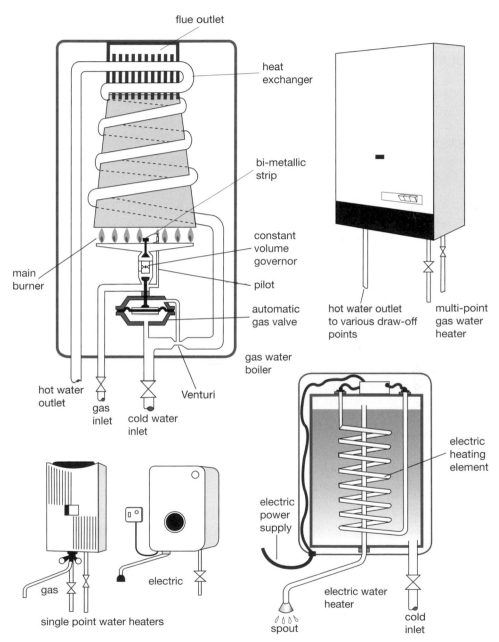

Instantaneous water heaters

INSTITUTE OF PLUMBING AND HEATING ENGINEERS (IPHE)

An independent organisation concerned with all aspects of the plumbing industry. It is concerned with the standards of good plumbing work and sees its prime role as the identification and promotion of competence, good standards and skilled professional work for the benefit of the public.

INSULATION
The application of a material to prevent the transmission of heat, electricity and sound. See THERMAL INSULATION.

INSULATOR   See CONDUCTOR.

INTEGRAL OVERFLOW
An overflow pipe which is built into the appliance, for example, see WASTE FITTING.

INTENSITY OF PRESSURE   See WATER PRESSURE.

INTERCEPTING TRAP (DISCONNECTING TRAP)
Usually found in older properties in the INSPECTION CHAMBER nearest the main sewer, often called a disconnecting trap because they disconnect the drain from the sewer. The prime function of an intercepting trap is to prevent sewer gas from entering the house drains, but nowadays, because drainage systems are better designed and sewer gases seldom cause trouble, in most cases an intercepting trap is omitted. Often in the past they proved a common cause of blockages due to the collection of debris in the trap. See also ANTI-FLOOD INTERCEPTOR.

(See illustration opposite)

INTERLOCK DEVICE
A device that monitors the safe operation of an appliance and causes it to shut down if a dangerous situation may result. Interlock devices generally require manual resetting. See BOILER INTERLOCK.

INTERMEDIATE PRESSURE STAGE
The section of LPG pipe work between a FIRST STAGE REGULATOR sited at a bulk tank or cylinder and a SECOND STAGE REGULATOR usually located at the entrance of the building. For example, see BULK TANK.

INTERMITTENT PILOT
A gas PILOT FLAME that does not burn continuously and will only be alight when heat is called for. When the boiler control calls for heat, a spark ignition sequence is initiated which allows gas through to the pilot. When this is established, the main burner will ignite.

INTERNAL CORNER   See EXTERNAL AND INTERNAL CORNERS.

INTERNAL DORMER   See DORMER.

INTERPOSED CISTERN
A cold STORAGE CISTERN which separates a cold DISTRIBUTION PIPE from the MAINS supply pipe and therefore the cistern acts as a backflow prevention device thus protecting the mains water supply from BACK-SIPHONAGE of contaminated liquids.

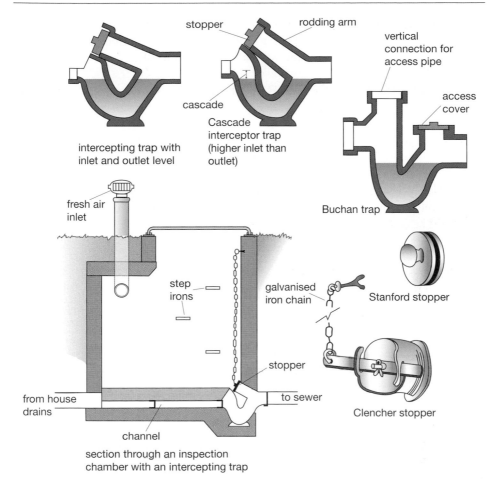

stopper   rodding arm

vertical connection for access pipe

access cover

cascade

intercepting trap with inlet and outlet level

Cascade interceptor trap (higher inlet than outlet)

Buchan trap

fresh air inlet

step irons

galvanised iron chain

Stanford stopper

stopper

from house drains

to sewer

Clencher stopper

channel

section through an inspection chamber with an intercepting trap

Intercepting traps

INTRINSICALLY SAFE
A term used to indicate that no sparks will be produced as the result of using an electrically operated device.

INTUMESCENT VENT
A special air grill or vent which incorporates an arrangement of holes that are closed in the event of a fire to prevent the spread of smoke. Basically the vent works by causing the latticework to expand as a result of heat from a fire, filling the ventilation gaps.

INVAR
An iron/nickel alloy which has a low expansion rate and is used in thermostats and BIMETALLIC STRIPS to assist their operation. Invar is also used in some measuring tapes to give very accurate measurements irrelevant of the weather and temperature.

INVERT LEVEL
The lowest point of the internal surface of a pipe or channel when fitted horizontally.

INVERTED BRANCH

A branch fitting in which the branch is looking downwards, intended for the connection of a branch ventilating pipe. See BRANCH for example.

ION EXCHANGE WATER SOFTENER   See BASE (ION) EXCHANGE WATER SOFTENER.

IONS

A term used to indicate the positive or negative charged particles that flow through gases and liquids. As an example, should a simple 'primary cell' battery containing copper and zinc suspended in diluted sulphuric acid (H_2SO_4) be used, the copper would slowly destroy the zinc (electrolysis). In so doing it would convert the H_2SO_4 to positively charged hydrogen and negatively charged sulphur dioxide (i.e. a flow of ions). The current would be referred to as direct (DC).

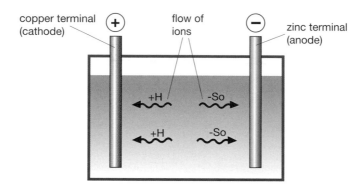

Direct current being generated

IRON

One of the most used and cheapest metals available. Iron makes up about 5% of the Earth's crust and is found in various minerals and ores. Iron is separated from the associated rock by smelting it in a continuously operated blast furnace together with coke and limestone. Iron trickles from the base of the furnace and is collected. During the smelting process much carbon is taken in with the iron and thus it is known as CAST IRON or pig iron. Further treatment is needed to remove the carbon content to produce WROUGHT IRON. Iron is known as a FERROUS METAL and is hardly attacked by dry air, but readily corrodes should any water be present.

Chemical symbol	Fe
Colour	Grey
Melting point	1535°C
Boiling point	3000°C
Coefficient of linear expansion	0.000011/°C
Density	7860 kg/m³

IRRIGATION SYSTEM

A watering system, designed to water plants or crops. These systems generally consist of a porous hosepipe laid above ground.

ISLE OF MAN KEY

A special three-headed key in which each of the three heads is recessed with a square hole of differing sizes. It is used to control the flow through a valve fitted with a square-headed spindle. For example, see LOCKSHIELD VALVE.

ISOLATION VALVE

Any type of valve, e.g. PLUG COCK, GATEVALVE, STOPCOCK, etc. which is fitted onto a pipe and is designed to prevent the flow of gas or liquid to or from an appliance or fitting, etc. It can be used to 'isolate' the required section for repair or service work. See also SERVICE VALVE.

ISOMETRIC DRAWING   See PROJECTION DRAWING.

I

JACK PUMP

The old village pump; the jack pump is a pump which lifts water up from a well. It is exactly the same as a lift pump except that with a lift pump the water is pumped up into a delivery pipe, whereas the jack pump discharges the water it has lifted out of a spout. See LIFT PUMP, for example.

JACK WELL

A SUMP or hole where the SUCTION PIPE and its FOOT VALVE from a pump are located.

JACUZZI

A design of whirlpool bath which incorporates a VENTURI tube in its pumping arrangement to give a water/air mix. See WHIRLPOOLS AND SPAS.

JAMNUT   See BACKNUT.

JOINDER JUNCTION

A branch fitting which is manufactured with the branch connection sealed off. It is designed to be installed in a drainage run where a connection will be made in the pipe at a later date. When the branch connection is required the closed end of the branch is cut off. See CONNECTIONS TO EXISTING DRAINS.

JOINT RAKER

Also called a plugging chisel, see CHISELS.

JOINTING COMPOUND

A compound used to assist the making of joints in pipelines. Several types will be found, including green, blue and white jointing compound, black graphite compound and red haematite compound. The white compound is the most common and is sold under several trade names such as Rocol Gasseal, Boss White, Hawk White, etc. It is used on water, gas and compressed air lines, although a check of its uses, by referring to the side of the tin is advisable, as some compounds are of no use with natural gas and may not be permitted on potable water supplies. The black graphite compound has a non-setting nature and is used on steam installations. The red haematite compound consists of a RED LEAD base and does tend to set hard when exposed to the air and is used for oil and most petroleum spirits. See also PTFE TAPE.

JOULE   See QUANTITY OF HEAT.

JUMPER (VALVE PLATE or JUMPER PLATE)

The piece of brass or plastic which holds and supports the washer in a screwdown valve. For example, see STOPCOCK.

JUNCTION BLOCK

A drainage fitting used to make the connection of a branch drain into an existing sewer. See CONNECTIONS TO EXISTING DRAINS.

JUNIOR HACKSAW   See HACKSAW.

J

KELVIN

The SI unit for temperature. The increments of this scale increase/decrease at the same rate of that used with the Celsius (centigrade) scale. The prime difference is that 0° Kelvin is the theoretical lowest possible temperature at which nothing can be colder. It is sometimes referred to as absolute zero, whereas 0° Celsius is only the freezing point of water (0°K = −273°C or conversely 273°K = 0°C). See also METRIC SYSTEM.

KEROSENE (CLASS C2, 28 SECOND FUEL)

A distillate grade fuel oil used for domestic oil burning appliances. This fuel is by far the most common oil used today for domestic oil burners. It is a clear oil consisting of approximately 84% carbon and 16% hydrogen. See also VISCOSITY.

KETTLING   See BOILER NOISES.

KEY

The roughening made on the surface of something to provide a better adherence to another surface or material when jointing.

KEY

A tool or device used to lift the lid off from inspection covers and manholes. See also AIR COCK KEY.

KILLED SPIRITS   See ZINC CHLORIDE.

KILOGRAM (kg)

A basic metric unit of mass and weight, being equal to a mass of platinum-iridium in the shape of a cylinder kept in the international bureau of weights and measures near Paris. This cylinder is almost equal to a mass of 1000 cubic centimetres of water (1 litre) at 4°C, at 1 bar pressure. The old British Imperial weight the pound is now defined as 0.45359237 kilograms exactly. See also METRIC SYSTEM.

KILONEWTON (kN)   See PRESSURE.

KILOWATT (kW)

A unit of power equal to 1000 watts which is about 1.34 horsepower. See WATT. See also BRITISH THERMAL UNIT.

KINETIC ENERGY
Energy that has been produced as the result of movement.

KING ROLL (CENTRE ROLL)
The intersection of ROLLS running in different directions located at the highest point on a flat roof. It is used on metallic roof coverings to allow for expansion. Rolls run with the fall of a roof and therefore do not normally meet, but should a flat roof finish at a high point and run in two or more directions a king roll may be required at this point.

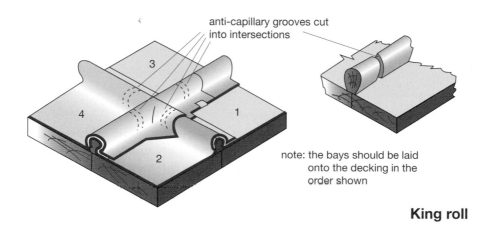

anti-capillary grooves cut into intersections

note: the bays should be laid onto the decking in the order shown

King roll

KITE MARK
A mark given to indicate British Standard approval and placed on all items manufactured to an approved quality standard and design.

KNEE BEND
A sharp bend used on CLAYWARE PIPES having little or no radius. See also KNUCKLE BEND.

KNUCKLE BEND
A bend with a very small radius, usually called a SHORT RADIUS BEND.

KNUCKLE JOINT   See WIPED SOLDER JOINT.

LADLE
A large spoon-shaped tool used to transport and pour molten lead from a melting pot.

LAGGING   See THERMAL INSULATION.

LAND DRAIN (FIELD DRAIN/SUBSOIL DRAIN)
Porous, perforated pipes laid in the ground to drain away, to a suitable position, the excess SURFACE WATER. Land drains could be used in situations where the ground becomes very waterlogged, therefore eliminating surface flooding and improving the stability of the ground surface. When laying any form of land drain the joints between the pipes are not made sound and the pipe is surrounded in brick rubble or clinker to allow water to soak easily into the pipe.

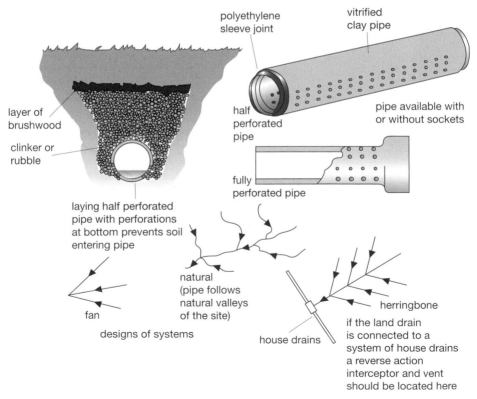

polyethylene sleeve joint

vitrified clay pipe

layer of brushwood

clinker or rubble

half perforated pipe

pipe available with or without sockets

fully perforated pipe

laying half perforated pipe with perforations at bottom prevents soil entering pipe

natural (pipe follows natural valleys of the site)

fan

designs of systems

house drains

herringbone

if the land drain is connected to a system of house drains a reverse action interceptor and vent should be located here

Land drains

LANTERN LIGHT

A raised frame which holds a series of windows and is fixed above the general level of flat roofs or forming the RIDGE of a pitched roof. Its purpose is to provide light and sometimes ventilation to the room below. See also SKYLIGHT.

LAP JOINT

A joint on sheet weathering materials in which the OVERCLOAK just lays on top of the UNDERCLOAK, used on steeply pitched roofs. When a lap joint is employed, it is essential to ensure that the higher sheet (the overcloak) lays over the lower sheet (the undercloak) a minimum vertical distance of 75 mm, see COVER FLASHING. See also NURALITE for lap joints made with this non-metallic roof material.

LAP WELDED JOINT   See LEAD WELDING.

LATENT HEAT (HIDDEN HEAT)

The amount of heat that would be required to bring about a PHYSICAL CHANGE of a substance, i.e. solid to liquid or liquid to gas. On heating up a substance the temperature usually slowly rises and this is known as sensible heat. When the temperature reaches the melting or boiling point a strange phenomenon occurs; although more heat may be applied, no rise in the temperature of the substance takes place until the whole of the substance has undergone a physical change and has been completely melted or boiled away. The amount of heat that had to be applied before the temperature of the substance began to rise again is known as the latent heat. See also QUANTITY OF HEAT and BOILING POINT.

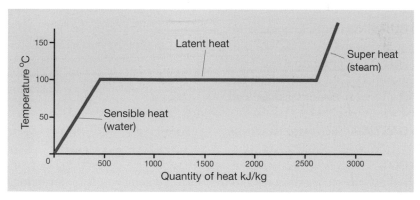

Water heated at atmospheric pressure

Latent heat

LAYBOARD

A timber board which is used to support lead sheet or pipe. A layboard may be found behind a chimney stack to support the BACKGUTTER which goes up under the roof tiles. For example, see BUILDING TERMINOLOGY.

LCC PATTERN PIPE

Name sometimes used to identify a socketed cast iron pipe used above ground. The pipe is identified by two raised rings at the top and bottom of the socket, for example, see CAULKED JOINT.

LDF
Abbreviation for LEAK DETECTION FLUID. See LEAK DETECTION SPRAY.

LEAD
A NON-FERROUS METAL produced in a variety of ways, but its most common form will be found supplied as pipe or in sheet form. The Latin name for lead is plumbum, this being the word from which the term 'PLUMBER' is derived, i.e. the worker of lead.

Chemical symbol	Pb
Colour	Blue-grey
Melting point	327°C
Boiling point	1744°C
Coefficient of linear expansion	0.0000293/°C
Density, cast lead	11300 kg/m^3
Density, milled lead	11340 kg/m^3

It will be seen that CAST LEAD is less dense than MILLED LEAD and thus it is easier to use when bossing, but due to the manufacturing process of cast lead it is used less nowadays. It is worth noting the boiling point of lead, which is 1744°C, at this temperature it changes from a liquid to a vapour therefore when carrying out the process of heating lead, such as in welding, it is advisable to ensure there is adequate ventilation to avoid breathing in any lead fumes. See also LEAD PIPES AND FITTINGS and LEAD SHEET AND ROOF COVERINGS.

LEAD BURNING   See LEAD WELDING.

LEAD BUTTON (LEAD DOME)
A purpose-made lead dome which fixes over a brass washer and hides the screwed fixing of a vertical section of sheet lead.

LEAD CASTING   See SAND CASTING.

LEAD DOT
A method used to secure sheet lead to masonry. To achieve a lead dot fixing, first a dovetail-shaped hole is cut in the brickwork, then once the lead has been laid a dot mould is placed over the hole and molten lead is poured into it filling the dovetail cavity and leaving a dome shaped head. There are two other types of dots, the soldered dot and the lead welded dot, which are often used to weather the fixings securing lead to vertical timber CHEEKS or CLADDING. See also LEAD BUTTON and SHEET FIXINGS.

(See illustration opposite)

LEAD FLASHING
See CHIMNEY FLASHINGS and COVER FLASHINGS.

LEAD-FREE SOLDER   See SOLDER.

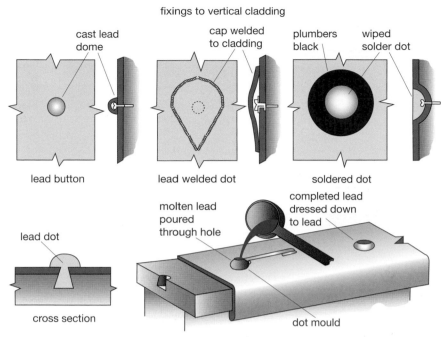

Lead and solder dots

LEAD KNIFE (DRAW KNIFE)
A knife with a thick curved blade which is used to cut sheet lead.

LEAD PIPES AND FITTINGS
The use of lead for water supply pipes is not allowed in new work nowadays not even for repair or replacement work, without water authority permission. All connections to lead pipes should be made with the use of COMPRESSION JOINTS designed specifically for this purpose. When carrying out repair work to existing lead mains where a new piece of material is required to be inserted, plastic pipe should be used, inserting a piece of copper pipe is no longer permitted, (see ELECTROLYTIC CORROSION). The old traditional method of connecting to lead piping was by the use of WIPED SOLDER JOINTS, but this method of jointing is now restricted to waste pipe connections and even then compression joints are generally used.

LEAD SAFE   See SAFE.

LEAD SHEET AND ROOF COVERINGS
Sheet lead is available in almost any size, although the standard width of a roll is 2.4 m with lengths up to 12 m. Sheets of this size are very heavy to handle therefore in many cases it is worth the extra charge of buying it in smaller (strip) widths. Sheet lead is available in a range of thicknesses and is colour coded accordingly (see p. 240).

Lead is highly MALLEABLE and lead BOSSING can be carried out with ease. Due to its low TENSILE STRENGTH and lack of ELASTICITY lead will CREEP therefore correct installation of the material is essential. Lead can very easily be welded together, for examples, see LEAD WELDING. On large roofs, allowance for expansion must be

British Standard thickness of sheet lead

BS code no.	Thickness mm	Weight kg/m^2	Colour code
3	1.32	14.97	Green
4	1.80	20.41	Blue
5	2.24	25.40	Red
6	2.65	30.05	Black
7	3.15	35.72	White
8	3.55	40.26	Orange

Recommended thicknesses of sheet lead

Location	BS code no.
Damp proof course, soakers	3–4
Small flat roofs, dormer cheeks, chimney flashings, cover flashings, and valleys	4–5
Large flat roofs, parapet gutters, and tapering valleys	5–6 or 7

made and the method of jointing the sheets together varies depending upon the circumstances and specifications. But in general, for joints which run across the flow of water (transverse joints) the method of jointing is either by drips, or welts. Welts should only be used on steeply pitched roofs or wall cladding. For joints which run in the same direction as the flow of water wood-cored rolls, hollow rolls or standing seams should be used. Wood-cored rolls should be used on shallow pitched roofs to prevent any flattening of the joint if trodden on. Large roofs are divided into a series of smaller areas called bays and the size of a bay will depend upon the thickness of the sheet lead and its location, e.g. flat or pitched roofs, or wall cladding. The lead is divided into bays to overcome the problems caused by thermal movement. See also DOUBLE LOCK WELT, SINGLE LOCK WELT, STANDING SEAM, HOLLOW ROLL and SHEET FIXINGS.

(See illustration opposite)

LEAD SLATE   See SLATE PIECE.

LEAD TACK (TINGLE)
The fixing clip used on sheet lead roof work. There are different types of lead tacks such as a bale tack or secret tack. For examples, see SHEET FIXINGS. The term lead tack is also used to mean a fixing lug which has been welded onto a piece of lead pipe for fixing it to a wall.

LEAD WEDGES   See WEDGES.

LEAD WELDING (LEAD BURNING)
Lead can easily be welded together with either oxy-acetylene welding equipment or propane gas, although propane gas welding is limited to simple butt and lap joints. Lead welding is one type of AUTOGENOUS WELDING in which the filler rod used is of

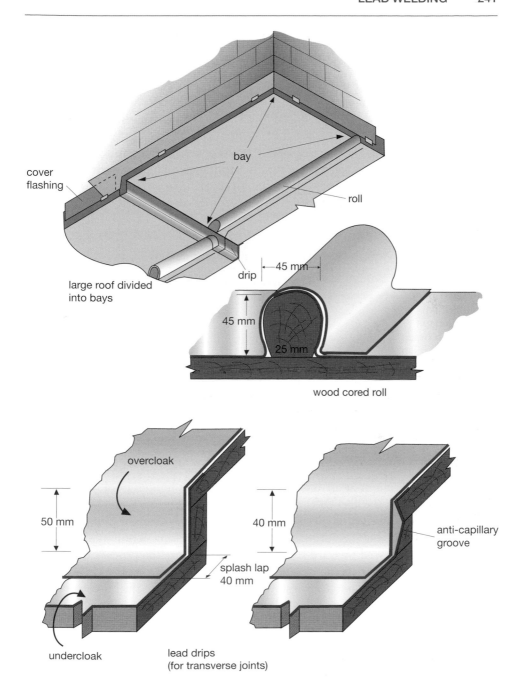

bay

cover flashing

roll

large roof divided into bays

drip

45 mm

45 mm

25 mm

wood cored roll

L

overcloak

50 mm

40 mm

anti-capillary groove

splash lap 40 mm

undercloak

lead drips (for transverse joints)

Allowance for expansion on lead roof coverings

the same material as that to be welded. Therefore the filler rods are generally made from strips of sheet lead 6–8 mm wide, cut from waste. Some plumbers cast their filler rods in a special mould, this produces a thicker filler rod and therefore saves a lot of stops and starts in a long run of welding. One of the prerequisites to successful welding is cleanliness of the filler rod and the surfaces to be welded. The correct nozzle must be chosen and adjusted to a neutral flame (see OXY-ACETYLENE FLAME).

The pressure at both oxygen and acetylene regulators should be adjusted to 14 kN or 0.14 bar (2–3 lb/in^2).

There are several techniques which can be adopted in order to weld successfully. The finish weld pattern can be either of a herringbone design, which is produced with a larger flame and where the welder has to work very fast, or the weld could be a thumb nail design, which is produced by a smaller flame. A finished weld should look uniform in size and shape and it is essential that there is sufficient PENETRA-TION, equally there should be no UNDERCUTTING or overlapping to the edge of the weld. The method of welding used generally varies depending upon the joint being carried out but basically the LEFTWARD WELDING TECHNIQUE is adopted in which the filler rod is applied as described below except in vertical welding in which no filler rod is used.

Size of nozzle for oxy-acetylene lead welding

Type of weld	Thickness of lead BS code no.	Nozzle size Model 'O' blowpipe
Flat butt or lap	3–5	2–3
Flat butt or lap	over 6	4
Inclined butt or lap	3–6	2–3
Inclined butt or lap	over 7	4
Vertical weld	3–5	1
Vertical weld	over 6	2–3
Lead to brass	over 5	3–4

To carry out a butt weld, first the meeting surfaces are shaved clean approx-imately 6 mm wide on each sheet, then a TACK WELD is applied at several intervals along the joint, this prevents the joint opening due to expansion. The welding nozzle is held close to the joint at an angle of about 60° and a molten pool is allowed to become established, the filler rod is introduced close to the nozzle and the blowpipe is slightly raised to melt off a piece of the filler rod which drops and fuses with the molten pool. The flame is returned to the pool in a stroking action as shown. The blowpipe is then moved forward to a new position where the process is repeated, slowly moving along the sheet.

The lap joint is carried out using a similar welding process to the butt joint except that one sheet (the overcloak) is lapped on top of the sheet to be joined by 25 mm. With this joint the molten pool is not allowed to penetrate right through to the underside of the undercloak and also two runs of welding are required to give full support to the weld. To prepare this joint, the overcloak is cleaned on both sides and placed on the cleaned undercloak and tacked in position. The first weld is made joining the two sheets; this causes a certain amount of undercutting to the overcloak therefore a second weld is required as reinforcement. The second weld is made on top of the first run leaving about 3–4 mm of it exposed.

With vertical welding no filler rod is used and the joint is prepared as for lap weld-ing. To weld the joint the overcloak is fused with the undercloak and the flame is removed to allow the weld to solidify. The nozzle is again held close to the over-cloak and as the lead begins to melt, the nozzle is circled round like a number six, finishing off at the undercloak, a molten lead bead will follow the flame and when it becomes fused with the under cloak the flame is again removed. This process is continued up the joint. See also INCLINED LEAD WELDING, ANGLE WELDED JOINT and WELDING DEFECTS.

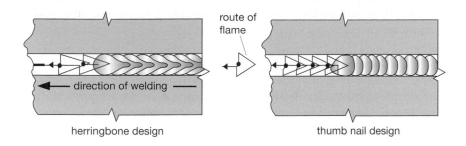

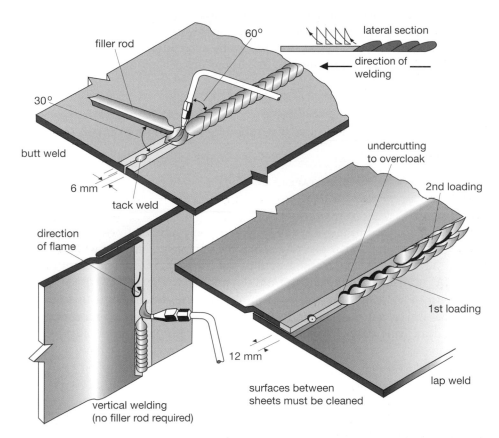

Lead welding

LEAD WOOL (SPUN LEAD)
Strands of lead fibre sold in bundles and used to make CAULKED JOINTS to cast iron pipes in damp conditions.

LEADED LIGHTS
Windows which consist of lots of small square or diamond shaped panes of glass which are held into the frame by 'H'-shaped pieces of lead called CAMES.

LEAK DETECTION SPRAY (LEAD DETECTION FLUID OR LDF)

A soap solution used to assist in finding leaks in pipe work such as gas supplies or sanitary pipe work. By maintaining pressure within the pipe and washing or spraying the joints with this soap solution, bubbles will form at any leaking connection as the gas or air escapes.

LEFTWARD WELDING TECHNIQUE

A welding technique in which the welding is started at the right hand side of the joint and the blowpipe moves slowly forward moving from side to side at the desired rate. The filler rod is moved progressively in front of the blowpipe. See LOW CARBON STEEL WELDING and LEAD WELDING. See also RIGHTWARD WELDING TECHNIQUE.

LEGIONELLA

Bacteria that grow in the warm water that is environmentally associated with water-cooling towers and showers. If ambient conditions are suitable, should the bacteria be present and the mist created breathed in by susceptible persons, a potentially fatal infection, similar to pneumonia, can be contracted. This infection is referred to as Legionnaires' disease.

LET-BY

Term used to indicate that a valve is not shutting off fully, for example, see TIGHT-NESS TESTING.

LEVEL INVERT TAPER   See TAPER.

LEVER

A tool/machine used to move heavy loads with only a little effort. A simple lever might consist of an iron bar pivoting on a piece of timber. The lever will only work to your advantage as long as the length of the iron bar from the fulcrum to your hands (A) is longer than the length from the fulcrum to the load (B). By extending the length of 'A' less effort will be required to move the load, this is known as a 'mechanical advantage'. The principle of leverage has many applications from moving dead loads and pulling nails out of timber to the operation of ballvalves, etc.

The amount of effort required to move a load is found by striking a balance between the load and effort. To achieve this the effort, E, multiplied by A must equal the load multiplied by B. Therefore in the example shown

$$E \times A = L \times B$$

By transposition of the formula

$$E = \frac{L \times B}{A} = \frac{1000 \text{ kg} \times 0.3}{2} = 150$$

the effort required would be 150 kg.

(See illustration opposite)

LEVER-HANDLED TAP

A quick action tap such as an ELBOW ACTION TAP.

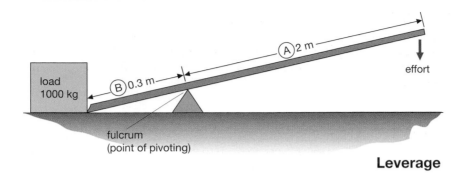

Leverage

LEVER LOCKING STOPPER   See CLENCHER STOPPER.

LEVER WRENCH (SELF-GRIP WRENCH/MOLE GRIP)
A tool in which the jaws can be made to lock on to all sorts of objects, the clamping of the jaws is made by simply squeezing the handles together. For example, see WRENCH.

LIFT OFF
A situation where the flame from a gas burner jet is burning some distance away from the burner itself. It is caused by the gas VELOCITY in the pipe being greater than the burning velocity of the fuel. See GAS BURNER; see also LIGHT BACK and FLAME RETENTION.

LIFT PUMP
A pump designed to lift water out of a well or SUMP HOLE. The height to which the water can be lifted is governed by atmospheric pressure. The pump works as follows:

(1) On the upward stroke the non-return valve A opens and water is sucked into the body of the pump due to a partial vacuum being formed.
(2) On the downstroke valve A closes and the non-return valve B opens allowing water into the area above this valve.
(3) On the second upward stroke more water is sucked into the body and that already above valve B is lifted to discharge out of a spout in the case of a JACK PUMP or is forced through a third non-return valve C into a delivery pipe. The process is repeated slowly lifting the water. See also FORCE PUMP.

(*See illustration over.*)

LIFTING KEY
The tool used to assist the lifting of INSPECTION COVERS, etc. It is a T shaped piece of metal with a bent lug on one end for hooking into a special slot in the cover.

LIGHT BACK (STRIKE BACK)
Light back is the opposite to flame LIFT OFF and is a situation that occurs when the gas VELOCITY is slower than the BURNING VELOCITY of the gas and as a result the flame burns inside the burner itself. Due to insufficient air being supplied at this point complete combustion of the fuel cannot take place which results in the production of CARBON MONOXIDE gas. Light back can occur in AERATED BURNERS and

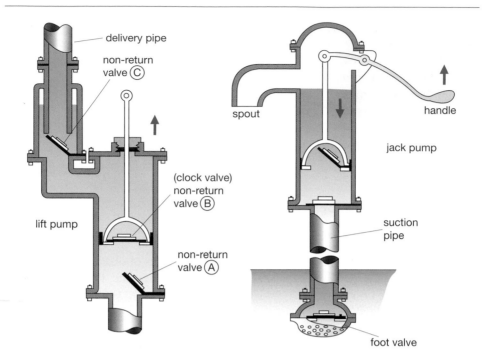

delivery pipe

non-return valve Ⓒ

spout

handle

jack pump

(clock valve) non-return valve Ⓑ

lift pump

suction pipe

non-return valve Ⓐ

foot valve

Lift and jack pumps

oxy-acetylene welding equipment, although light back in oxy-acetylene welding is generally known as FLASHBACK or BACKFIRE.

LIME SODA PROCESS   See CLARKS PROCESS.

LIMITS OF FLAMMABILITY
The amount of gas mixed with air to produce a flammable mixture, for example, butane gas will burn with 1.9%–8.5% of gas in a volume of air. Anything less than 1.9% will be too weak and anything more than 8.5% will be too rich to support combustion.

Limits of flammability

Gas	Percentage of gas in air	
	Minimum	Maximum
Butane	1.9	8.5
Propane	2	11
Natural	5	15

LINDEWELDING
A special welding process sometimes adopted to weld steel piping. The process increases the welding speed considerably in comparison to the LEFTWARD and RIGHTWARD WELDING TECHNIQUES using a NEUTRAL FLAME. The lindewelding

technique is based on the ability of the steel to absorb carbon produced by a carbonising flame (see OXY-ACETYLENE FLAME). When the steel has a greater portion of carbon in its make-up, its melting temperature is reduced. This gives rise to the rapid melting of the surfaces along the weld zone. To weld successfully using this technique, special filler rods are used which contain silicon and manganese.

LINEAR EXPANSION   See COEFFICIENT OF THERMAL EXPANSION.

LINEAR FLOW MANIFOLD   See MICRO-BORE.

LINK CUTTER (CHAIN CUTTER)   See PIPE CUTTER.

LINT
Term used to mean dust, carpet fibres, hairs, fluff, etc. that collects on the surface of grills, filters, etc.

LINT ARRESTOR
The mesh, gauze or filter that is found in the air supply to a burner. It is designed to prevent any of the dust, fluff or fibres collecting that may prevent an adequate air supply used in the combustion process reaching the flame.

LIP TEE   See TONGUE TEE.

L

LIQUEFIED PETROLEUM GAS (LPG)
Gases such as BUTANE and PROPANE which are given off from crude oil during the refining process and are used as fuels. These gases are a mixture of carbon and hydrogen. When they are subjected to low pressures they change to a liquid and are stored in a steel container, which has a left handed thread for the connection of the regulator. When the gas cylinder valve is opened the pressure within the liquid is forced out and once the pressure has been released the liquid reverts back to gas. Liquefied petroleum gases like natural gas are odourless and therefore during the production process a chemical is added to give it a distinctive smell, this acts as a warning should there be a leak. Both butane and propane are heavier than air therefore when working in trenches or basements special care should be taken to avoid any leaked gas accumulating, resulting in a possible explosion. These gases are sold by weight and in liquid form therefore it is possible to tell roughly how much gas is left in the cylinder by weighing it and subtracting from this figure the weight of an empty cylinder (1 kg of liquid = approximately 0.4 m^3 of LPG). Sometimes frost will appear on the outside of cylinders when the AMBIENT AIR temperature is low and large quantities of gas are taken from the cylinder; this is due to the fact that in converting the liquid petroleum to its gaseous state heat is required (just like heat being required to change water to steam) thus this heat is taken from the air surrounding the cylinder. To overcome this problem, remove the cylinder to a warmer area, or wrap the cylinder in sacking or blankets, etc., never subject the cylinders to direct heat.

The carriage and storage of LPG may be subject to legal requirements in addition to the general requirements laid down in the Health and Safety at Work Act

1974 – for advice the guidance notes published by the Health and Safety Executive should be sought.

LIQUID EXPANSION TYPE THERMOSTAT   See THERMOSTAT.

LIQUID VAPOUR FLAME FAILURE DEVICE (VAPOUR PRESSURE FSD)

A device designed to open and close the valve that serves the main burner of gas burning appliances automatically. Several designs can be seen but the most common is found within a gas oven and works as follows.

(1) When the thermostat is calling for heat, gas is allowed to flow into the valve and a small amount is allowed to by-pass the valve and discharge into the oven.
(2) The by-pass gas is then ignited, usually with a piezo-electric spark. The small flame generated then plays on to a phial that is filled with a VOLATILE FLUID. The fluid expands and causes the bellows chamber to expand so that it pushes against the spring and forces the valve open. A greater volume of gas can now flow into the oven, allowing the flame to increase and the oven come to up to full heat.
(3) When the oven reaches the required temperature, the gas to the valve reduces in volume back to the simmer or by-pass rate.

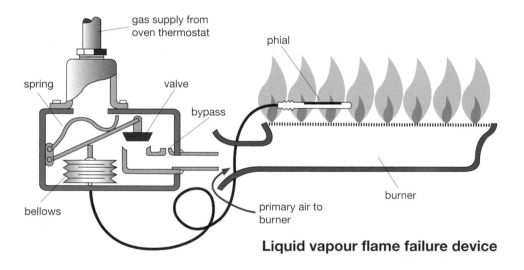

Liquid vapour flame failure device

LITRE

The unit of volume in the metric system, being equal to 1000 cubic centimetres (1000 cm³). One litre of pure water weighs almost exactly one kilogram at 4°C at standard atmospheric pressure (1 bar).

LOADING UNIT

A factor which is used to assist in finding the recommended design flow rate to various SANITARY APPLIANCES. A loading unit takes into account the flow rate at an appliance and its probable usage. By using the table below to find the total loading units required and by cross-reference to the graph, the design flow rate can be directly read.

Sanitary appliance	Loading units
WC cistern	2
Wash basins	1.5
Baths	10
Sinks	3
Showers	3

Example

Find the design flow rate for a cold distribution pipe serving 10 WCs, 10 wash basins, 5 baths and 5 sinks in a block of flats.

$$10 \text{ WCs} \times 2 = 20$$
$$10 \text{ basins} \times 1.5 = 15$$
$$5 \text{ baths} \times 10 = 50$$
$$5 \text{ sinks} \times 3 = 15$$
$$\text{Total} = 100$$

The design flow rate for 100 loading units from the graph opposite reads 1.3 litres per second. See also PIPE SIZING (hot and cold water supplies).

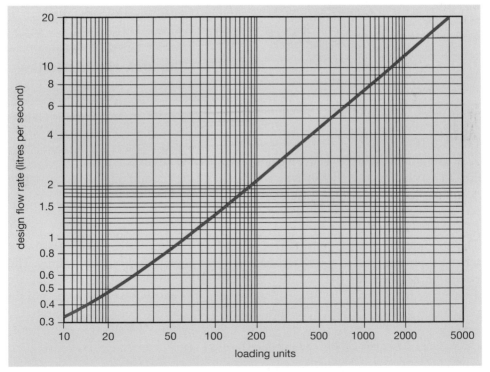

Graph used to find the required flow rate of water to an installation

LOCALISED HOT WATER SUPPLY

A method of hot water supply where the water is heated locally to its needs. Most methods of localised hot water supply use INSTANTANEOUS WATER HEATERS which have the advantage of only heating the hot water as and when it is required. Also localised hot water overcomes the problem of excessively long draw off points from a CENTRALISED HOT WATER SUPPLY SYSTEM. See also STORAGE WATER HEATER.

LOCK-SHIELD HEAD

A cap or device fitted to stopvalves, radiator valves, etc. to prevent unauthorised persons tampering with the flow of water. The valve is adjusted by use of a special key or simply by removing the lock-shield head and using a spanner to turn the spindle.

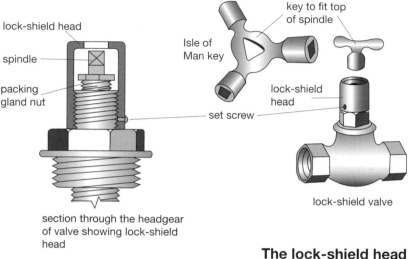

section through the headgear of valve showing lock-shield head

The lock-shield head

LOCKNUT   See BACKNUT.

LOCKOUT

A situation in a control system, e.g. an oil or gas burner or fire protection system, etc., in which the power supply is shut down to a safety condition. No restart can be accomplished without some form of manual intervention.

LOCK UP

The condition of the gas REGULATOR, located at the incoming supply gas meter, when the pressure within the system has risen to a point and closed off its inlet. This condition is caused by the build-up of pressure within the installation as all the appliances are closed. When they are re-opened, the gas will be drawn off and the regulator will itself re-open, aiming to maintain the correct WORKING PRESSURE.

LONDON PATTERN SINK   See SINK.

LONG AND SHORT HOPPER WC PAN

Early designs of trapped WC PANS made in the late nineteenth century. The long hopper consisted of a conical shaped bowl (hopper) connected to a trap. The pan was flushed by a jet of water which twirled itself down to the trap in a thin spiral of water. Due to the length of the hopper, by the time the water reached the trap it had little energy left to carry the contents of the trap with it; also with the size of the area to be cleaned the sides soon became soiled. A later model, 'the short hopper' proved to be an improvement although it was only when a method of flushing the water around the rim of the hopper was found that improvement was really made.

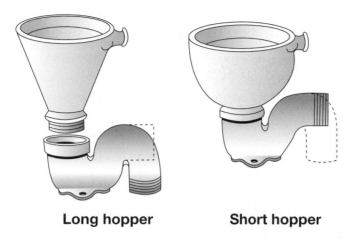

Long hopper   **Short hopper**

LONG BRANCH (LONG TAIL BRANCH)
A branch fitting in which the length of the spigot is 900 mm long. This type of branch fitting proves useful when making connections to a range of WCs. The spigot may be cut to give shorter centres if necessary, see sketch.

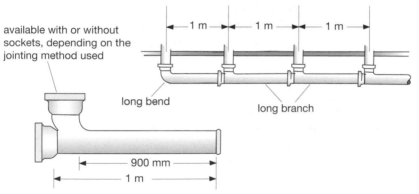

Long branch

LONG HANDED DUMMY   See DUMMY.

LONG RADIUS BEND   See BEND.

LONG STRIP ROOFING
A method sometimes employed, using STANDING SEAMS, to cover roofs with the harder metallic sheet materials such as copper, stainless steel and zinc. Long strip roofing has been carried out on the continent for many years but it is only since about 1957 that it has been adopted in this country. Long strip roofing differs from the traditional methods of laying the material in that no CROSS WELTS and DRIPS are used, thus a substantial saving in the laying cost is made. This design of covering can be used on roofs with a pitch ranging from 2.5° to 45° and the individual bays can be up to 8.5 m long and 0.6 m wide compared with the maximum 1.8 m long bay

used in the traditional practice. In order to achieve such long bays two important points must be observed.

(1) The standing seams are to be formed with a gap at their base to allow for thermal movement. Should the roof be very long it is recommended that an expansion type of BATTEN ROLL be fitted every twelve bays to facilitate both transverse and longitudinal movement.

(2) Special expansion clips must be incorporated with the standing seams which are to provide both a secure fixing and allow for lengthwise movements of the sheet. See sketch.

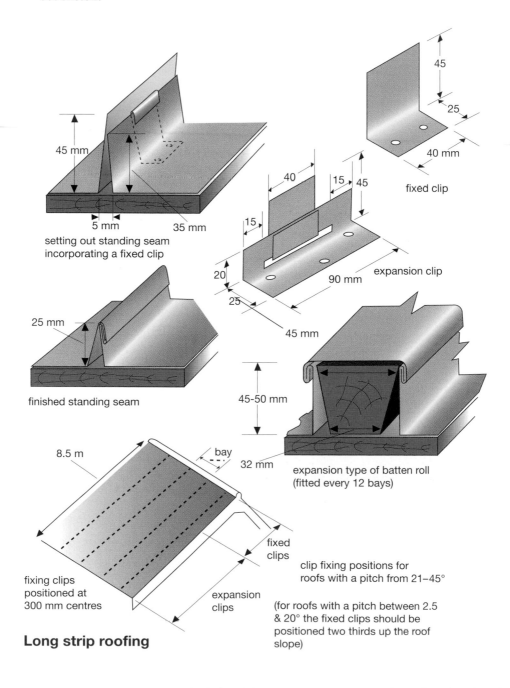

45

25

40 mm

fixed clip

45 mm

5 mm   35 mm

setting out standing seam
incorporating a fixed clip

40

15   45

15

20

25

90 mm

45 mm

expansion clip

25 mm

finished standing seam

45-50 mm

32 mm

expansion type of batten roll
(fitted every 12 bays)

8.5 m

bay

fixed
clips

clip fixing positions for
roofs with a pitch from 21–45°

fixing clips
positioned at
300 mm centres

expansion
clips

(for roofs with a pitch between 2.5
& 20° the fixed clips should be
positioned two thirds up the roof
slope)

Long strip roofing

The method of laying the material to this design of roof can be carried out in the traditional fashion using conventional hand tools, or special power tools are available to assist the forming of the standing seams with great speed.

LONGSCREW (CONNECTOR)

A joint used to join two pieces of LOW CARBON STEEL tube, neither of which can be rotated. It has a male thread at each end, one being sufficiently long enough to accommodate a backnut and socket. To make the joint, first the backnut and socket are wound along the long parallel thread, the short tapered thread is then connected onto one of the pipe ends using another socket, this completes one of the connections. The other end is made by winding back the socket from the parallel thread to connect onto the second pipe. Finally, the backnut is wound up to the socket trapping a GROMMET between the two mating surfaces, thus making a water-tight seal, see sketch. See also UNION CONNECTOR.

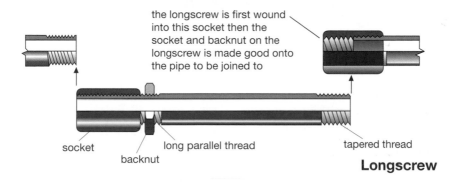

the longscrew is first wound into this socket then the socket and backnut on the longscrew is made good onto the pipe to be joined to

socket   long parallel thread   tapered thread

backnut

Longscrew

LOOP VENT   See BRANCH VENTILATING PIPE.

LOUVRE

An assembly of fixed or pivoted vanes contained in a frame that is designed to permit or restrict the passage of light or air flow. Note that in the case of combustion ventilation the louvres must not be allowed to be adjusted and air passage must be maintained at all times.

LOW CARBON STEEL (MILD STEEL)

Iron containing up to 0.2% of carbon. It is extensively used throughout the building industry for all types of structural work, fixtures and fittings, etc. See LOW CARBON STEEL PIPES. See also STEEL.

LOW CARBON STEEL PIPES AND FITTINGS

LOW CARBON STEEL pipes are available either painted black or GALVANISED. Black iron pipes should be used for hot water heating systems and gas supplies only, if it were used where fresh water is continously being drawn off through the pipeline it would soon become liable to corrosion problems. LCS pipes are manufactured in three grades and colour coded accordingly. These being heavy gauge, red; medium gauge, blue; and light gauge, brown. Only heavy gauge tube is permitted to be used underground, which must also have external protection from the soil. All grades of pipe are permitted for use above ground although light gauge is restricted to the installation of DRY PIPE SPRINKLER SYSTEMS only. The method of jointing LCS pipes is usually by the use of screwed joints which are made tight by the use of either PTFE TAPE or JOINTING COMPOUND and HEMP. Black iron pipes can also be jointed by welding; one should not attempt to weld galvanised pipes as the zinc coating would be burnt off, also the fumes created could damage your health. Threads are cut into the pipes using STOCKS AND DIES and an assortment of fittings are used to join to them. The fittings could be made from either malleable cast iron or steel. Steel fittings tend to be stronger, although more expensive than those of malleable cast iron. See also UNION CONNECTOR and LONGSCREW.

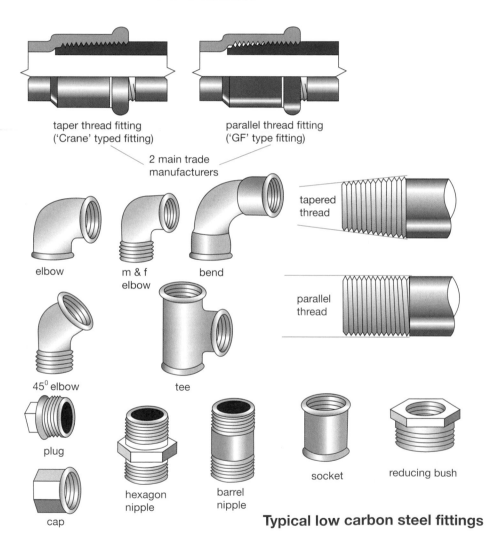

taper thread fitting
('Crane' typed fitting)

parallel thread fitting
('GF' type fitting)

2 main trade
manufacturers

elbow

m & f
elbow

bend

tapered
thread

parallel
thread

45⁰ elbow

tee

plug

cap

hexagon
nipple

barrel
nipple

socket

reducing bush

Typical low carbon steel fittings

LOW CARBON STEEL WELDING

LOW CARBON STEEL can easily be welded together with either oxy-acetylene or electric arc welding equipment. Low carbon steel is one type of autogenous welding in which the filler rod used is the same material as that being welded. For electric arc welding see ARC WELDING. Oxy-acetylene welding is carried out by first ensuring that the edges of the sheet to be joined are free from any dirt, scale or grease, etc. The joint is then prepared at the edges as shown opposite. The correct nozzle must be chosen and adjusted to a neutral flame (see OXY-ACETYLENE FLAME). The pressure at the regulators should be adjusted as shown below.

Leftward welding technique

Thickness mm	Preparation	Gap allowance	Nozzle size	Approximate pressure bar
up to 3	square	3 mm	5–7	0.14
3–5	40° bevel	3 mm	7–10	0.21
over 5	40–45° bevel	3–5 mm	10–18	0.28

Rightward welding technique

Thickness mm	Preparation	Gap allowance	Nozzle size	Approximate pressure bar
up to 3	square	3 mm	7–10	0.21
3–5	square	3 mm	10–13	0.28
over 5	30° bevel	3–5 mm	13–25	0.3

Note The nozzle size specified is for a DH type blowpipe.

There are two basic welding techniques which can be adopted in order to weld successfully these being either the leftward or rightward methods, but see also LINDEWELDING. With the LEFTWARD WELDING TECHNIQUE the filler rod precedes the blowpipe and the weld progresses from right to left with the blowpipe nozzle pointing in the direction of the unwelded surfaces, see sketch. Conversely with the RIGHTWARD WELDING TECHNIQUE the welding progresses in the opposite direction, i.e. from left to right. In order to achieve this and weld successfully the blowpipe nozzle is directed into the completed weld at a much lower angle, also with this technique the blowpipe flame precedes the filler rod along the weld. The rightward technique can have several advantages over the leftward technique in so far as less preparation is required in preparing the edges to be joined, see tables above. For example, metal up to 5 mm thick can be welded without grinding the edges to a bevel, this also leads to a saving in the amount of filler rod used. Also with the rightward technique a larger nozzle is used which leads to an increased welding speed and finally by working in front of the welding flame one can observe the melting of the sides and root of the weld to ensure good penetration. See also WELDING DEFECTS and VERTICAL WELDING.

(See illustration over.)

LOW PRESSURE CUT OFF
A special gas valve designed to prevent gas entering a system of pipe-work until it has been proved that there are no open ends. These devices are typically found in school science labs where students may have interfered with the gas taps and left them in the open position.

(See illustration on p. 257)

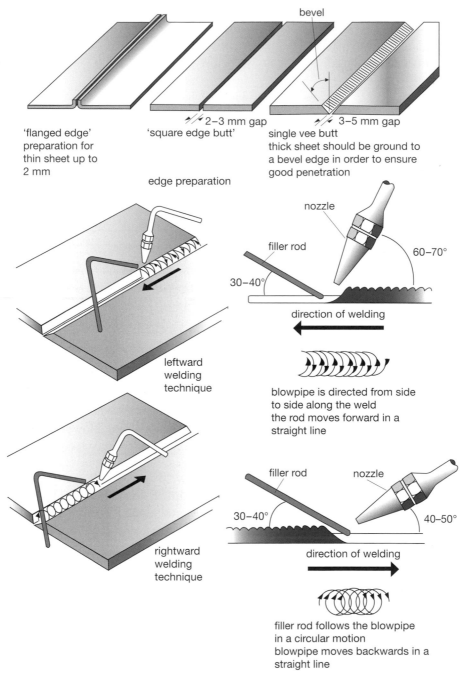

bevel

'flanged edge'
preparation for
thin sheet up to
2 mm

2–3 mm gap
'square edge butt'

3–5 mm gap
single vee butt
thick sheet should be ground to
a bevel edge in order to ensure
good penetration

edge preparation

leftward
welding
technique

nozzle

filler rod

30–40°

60–70°

direction of welding

blowpipe is directed from side
to side along the weld
the rod moves forward in a
straight line

rightward
welding
technique

filler rod   nozzle

30–40°

40–50°

direction of welding

filler rod follows the blowpipe
in a circular motion
blowpipe moves backwards in a
straight line

Low carbon steel welding

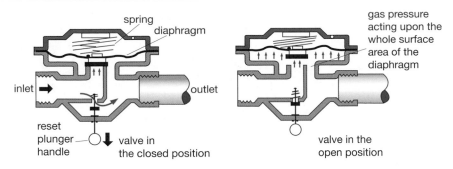

Low pressure cut-off device

The valve shown works as follows.

(1) Gas entering the pipework travels to the valve and stops as the valve is in the closed position. The gas pressure acting on the small surface area of the valve outlet is insufficient to push the valve from its seating.
(2) With the reset plunger held in the open position, gas is allowed to enter into the protected supply pipework. Should all open ends or appliances within this system be closed the pressure will build up. This, in turn, allows the gas pressure to act over the whole surface area of the weighted or sprung loaded diaphragm; it then overcomes the pressure acting downwards. Thus the valve opens.
(3) Once opened, the appliances can be operated. The pressure within the system continues to act upon the whole surface of the diaphragm and the valve remains open.
(4) Should the gas supply be isolated and shut off and the pressure released from the system, the valve will drop and seal off the system.

LOW PRESSURE HOT WATER HEATING SYSTEM
A hot water heating system which is open to the atmosphere via a FEED AND EXPANSION CISTERN and a VENT PIPE. See also HIGH PRESSURE HOT WATER HEATING SYSTEM.

LOWER EXPLOSIVE LIMITS   See FLAMMABILITY LIMITS.

LOWER FLAMMABILITY LIMITS   See FLAMMABILITY LIMITS.

LPG   Abbreviation for LIQUEFIED PETROLEUM GAS.

LUMINOUS FLAME   See NON-AERATED FLAME.

M AND F ELBOW (STREET ELBOW)

Abbreviation for an elbow which has a male thread on one end and a female thread on the other. See LOW CARBON STEEL PIPES AND FITTINGS.

MACERATOR PUMP   See WC MACERATOR PUMP.

MACHINE CAST LEAD SHEET

Sheet lead which is produced by a continuous casting process in which a rotating water cooled drum is allowed to make contact with the surface of molten lead. On contact the lead solidifies and is subsequently wound onto rolls. The thickness of the lead is regulated by the temperature and speed of the drum and is available in codes 3, 4, 5 and 6.

MACKINTOSH SINK

A SANITARY APPLIANCE found in some hospitals consisting of a shallow sink with a SLOP HOPPER at one end. It is used for the cleansing of mackintosh (waterproof) sheets.

MAG VALVE   See SOLENOID VALVE.

MAGNA BATH   See BATH.

MAGNESIUM

The lightest structural metal available, it is a silvery white colour. Magnesium is hardly used in industry except where lightness is essential (being only two thirds the weight of aluminium) and for this reason it is found extensively used in the aerospace industry. Sometimes a block of magnesium is placed in the cold storage cistern to give CATHODIC PROTECTION to the metal pipework, etc. in the system.

MAGNETIC VALVE   See SOLENOID VALVE.

MAGNETIC WATER CONDITIONER   See WATER CONDITIONER.

MAGNETICALLY OPERATED FLAME FAILURE DEVICE   See THERMOELECTRIC VALVE.

MAGNETITE

The black sludge which is often found in water filled central heating systems. Magnetite is the black rust or the black OXIDE of iron and is the result of corrosion within the system. Should too much of this magnetite build up within the system its efficiency will be lost, therefore it is essential to prevent this build up by the addition of a CORROSION INHIBITOR.

MAGS WELDING   See METAL ARC GAS SHIELDED WELDING.

MAGUIRE'S RULE
A table identifying the suggested fall to be used in drainage systems which normally carry only intermittent flows of water as might be expected from one or two households, unlike a constant flow of liquid which might flow through a drain serving several dwellings.

Diameter of pipe mm	Recommended fall
100	1 in 40
150	1 in 60
225	1 in 90

An example of a fall of 1 in 40 would mean that for every 40 mm you would drop 1 mm, or for every 40 m you would drop 1 m, etc. See GRADIENT. See also DISCHARGE UNITS and CHEZY FORMULA.

MAIN EARTHING TERMINAL
The terminal block located near the electrical consumer unit where all connections to the supply earth are made, this includes the main equipotential bonding wires.

MAINS
A term often used to represent the incoming supply of water from the Water Authority. The term 'service main' is given to the pipe laid by the water authority giving a general supply of water to many dwellings. See also SERVICE PIPE.

MAINS TAPPING MACHINE   See UNDER PRESSURE MAINS TAPPING MACHINE.

MALE THREAD
An external thread cut onto rods and pipes with dies. A male thread screws into a FEMALE THREAD.

MALLEABILITY
The ability of a metal to be worked, e.g. bossed or hammered into a new shape without breaking. See BOSSING for an example. Most metals become more malleable when their temperature is increased.

MALLEABLE IRON (MALLEABLE CAST IRON)
Cast iron which has been heat treated by an ANNEALING process making it less brittle. Basically white cast iron is packed in sand or iron ore and the temperature is raised to 900°C and kept there for 2–3 days. The temperature is then slowly reduced, this releases some of the carbon and has the effect of producing a metal which has strength, DUCTILITY and resistance to shock. Malleable iron is often used to make low carbon steel pipe fittings.

M

MALLET
A tool shaped like a hammer but instead of a steel head, as found on hammers, a head made of either hard wood or rubber is used. A mallet is used to strike wooden tools to absorb some of the blow, and prevent damage.

MANDREL (MANDRIL)
A cylindrical shaped wooden tool which was driven through lead pipe to restore it to its true bore.

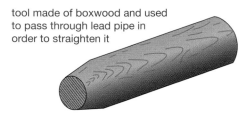

tool made of boxwood and used to pass through lead pipe in order to straighten it

Plumber's mandrel

MANHOLE
A term in common use, but it is very much being superseded by the term INSPECTION CHAMBER. The distinction between the two is very slight, therefore in this book the term inspection chamber will be used in preference to manhole.

MANIFOLD
A fitting or arrangement of pipework which is designed to branch lots of pipes from one main pipe. For examples of different manifolds see MICRO-BORE and COUPLING OF STORAGE CISTERNS.

MANIPULATIVE COMPRESSION JOINT   See COMPRESSION JOINT.

MANLID
Old fashioned term used meaning the ACCESS COVER opening into an enclosed vessel.

MANOMETER (WATER or 'U' GAUGE)
An instrument used to measure pressure and test for leaks within a pipe. It consists of a clear glass or plastic tube bent to form a 'U' shape. There are two basic types of manometer, small, for testing drainage and sanitary pipework and large, for testing gas installations which require higher readings. The tube is filled with coloured water and when pressure is applied to one leg of the tube the liquid is pushed back up the other leg. Manometers are supplied with their gauges registered as direct or indirect reading types. Both are shown in the sketch, the only difference is that with the direct reading type the scale goes up in twos whereas with the indirect type the gauge measurement in each leg has to be added to obtain a reading. For examples of manometers being put to use see AIR TEST and TIGHTNESS TESTING.

(See illustration opposite)

MANSARD ROOF (CURB ROOF)
A roof structure consisting of two pitches.

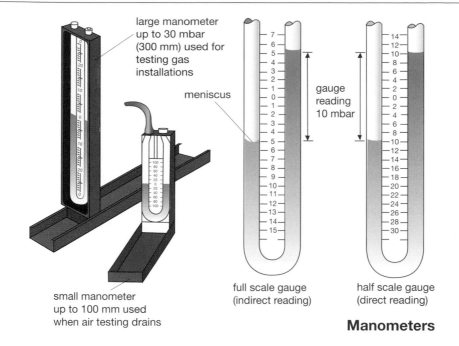

large manometer up to 30 mbar (300 mm) used for testing gas installations

meniscus

gauge reading 10 mbar

full scale gauge (indirect reading)

half scale gauge (direct reading)

small manometer up to 100 mm used when air testing drains

Manometers

MANUFACTURED GAS
Gas once commonly used in the gas distribution main. Manufactured gases are generally made from coal or oil. See TOWN GAS.

MASONRY DRILL
A type of drill used to make holes in brickwork and concrete. It has a tungsten carbide tip and can only be sharpened on a special grinding wheel. See also TWIST DRILL.

MASS
The amount of matter there is in an object. Weight can vary according to the gravitational pull.

MASTER GULLY
A trapped gully situated in a convenient position to receive SURFACE WATER from various RAINWATER SHOES. The purpose of using a master gully is to reduce the number of GULLY traps from rainwater connections in a COMBINED SYSTEM OF DRAINAGE.

(See illustration over.)

MAXIMUM INCIDENTAL or OPERATING PRESSURE   See GAS PRESSURE.

McALPINE RESEALING TRAP   See RESEALING TRAP.

MECHANICAL ADVANTAGE   See LEVER.

MECHANICAL SERVICES (MECHANICAL ENGINEERING SERVICES)
A term which is being used more and more as the years go by. It is an umbrella term for all the various trades such as plumbing, pipe-fitting, heating, ventilating, welding,

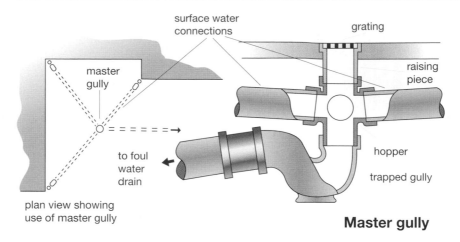

plan view showing
use of master gully

Master gully

electrical work, sanitation, drainage, gas fitting, metallic and some non-metallic roof coverings.

MECHANICAL THERMOSTAT

A thermostat that operates by the movement of a valve closing off the gas supply pipeline to an appliance as opposed to making or breaking an electrical contact. For example, see THERMOSTAT.

MECHANICAL VENTILATION

A method of ventilation designed to guarantee a given number of air changes per hour. A simple form of mechanical ventilation could consist of an extractor fan fitted to an external wall or a small pipe duct. See also NATURAL VENTILATION.

MEDIUM PRESSURE GAS REGULATOR

A pressure regulator sometimes found at the point of entry of the gas supply to the property. It is designed to take pressures up to 2 bar and as such will include an over-pressure shut-off device thereby ensuring such pressures cannot enter the property, which would prove dangerous.

MELTING POT

A cast iron vessel used on site to melt down lead for various purposes. It is just like a small cooking pot.

MENISCUS

The curvature of the water surface in a tube due to the COHESION of the water molecules, for example, see MANOMETER.

MERCURY VAPOUR FLAME FAILURE DEVICE

A LIQUID VAPOUR FLAME FAILURE DEVICE that used mercury as the VOLATILE FLUID.

MERINGUE DEZINCIFICATION

A form of DEZINCIFICATION OF BRASS in which hollow puff shaped moulds, of zinc corrosion products, are produced at the point where the destruction of the zinc in the brass is taking place.

METAL ARC GAS SHIELDED WELDING (MAGS WELDING)

An arc welding process, sometimes referred to as metal inert gas welding (MIG welding) which employs a separate supply of gas to shield the weld pool from atmospheric contamination. In comparison with manual arc welding (stick welding) MAGS welding has an advantage in that the filler rod is supplied via a continuous electrode and thus welding can prove to be faster. The gas is supplied to the weld zone to form a shield around the molten pool and thus it excludes the gases from the atmosphere which might otherwise cause OXIDATION. The gas used to shield the weld is determined by the metal being welded, for example argon and helium are used for NON-FERROUS METALS whereas for FERROUS METALS argon mixed with oxygen or carbon dioxide, or just 100% carbon dioxide is used, the latter process is often referred to as CO_2 welding. The steel electrode wire used has a thin coating of copper to prevent it from rusting and to reduce friction. See also ARC WELDING.

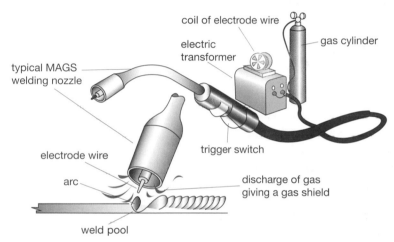

Metal arc gas shielded welding

METALLIC ROOF COVERINGS

Roof coverings in metal, the most common being lead, copper, aluminium and zinc, although we could add to this list a new material to this country, that is stainless steel.

METAPHOSPHATES

A type of crystal which is sometimes introduced into hot water systems in order to reduce scale in the pipework. See WATER CONDITIONERS.

METER   See GAS METER and WATER METER.

METER BY-PASS

A pipe which by-passes the meter into the property. Every meter by-pass pipe is fitted with a closed valve which has a water or gas authority seal (whichever is applicable). The purpose of a by-pass is to allow its temporary opening should the meter need to be removed, e.g. if the supply is completely shut down it could be dangerous. See GAS METER.

METER PIT

A chamber which is constructed below ground level to house a water meter. The chamber should be IMPERVIOUS to water and be fitted with a surface access cover for inspection purposes.

METER PRESSURE

The pressure reading at a gas meter. Two readings are generally taken, these being the standing pressure and the meter working pressure. These readings are taken with the use of a MANOMETER connected to the meter test nipple. The standing pressure is the pressure with no appliances alight and should be between 20 and 25 mbar, yet should not exceed 30 mbar. The working pressure should be 21 mbar ± 2 mbar with any gas appliance alight at full rate giving a gas flow of 0.5 m^3/h, which is equivalent to 5^1/$_2$ kW. In the case of a cooker, three hot plate rings are sufficient.

METHANE

A hydrocarbon consisting of one carbon atom and four hydrogen atoms (CH_4). It is found primarily within the mixture of NATURAL GAS. See NATURAL GAS for further reference.

METRIC SYSTEM

A set of units used internationally by most nations. It was introduced by the French National Assembly late in the sixteenth century and was adopted by the British in the early 1970s. One of the main characteristics of the system is its decimal nature; therefore the conversion between smaller and larger units is made by moving the decimal point to the left or right.

The SI system of units (Système International d'Unités), developed from the metric system, has been defined and recommended as the system of choice for scientific use worldwide. It should be noted that not all metric units are SI units. For example, hectare and litre are not SI units.

Metric equivalents

Quantity	Metric unit	Imperial unit
Length	metre (m)	yard (yd)
Length	millimetre (mm)	inch (in)
Area	square metre (m^2)	square yard (yd^2)
Area	square millimetre (mm^2)	square inch (in^2)
Area	hectare (ha)	acre
Volume	cubic metre (m^3)	cubic yard (yd^3)
Volume	cubic millimetre (mm^3)	cubic inch (in^3)
Capacity	litre (l)	gallon (gal)
Mass	kilogram (kg)	pound (lb)
Force	newton (N)	pound force (lbf)
Pressure	newton per square metre (N/m^2)	pound force per square inch (lbf/in^2)
Velocity	metres per second (m/s)	foot per second (ft/s)
Temperature	kelvin (K)	degree Fahrenheit (°F)
Temperature	degree Celsius (°C) [centigrade]	degree Fahrenheit (°F)
Heat	joule (J)	British Thermal Unit (BThU)
Power	kilowatt (kW)	BThU per hour (BThU/h)
Power	kilowatt (kW)	horse power (hp)

Conversion factors

1 in	= 25.4 mm	1 lb	= 0.4536 kg
1 yd	= 0.9144 m	1 lbf	= 4.448 N
1 in^2	$= 645.2 \text{ mm}^2$	1 lbf/in^2	$= 6894 \text{ N/m}^2$
1 yd^2	$= 0.8361 \text{ m}^2$	1 lbf/in^2	= 0.06894 bar
1 in^3	$= 16390 \text{ mm}^3$	1 ft/s	= 0.3048 m/s
1 yd^3	$= 0.7646 \text{ m}^3$	1 BThU	= 1055 J
1 acre	= 0.404686 ha	1 BThU/h	= 0.0002931 kW
1 gal	= 4.536 l	1 hp	= 0.7457 kW

To convert metric to Imperial divide by the conversion factor. To convert Imperial to metric multiply by the conversion factor.

Examples

$$10 \text{ m}^2 = 10 \text{ divided by } 0.8361 \ = 11.96 \text{ yd}^2$$
$$75 \text{ kg} = 75 \text{ divided by } 0.4536 \ = 165.34 \text{ lb}$$
$$75 \text{ gal} = 75 \text{ multiplied by } 4.546 = 340.95 \text{ l}$$

Converting temperature scales

$$0°C \quad = 273.15 \text{ K}$$
$$°C \text{ to } °F = °C \times 9 \div 5 + 32 = °F$$
$$°F \text{ to } °C = (°F - 32) \times 5 \div 9 = °C$$

Examples

$$21°C \times 9 \ = 189 \div 5 \ = 38 + 32 = 70°F$$
$$70°F - 32 = 38 \times 5 \ = 190 \div 9 = 21°C$$
$$15°C \quad = 273 + 15 = 288 \text{ K}$$

The following table shows how the metric units can be prefixed to increase or decrease their value:

Prefix	Symbol	Calculation factor
atto	a	divided by 10 18 times
femto	f	divided by 10 15 times
pico	p	divided by 10 12 times
nano	n	divided by 10 9 times
micro	μ	divided by 10 6 times
milli	m	divided by 10 3 times
centi	c	divided by 10 2 times
deci	d	divided by 10 1 times
deca	da	multiplied by 10 1 time
hecto	h	multiplied by 10 2 times
kilo	k	multiplied by 10 3 times
mega	M	multiplied by 10 6 times
giga	G	multiplied by 10 9 times
tera	T	multiplied by 10 12 times
peta	P	multiplied by 10 15 times
exa	E	multiplied by 10 18 times

Example

$$1 \text{ millimetre} = 0.001 \text{ metre}$$
$$1 \text{ decimetre} = 0.1 \text{ metre}$$
$$1 \text{ kilometre} = 1000 \text{ metres}$$

MICRO-BORE (MINI-BORE)

A HOT WATER HEATING SYSTEM using pipes as small as 6 mm. The main advantages of this type of system are the ease in which it can be installed, the reduced cost of installation and the faster pump velocities which can be used giving a quicker heat up period.

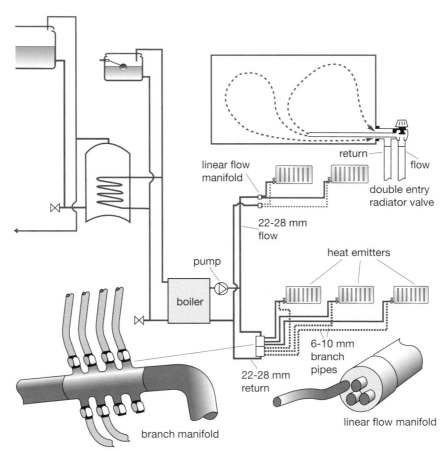

Micro-bore heating system

MICRO-SWITCH

A small electrical switch found in components such as a pressure switch or motorised valve. They are generally designed to operate automatically in response to the component on which it is installed, thereby allowing an electrical current to pass onto the next stage.

MID-POSITION VALVE   See MOTORISED VALVE.

MIG WELDING   See METAL ARC GAS SHIELDED WELDING.

MILD STEEL   See LOW CARBON STEEL.

MILLED LEAD
The bulk of sheet lead produced today. Milled lead is produced by passing a slab of lead to and fro between steel rollers which press it until the required thickness is obtained.

MINI-BORE   See MICRO-BORE.

MINIATURE CIRCUIT BREAKER (MCB)   See CIRCUIT BREAKER.

MINUETE VALVE
The trade name for one design of BALLVALVE, being a quarter turn valve fitted in the pipeline to isolate an appliance.

M.I.P.   See GAS PRESSURE.

MIRROR TEST   See REFLECTION TEST.

MITRED HIP or VALLEY   See CLOSE CUT HIP OR VALLEY.

MIXER TAP
A tap in which hot and cold water are delivered through a common spout. There are two basic designs of mixer taps, those in which the hot and cold water is mixed in

M

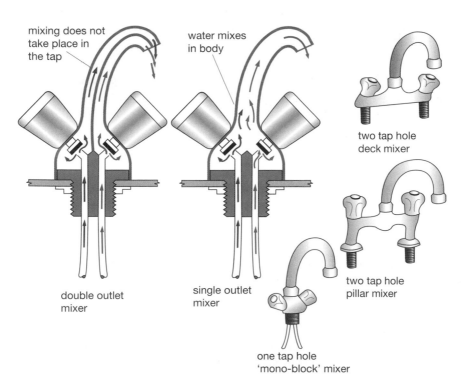

Mixer taps

the tap body or spout (single outlet mixer) and those in which the hot and cold water does not actually mix until it is discharged from the outlet nozzle (double outlet mixer). The first type described is only suitable for use when the water from the hot and cold supplies are at equal pressures, failure to observe this could result in the water having the greater pressure flowing back down the other feed pipe. This results in unsatisfactory water flows and temperatures discharging from the nozzle outlet and can lead to excessive pressure being exerted in low pressure systems causing damage; also there is a greater risk of contamination from stored water supplies to the mains supply. Many mixing taps are fitted with a swivel outlet. This allows the water to be discharged at various positions and proves useful with modern sinks which often have more than one bowl. One major draw back to mixer taps fitted with a swivel outlet is the fact that the neoprene 'O' ring used to ensure a watertight seal does tend to wear out and therefore requires frequent replacements. See also SPRAY MIXING TAP.

MIXING VALVE
A valve in which hot and cold water from separate supplies mix together. See MIXER TAP and SHOWER MIXING VALVE. See also THERMOSTATIC MIXING VALVE.

MODIFIED INDIRECT COLD WATER SUPPLY
An INDIRECT SYSTEM OF COLD WATER SUPPLY in which the wash basins and the kitchen sinks are supplied directly from the mains supply and all other sanitary appliances are supplied from a cold water storage cistern. With a system such as this, drinking water is available at the basins as well as the sink which is desirable especially when cleaning your teeth, etc.

MODIFIED ONE PIPE SYSTEM   See SECONDARY VENTILATED STACK SYSTEM.

MODIFIED SINGLE STACK SYSTEM   See SECONDARY VENTILATED STACK SYSTEM.

MODULAR BOILER
A boiler consisting of three or more identical boilers (or modules) installed together that are designed to share the same heating load.

MODULAR FLUE (COMMON FLUE SYSTEM)
A flue system in which several appliances connect into the same flue, such as from a modular boiler arrangement.

MODULATING BURNER
A burner designed to operate at varying high and low settings, as compared with the 'full on' or 'off' function of the more traditional design of burner found in conventional boilers with cast iron heat exchangers. Modern boiler heat exchangers have improved efficiency and a greater heat transfer can be achieved, down to about 25% of the full gas rate therefore allowing for a reduced flame size. Modulating burners tend to eliminate burner ignition noises and reduce the tendency for corrosion conditions on the heat exchanger surface.

MODULATING VALVE   See COMPENSATING VALVE.

MOISTURE BARRIER   See VAPOUR BARRIER.

MOLE GRIP   See LEVER WRENCH.

MOLE PLOUGH
A device which bores a hole/tunnel through the ground and pulls a plastic or copper tube. The mole plough consists of a long vertical blade about 10 mm thick which slices through the soil. At the bottom of the blade is fixed a pointed cylindrical cutter or 'mole'. First the start and finish holes are dug and the blade with its mole is adjusted to the required depth and positioned in the starting hole. The plough is then pulled by a tractor to the second hole boring the hole and pulling the tube with it. Mole ploughing is generally limited to ground which is free from obstructions.

MOLECULE
The smallest portion of a substance capable of independent existence, made up from two or more atoms. For example, many thousand water molecules go to make one droplet of WATER.

MOLESKIN   See WIPING CLOTH.

MOMENTUM   See TRAP SEAL LOSS.

MONKEY'S TAIL   See RUNNING ROPE.

MONKEY WRENCH
An American term meaning a type of adjustable spanner in which the jaws are at right angles to the handle. The term monkey wrench was adopted on account of its inventor's name Charles Moncky.

MONO-BLOCK TAP
A mixer tap which only uses one hole through the sink top or work surface. For example, see MIXER TAP.

MONODRAFT FLUE
A design of flue which allows the products of combustion to pass up through a central flue route and the air supply to pass down to the appliance via a concentric void surrounding the flue. For example, see BALANCED COMPARTMENT.

M.O.P.   See GAS PRESSURE.

MOTORISED VALVE
A valve fitted with an electrically operated motor on top to open or close the valve automatically. The power switching on the supply to operate the valve is regulated by either a thermostat or a time clock. There are two basic types of motorised valve, a zone valve or a diverter valve. The zone valve simply opens and closes the waterway and is fitted in a straight run of pipe, whereas a diverter valve is fitted at a 'tee' connection and would send the flow of water either one way or the other, this valve

can be wired to give priority to either the domestic hot water or the heating circuit. Some diverter valves are designed to have a midway position allowing water to flow in both directions at the same time.

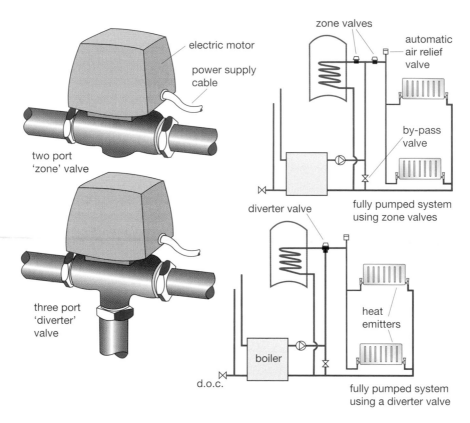

Motorised valves

MOULD BURNING

A method which can be adopted to lead weld vertically on thick sections of sheet lead. A small semi-circular mould is offered up to the jointing surfaces and the lead is introduced via a filler rod filling the mould and fusing the two surfaces. The section just welded is allowed to cool and the mould is moved up the joint to a new position for welding. When joining thick sheet it is advisable to bevel the jointing edges to ensure good penetration to the weld, see sketch.

(See illustration opposite)

MOUNTINGS

Term used to mean various accessories, for example, see BOILER MOUNTINGS.

MUD GULLY   See YARD GULLY.

MULSIFYRE SPRINKLER SYSTEM (OILFYRE SPRINKLER SYSTEM)

A sprinkler system designed to discharge a fine jet of water through a series of projector nozzles. The Mulsifyre system proves particularly useful in buildings where

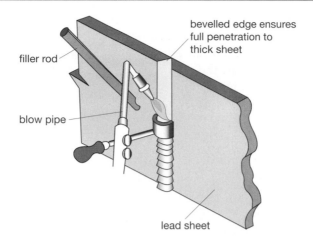

Mould burning

oil and paint fires can be a problem. Should the projector nozzles open droplets of water are discharged onto the fire and thus when the water emulsifies or mixes with the oil it cools and the resulting mixture will not support combustion. The Mulsifyre system works as follows:

(1) The air pipe to the SPRINKLER HEADS is charged with compressed air, this in turn holds the valve to the projector nozzles closed.
(2) Should a sprinkler head detector head burst due to excessive heat the air pressure is released from the supply valve and water is allowed to travel to all the nozzles and spray out water. See also SPRINKLER SYSTEM.

M

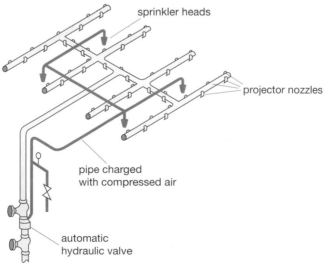

Mulsifyre sprinkler system

MULTI-BRANCH FITTING
A branch fitting with more than one branch connection, it is sometimes referred to as a group connector. See also COLLAR BOSS.

MULTI-FUNCTION GAS VALVE

A component part of most modern domestic gas boilers. The multi-function gas valve is really a series of component valves all incorporated in the one unit. Found within the gas valve shown are the GAS GOVERNOR, a THERMOELECTRIC VALVE with its connecting THERMOCOUPLE and the SOLENOID VALVE which is operated by the THERMOSTAT and allows gas to the main burners. The multi-function gas valve shown works as follows:

(1) Gas enters the valve and passes through the governor to reduce its pressure to that acceptable for the boiler.
(2) The boiler thermostat is switched to zero.
(3) The large button on the valve is depressed and held down, this opens the thermoelectric valve and allows gas through to the pilot jet; simultaneously the smaller button is depressed several times, this operates a spark to ignite the pilot flame.
(4) Once the pilot has been alight for 20–30 seconds the flame playing onto the thermocouple energises the thermoelectric solenoid valve, thus holding the valve open, therefore the pressure can be released from the large button.
(5) The main flow of gas can now continue to the next solenoid valve which only opens and allows gas to the main burners when the boiler thermostat is turned up and calls for heat, this electrically energises this valve causing it to open.

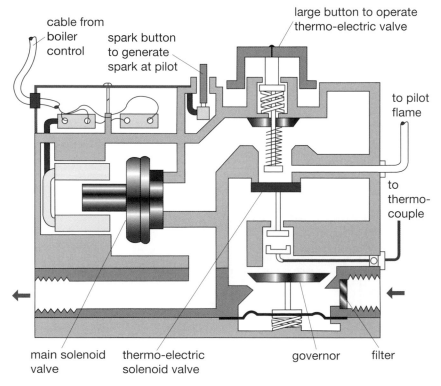

Multi-function gas valve

MULTI-JET SPRINKLER SYSTEM   See SPRINKLER SYSTEM.

MULTI-METER

A meter designed to measure an assortment of electrical readings, including: voltage, amperes – direct and alternating current, resistance and continuity. The dial is set on the desired location/setting and, with the probes inserted into the correct ports, readings can be taken. Using a multi-meter does away with the need to have several different meters, e.g. voltmeter, ammeter, ohmmeter, etc.

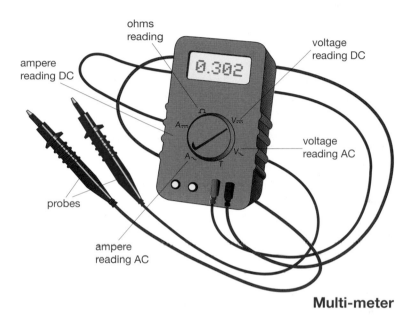

Multi-meter

M

MULTI-POINT   See INSTANTANEOUS WATER HEATER.

MULTI-QUICK

The trade name for a well known WC pan connector. For example, see WC SUITE.

MUNZING RING (MUNSON RING)   See PIPE SUPPORT.

N

NATURAL DRAUGHT BURNER   See ATMOSPHERIC BURNER

NATURAL DRAUGHT FLUE
A flue which removes the products of combustion from an appliance by means of CONVECTION CURRENTS, see OPEN FLUE and BALANCED FLUE.

NATURAL GAS
Most of the gas used in this country as a fuel in the home is now natural gas. Natural gas is a mixture of hydrocarbons, mainly methane and ethane, that form beneath the Earth's surface often occurring in association with oil. The gas is trapped below an impermeable strata of rock, salt or shale and extracted by drilling and pumping. The composition of natural gas is given in the table, note that there is no hydrogen and carbon monoxide present in the gas unlike the old town gas once used in this country. The percentage given can vary depending upon the underground gas field.

Composition of natural gas

Constituents	Approximate percentage
Methane	94.0
Ethane	3.0
Nitrogen	1.3
Propane	0.7
Butane	
Carbon dioxide	
Helium	1.0
Other hydrocarbons	

Natural gas is odourless in its natural form therefore a chemical scent is added to give it a distinctive smell, this acts as a warning should there be a leak. See also TOWN GAS.

NATURAL VENTILATION
Natural ventilation is by far the most common method used to achieve the required AIR CHANGES in a room. Natural ventilation includes windows which can be opened, air bricks and openable ventilators which open to the external air. Natural ventilation is generally based on CONVECTION CURRENTS; fresh cool air is admitted at low level which results in the VITIATED AIR being forced out at high level. In general all habitable rooms, kitchens, bathrooms and WC compartments should have at least one ventilation opening of at least $1/20$ th of the floor space; if this cannot be achieved a means of mechanical ventilation should be installed.

NEAT FLAME   See NON-AERATED FLAME.

NEEDLE VALVE
A valve designed for use where a fine adjustment is required. The valve basically consists of a tapered/conical-shaped valve that draws away from its seating, allowing the gas or liquid to flow with a slowly increasing volume.

NEOPRENE
The trade name for a synthetic rubber product. Neoprene is possibly the most common synthetic rubber used in the plumbing industry, being used extensively in the manufacture of 'O' RINGS. It is resistant to the ultraviolet rays of the sun, and is also flame resistant, being self-extinguishing.

NEUTRAL FLAME
A flame which burns with equal amounts of oxygen and acetylene at the welding blowpipe, recognised by a small blue rounded inner cone at the welding tip, for example, see OXY-ACETYLENE FLAME. A neutral flame is used when welding metals such as lead, copper, low carbon steel and cast iron.

NEUTRAL POINT (DATUM POINT)
The neutral point is generally regarded as the cold feed entry position in a HOT WATER HEATING SYSTEM. It is at this point that any pressure change brought about by a pump being switched on and off is negligible. Therefore when installing any pipework layout the entry point for the inlet of cold water from the feed and expansion cistern must be carefully designed so that the neutral point is at a position where the vent pipe can also be connected, thus ensuring that there will also be no adverse positive or negative pressures along this pipe causing water to be pumped over the vent pipe into the cistern or air to be drawn into the system. In central heating systems, which are fully pumped, it is generally advisable to run the cold feed and vent pipe to a position within 150 mm of each other in the system, making this the neutral point. To achieve this, a DE-AERATOR sometimes is used.

(See illustration over.)

NEUTRAL WATER
A supply of water which is said to be neither ACIDIC nor ALKALINE, for classification, see pH VALUE and WATER.

NEWTON   See PRESSURE.

NIPPLE
A short piece of low carbon steel or brass pipe with a male thread at each end. See BARREL NIPPLE and HEXAGON NIPPLE.

NIPPLE (PRESSURE POINT)
The pressure test position on a gas supply pipe, for example, see TIGHTNESS TESTING.

NITROGEN
A colourless, odourless gas which is chemically inactive and forms about 78% of the Earth's atmosphere. Nitrogen is an important element being vital to all living organisms.

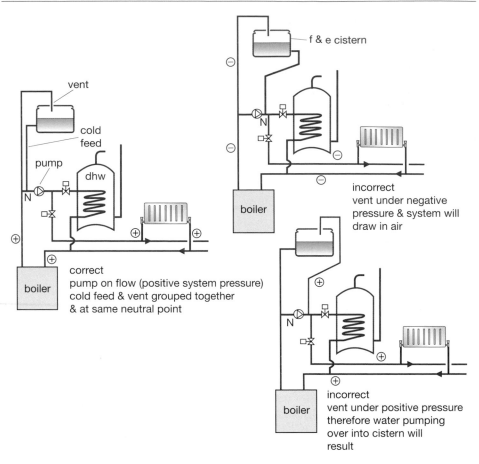

vent

cold feed

pump

dhw

N

boiler

correct
pump on flow (positive system pressure)
cold feed & vent grouped together
& at same neutral point

f & e cistern

N

boiler

incorrect
vent under negative
pressure & system will
draw in air

N

boiler

incorrect
vent under positive pressure
therefore water pumping
over into cistern will
result

The neutral point

NOBLE GASES (INERT GASES)

A family of six chemical elements, namely argon, helium, krypton, neon, radon and xenon. The term noble refers to the extreme lack of chemical reactivity of these gases, for example, they very rarely form compounds with other atoms. Most of these gases are used in some form of commercial application, neon for example is used in some types of fluorescent lighting and argon is used in the ARGON ARC WELDING process, the argon being used to prevent the oxygen in the air getting to the weld causing OXIDATION.

NOBLE METAL

Any of several metallic elements that have good resistance to OXIDATION. The collection of noble metals is not strictly defined but includes metals such as silver, platinum and gold. The term noble metal is also sometimes used to describe the difference between metals on the ELECTROMOTIVE SERIES, the less noble metal (the anode) being the one which is destroyed by another more noble metal (the cathode).

NOISES IN COLD WATER PIPEWORK

For the most common causes of noise in cold water pipework, see WATER HAMMER. But noises can also be caused due to the VELOCITY of water flowing through a pipe.

Pipes such as copper and stainless steel, being thin-walled, tend to transmit, by vibration, the sound caused by the passage of water through their surfaces and when these pipes are fixed on hollow surfaces such as stud partition walls the partition acts as a sounding box, magnifying the noise created. Therefore fixing to such surfaces should be avoided. Lagging the pipes and inserting a piece of rubber behind the fixing clip often cuts down these noises. See also NOISES IN HOT WATER PIPEWORK.

NOISES IN HOT WATER PIPEWORK
Most noises are caused as a result of insufficient room being left for expansion and contraction of the pipework due to thermal movement caused by the water heating and cooling. When running pipes through walls they should be sleeved, equally when running the pipes under floor boards it is essential to ensure that the notches cut in the joists are large enough for the pipe allowing for a bit of free movement, it is also a good idea to sit the pipe on a piece of felt. Allowing for movements should also be considered when fixing your pipework. Another noise problem in hot water pipework can be encountered if air becomes entrapped; therefore when designing and installing a system make sure that the air can escape via a VENT PIPE or AIR RELEASE VALVE. Pumps can sometimes give problems either because they are causing the pipe to vibrate and cause a continued tapping or because their velocity is too fast for the size and design of pipework. See also BOILER NOISES and EXPANSION JOINT.

NOMINAL CAPACITY
The capacity of water a cistern could hold if filled right up to the top. In reality the cistern used in a building is never filled with this much water because a ballvalve and overflow pipe are fitted in the top section of the cistern. See also ACTUAL CAPACITY.

NON-ACTIVE FLUX (INACTIVE FLUX)   See FLUX.

NON-AERATED BURNER
A burner in which all the air used to support combustion is mixed with gas as it leaves the burner nozzles, unlike an AERATED BURNER.

NON-AERATED FLAME (LUMINOUS FLAME/NEAT FLAME/POST AERATED FLAME)
The name given to a flame which burns at the point of combustion when using a NON-AERATED BURNER. The flame is quiet and stable and it will not LIGHT BACK into the burner jet. See also AERATED FLAME.

NON-COMBUSTIBLE
A material that will not readily burn.

NON-CONCUSSIVE VALVE (SPRING LOADED VALVE)
A self-closing valve or tap. Non-concussive valves are used on drinking fountains and public showers and are found on wash basins in public toilets and schools to conserve water. Non-concussive taps fitted to basins are generally designed so that the head has to be depressed to permit the valve to open, upon releasing the pressure the valve should slowly shut off. It is essential when fitting or repairing these valves to ensure that they are regulated to close slowly as the manufacturers' instructions, failure to do so could result in WATER HAMMER and an unsatisfactory operation of

the valve. The term non-concussive valve is also sometimes used to mean any tap in which its opening or closing does not cause water hammer.

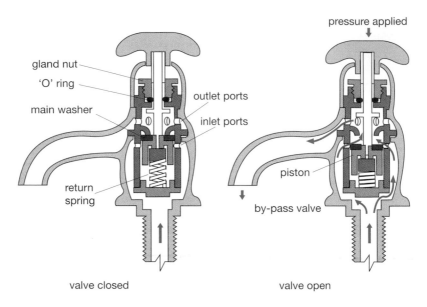

valve closed   valve open

Non-concussive tap

NON-FERROUS METAL
A metal or ALLOY containing no, or only the smallest trace of, iron, i.e. copper, lead, aluminium, zinc, brass, etc.

NON-MANIPULATIVE COMPRESSION JOINT   See COMPRESSION JOINT.

NON-METALLIC ROOF WEATHERING
Any roofing material which consists of anything except metal. This includes all the various slates, tiles and felts used. When the plumber talks of a non-metallic roof weathering material for a flat roof, 'Nuralite' sheet is often what they are referring to (Nuralite being the trade name for an asphalt bonded fibre sheet of laminar construction), but one should not jump to the conclusion that this is what is being referred to as non-metallic refers to any kind of bitumastic roofing felt. See NURALITE and NUTEC NURALITE.

NON-RETURN VALVE (REFLUX VALVE)
A valve which allows liquid to flow in one direction only. The valve is only opened when there is a flow of water, thus if the flow stops or there is a BACKFLOW of water the valve closes. There are several types of non-return valves which are sometimes called different names to identify them from each other. The two types shown are generally referred to as the check valve and the flap valve.

(See illustration opposite)

NON-STORAGE COLD WATER SUPPLY
A DIRECT SYSTEM OF COLD WATER SUPPLY.

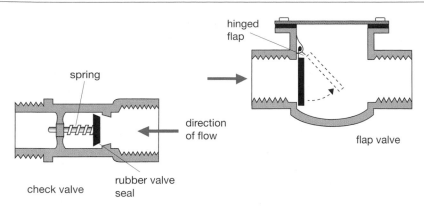

spring

direction
of flow

check valve

rubber valve
seal

hinged
flap

flap valve

Non-return valves

NORTH LIGHT
A pitched roof sometimes used in factories which has one steep and one shallow slope. The steeper side is normally glazed and faces north.

NOTICE PLATE   See CHIMNEY PLATE.

NOZZLE (SPOUT)
The outlet point from a tap. See BIB TAP.

NOZZLE   See WELDING TIP.

NURALITE
A non-metallic roof weathering material, 'Nuralite' is a trade name for an asphalt bonded asbestos roofing sheet of laminar construction. Nuralite sheet is available in various forms, although the standard sheet size is 2400 mm × 900 mm. It weighs about 2.45 kg/m². Nuralite is easy to cut with either tin snips or a knife and is usually worked by heat from a blowlamp, this softens the material for a period to allow one to mould it to the required shape. Once in its softened state only two details need to be mastered, these are the internal and external corners; the internal being constructed forming a DOG EAR as shown.

The jointing of Nuralite is very simply made by one of three basic methods, the D12C joint, the lap joint and the DE-LAMINATED JOINT. The third jointing method, the de-laminated joint is generally restricted to external corners only. The lap joint is normally restricted to sloping roofs and when making this joint a 75 mm lap would be required with the application of a No. 3 jointing compound. When laying Nuralite roofing sheet to roofs, the sheets are generally just laid onto the decking with a 2 mm gap between each sheet. One side of the Nuralite is smooth compared to the other and it is this smooth side that should be uppermost, exposed to the atmosphere, due to its superior water-shedding properties. The Nuralite should be fixed to the decking with either a No. 10 or No. 30 jointing compound as specified below. The edge of the sheet should be nailed at 300 mm centres, 12 mm from the edge of the sheet. Once the sheets have been laid and all the edges and sides completed the jointing method D12C is employed to make the joint water tight.

Nuralite jointing compounds	
Number 1	A bitumastic welding block used for de-laminated joints.
Number 3	A bitumastic block compound used for lap joints.
Number 10	A cold bitumastic undersheet adhesive used on roofs up to 40° pitch
Number 30	A hot bitumastic undersheet adhesive used on roofs under 15° pitch or where high wind pressures are likely.
D12C	A 100 mm wide, heat applied jointing strip.

Nuralite can be laid in the more traditional methods of ROLL CAP ROOFING as would be employed on zinc roofs but this method is only usually carried out for reasons of appearance. There are a range of preformed flashings which are available to meet many common applications and for more details contact the Nuralite company in Rochester, Kent. See also NURALITE FX SHEET and NUTEC NURALITE.

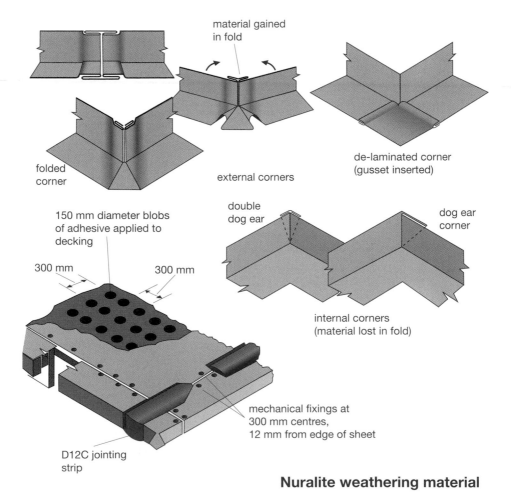

material gained in fold

folded corner

external corners

de-laminated corner (gusset inserted)

150 mm diameter blobs of adhesive applied to decking

300 mm   300 mm

double dog ear

dog ear corner

internal corners (material lost in fold)

mechanical fixings at 300 mm centres, 12 mm from edge of sheet

D12C jointing strip

Nuralite weathering material

NURALITE FX SHEET

The notes for Nuralite FX are the same as those for NURALITE (standard sheet) except that the FX sheet is much easier to fabricate especially when forming the DE-LAMINATED JOINT. The FX sheet also has other advantages over the standard Nuralite in that it will withstand more readily the movement experienced in new building work, it also tends to have a better resistance to general wear and tear. See also NUTEC NURALITE.

NUTEC NURALITE
The notes for 'Nutec' are the same as those for NURALITE (standard sheet) except that with Nutec the material does not contain any asbestos at all and is ideal in areas where asbestos is prohibited. Nutec NURALITE FX SHEET is also available.

NYLON
A strong, rotproof THERMOPLASTIC, used more and more in the manufacture of valves, valve seatings and washers, etc.

N

O GEE GUTTER   See EAVES GUTTER.

O RING
A round rubber ring usually made from NEOPRENE which comes in many sizes. O rings are found in mixer taps to allow a swivel spout to turn without water leaking from the joint; O rings are used in shower mixing valves to allow the spindle to move in and out opening and closing the valve; large O rings are also used to make certain types of push fit joints.

OBLIQUE DRAWING   See PROJECTION DRAWING.

OBSTRUCTION TEST TO DRAINS (BALL TEST)
A test carried out in which a ball 13 mm less than the diameter of the pipe is rolled through a drain, should there be no obstruction the ball will roll freely through, for example, see ALIGNMENT TEST.

OCCUPIED ZONE
Any space in which people linger for a 'significant time'. Generally if we were referring to an occupied room we would be indicating a typical height of 1.8 m, and bounded by a perimeter of 150 mm from the walls and not necessarily the total room volume which may have ceiling heights of over 3 to 4 metres.

OFFSET (SWAN NECK)
A fitting or formation of pipework used to join two pipes running in the same direction but not in line, for example, see PIPE BENDING TECHNIQUE.

OHM   See RESISTANCE.

OHM'S LAW
The relationship between current (amperes), voltage (volts) and resistance (ohms). Presuming there are no temperature changes, Ohm's Law allows one to find the missing electrical value, should the other two values be available.

Ohm's law is written as amps × ohms = volts

Which transposes to volts ÷ ohms = amps

or volts ÷ amps = ohms

Example

A cable of 0.043 Ω resistance carries a current of 140 A. What will be the voltage drop in the cable?

$$\text{amps} \times \text{ohms} = \text{volts}$$

$$\text{Voltage drop} = 140 \times 0.043 = \underline{6.02 \text{ volts}}$$

OIL FILTER

A filter which is fitted in the oil line from the oil storage tank to an oil burning appliance. It is designed to prevent any debris in the stored oil from blocking the fine jets on the burner.

OIL FIRED BOILER

Large volumes of oil stored in bulk will not easily burn therefore it is necessary to break the liquid up into a form of fine mist or vapour, thus when mixed with the correct amount of air a combustable mixture is produced. There are two types of oil fired boilers, the ATOMISING and VAPORISING BURNERS. Generally oil fired boilers are FREESTANDING and the flue can be either of the BALANCED FLUE type or a vertical chimney. When installing or servicing these boilers it is essential that a COMBUSTION EFFICIENCY TEST is carried out to ensure complete combustion of the fuel. See also OIL STORAGE and FLUE.

OIL STORAGE AND SUPPLY

There are two domestic grades of oil used for OIL FIRED BOILERS, class C2, 28 second fuel (kerosene or light oil) and class D, 35 second oil (gas oil or domestic fuel oil). The 35 second gas oil is slightly thicker and tends to wax up in winter, it is distinguished by its red colour. Generally the oil is stored in a large rectangular tank with a minimum capacity of 1250 litres, however in the interest of convenience and saving money on oil deliveries larger tanks are usually installed. When installing oil storage tanks a little thought must be given to its siting as it can be unsightly. The oil delivery tanker must be able to drive without 30 metres of it otherwise a 50 mm nominal bore pipe would be required to extend the oil inlet feed to such a position. A vent pipe must be fitted as shown being at least equal in diameter to the fill pipe and its outlet must be fitted with a BALLOON. The tank should be fitted on a support to give a minimum HEAD to the burner of at least 300 mm, otherwise a pump may be required; equally the head should not exceed 3 m. When siting the tank it should not be fitted on a level support but be made to slope one way, this ensures complete drainage of the vessel via a drain valve when required. The oil outlet to the burner should be taken from the opposite end away from the slope slightly above the base of the tank, and on this pipe an isolation valve should be fitted, see sketch. If the oil tank is sited externally it should be at least 1.8 metres from the building and preferably above ground. Should the tank be installed closer to the building than this or fitted internally a fire resisting wall must be provided. Generally fully annealed copper tube (grade table Y) is used for oil supply lines to the burner and the joints used should be of the MANIPULATIVE COMPRESSION type only. Along the supply pipe to the burner an oil filter and FIRE VALVE should be fitted as shown. Normally the oil supply line is of the one-pipe feed but if the minimum head of 300 mm cannot be achieved, a two-pipe feed is employed or a one pipe feed which incorporates a DE-AERATOR thus ensuring the burner is always primed with oil.

(See illustration over.)

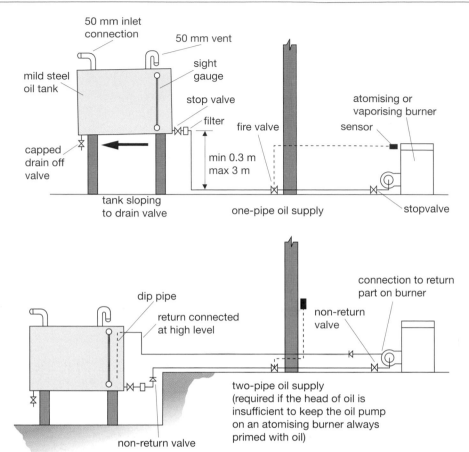

Oil storage and supply

OIL TANK   See OIL STORAGE AND SUPPLY.

OILFYRE SPRINKLER SYSTEM   See MULSIFYRE SPRINKLER SYSTEM.

OLIVE   See COMPRESSION RING.

ONE PIPE CIRCULATION (SINGLE PIPE CIRCULATION)
A form of PARASITIC CIRCULATION in which the circulatory flow of hot water, by CONVECTION CURRENTS, takes place in the same pipe. When installing a hot storage vessel it is advisable to run the vent or hot draw off connection horizontal for at least 450 mm, this would prevent one pipe circulation and its subsequent waste of heat.

(See illustration opposite)

ONE PIPE SYSTEM OF ABOVE GROUND DRAINAGE
The name given to the first attempts of discharging soil and waste water down the same discharge stack. Prior to the one pipe system the soil water from WCs, etc. did not mix with the WASTE WATER from baths and basins, etc. until it met at ground level in the drain (see TWO PIPE SYSTEM). Today, however, the term one pipe system is seldom used although strictly speaking it is still installed, see SANITARY PIPEWORK.

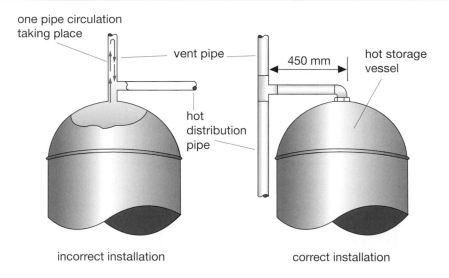

one pipe circulation taking place

vent pipe

450 mm

hot storage vessel

hot distribution pipe

incorrect installation

correct installation

One pipe circulation

ONE PIPE SYSTEM OF CENTRAL HEATING

A type of hot water heating system sometimes employed reducing the installation costs, due to less pipework being required; but equally these systems can have several disadvantages over the more expensive two pipe systems. Firstly, the first HEAT EMITTER in the system passes its cooler water back into the main FLOW PIPE thus resulting in the heat emitters at the end of the heating circuit being cooler than those at the beginning, therefore careful BALANCING of the system is essential. Secondly, due to its design, the pump only forces water around the main flow pipe and not through the individual radiators, these are only heated by CONVECTION CURRENTS, therefore the heat emitter used must offer only a very little resistance to the natural flow of water rising. The one pipe system can be completely designed to work by GRAVITY CIRCULATION although to do so requires larger pipes and a design of system which will offer no resistance to convection currents, therefore in most cases a pump is used to assist the circulation of water. Basically there are four designs of one pipe circulation systems, the drop, the ladder, the parallel and the ring systems. Each is shown in the sketch and the layout chosen usually depends upon the space available to run the pipework. In the case of the drop system the venting of the heat emitters is not required as air can escape from the system via the flow pipe. See also HOT WATER HEATING SYSTEM and TWO PIPE SYSTEM OF CENTRAL HEATING.

(See illustration over.)

ONE PIPE SYSTEM OF OIL SUPPLY   See OIL STORAGE AND SUPPLY.

ONION

The hole which is formed between two pieces of metal which are being welded. See VERTICAL WELDING.

OPEN-ENDED SPANNER   See SPANNER.

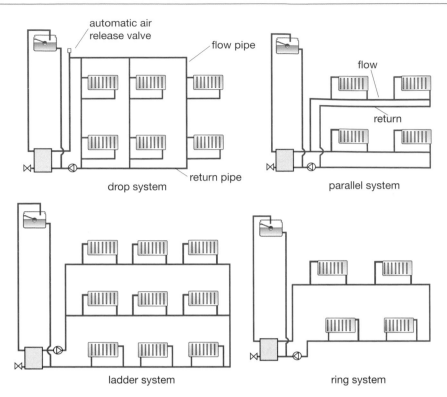

automatic air
release valve

flow pipe

drop system

return pipe

flow

return

parallel system

ladder system

ring system

One pipe central heating systems

OPEN FLUE (CONVENTIONAL FLUE)

A flue pipe or chimney that discharges its products of combustion to the outside environment in a safe position. The air used for combustion is taken from within the building, therefore with this type of flue arrangement it is essential to ensure that adequate ventilation is available. Most conventional flues work on the principle of convection currents. The hot products of combustion rise up the flue due to the cold air surrounding the base of the appliance. Because flues of this type rely on convection currents to make them work, the number of bends should be limited and, if used, should not be less than 135°. The up-draught of gases within the flue can be increased by extending the height of short flues and, equally, should the flue be too long, say, over 10 metres, the draught may become weaker and mechanical means may be required to assist the extraction of gases. A good height for a flue is usually between 6 and 9 metres. The termination of a flue should be at a position that is not adversely affected by wind pressure. See also BALANCED FLUE, FLUE, FLUE PIPE and VENTILATION GRILLE.

(See illustration opposite)

OPEN FRONT WC SEAT (OPEN RING SEAT/CUTAWAY SEAT)

A hinged WC seat in the shape of a horseshoe with a gap at the front. Designed for use in schools, factories, etc. to prevent careless splashes fouling its surface. For example, see WC SEAT.

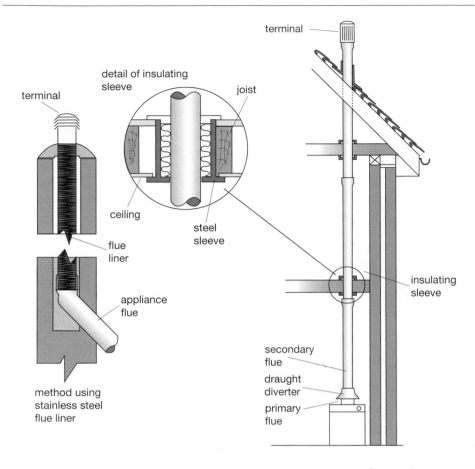

Open flue

O

OPEN VENTED SYSTEM
A hot water system that includes a storage cistern, for examples, see CENTRALISED
HOT WATER SUPPLY and HOT WATER HEATING SYSTEM.

OPERATING PRESSURE (BURNER PRESSURE)
The gas pressure reading to which a manufacturer requires a gas-burning appliance
to be set. Sometimes manufacturers give two setting pressures, indicating a maximum
and minimum range setting. The gas-commissioning engineer will set the appliance
accordingly within this range to suit the desired output for the appliance depending
upon the design requirements.

(See illustration over.)

OPSO/UPSO   See OVER PRESSURE/UNDER PRESSURE SHUT OFF.

OPTIMISER
A system of commercial heating control that considers the outside temperature. To
understand its principle you should review the entry relating to the COMPENSATING
VALVE. However, the optimiser also considers the time of day at which the heating is

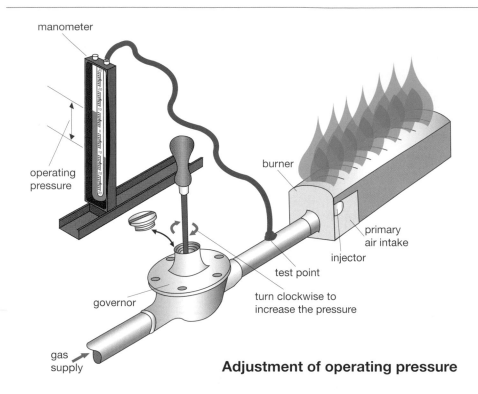

Adjustment of operating pressure

to be brought on. For example, during a cold night the optimiser will allow a longer period for the building to warm up than on a mild night. Thus, instead of using a fixed time clock to bring on the heating every morning, the time at which the heating comes on varies.

ORIFICE
The word orifice means an opening, but this term is sometimes used to represent the SEATING of a BALLVALVE.

ORTHOGRAPHIC PROJECTION   See PROJECTION DRAWING.

OSCILLATION
Name sometimes used to identify ballvalve murmur, see WATER HAMMER.

OUTFALL
The discharge point of a pipe conveying soil or surface water.

OVERCLOAK
When referring to a sheet weathering material, the overcloak is the sheet which lies on top of the next sheet thus ensuring a watertight joint. For one example of an overcloak, see LEAD SHEET AND ROOF COVERINGS.

OVER PRESSURE/UNDER PRESSURE SHUT-OFF (OPSO/UPSO)
A device fitted to an LPG installation to prevent excessively high or low gas pressures entering the domestic supply. In each case should the pressure be incorrectly

regulated, due to a malfunction, the relevant diaphragm within the valve lifts against its spring setting to close the supply. Manual re-setting will be required in each case. The OPSO re-set is sealed to prevent unauthorised re-setting. See also BULK TANK.

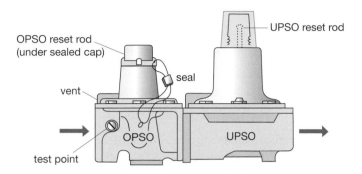

Over pressure/under pressure shut off

OVERFLOW PIPE
A pipe connected to any vessel or chamber containing water, its purpose is to remove excess water should the level rise too high. An overflow pipe could also be a WARNING PIPE, in which case it must be discharged in a position where it is readily seen. To ensure cold drafts do not blow up the overflow pipe connected to storage cisterns, possibly freezing the ballvalve and water in winter, it is advisable to turn down the overflow pipe connection into the water at the cistern connection. See STORAGE CISTERN.

OXIDATION
The combination of a substance with oxygen, usually forming an OXIDE of that substance. Some examples of oxidation include the reaction of iron with oxygen forming rust, which is basically an iron oxide. Clean pieces of copper or lead would soon tarnish if left exposed to the atmosphere, this thin coating of tarnish is an oxide coating. Oxidation also occurs with other substances as well as metals for example methane (which is a compound of carbon and hydrogen) when combined with oxygen forms carbon dioxide and water (hydrogen oxide). The process of oxidation is usually speeded up if a flame or heat is applied and it is for this reason that special precautions must be taken when soldering or welding metals. The use of heat causing oxidation to a metal can also be put to use when using a cutting blowpipe, see CUTTING BLOWPIPE.

OXIDE
A compound in which an element is combined with oxygen. All metals form an oxide coating, such as rust on FERROUS METALS and PATINA on NONFERROUS METALS. It is not only metals which form oxides and oxide coatings, see OXIDATION.

OXIDISING FLAME (OXIDIZING FLAME)
A flame which has an excess of oxygen at the welding blowpipe. This tends to show as a sharp pointed bright inner cone at the welding tip (see OXY-ACETYLENE FLAME). A slightly oxidising flame is used when welding brasses, bronze welding or BRAZING.

It is designed to prevent the zinc which melts at a low temperature from changing to a vapour and escaping from the weld; the excess oxygen mixes with the zinc to form a zinc oxide which has a much higher melting temperature, see BRONZE WELDING.

OXY-ACETYLENE FLAME

A process in which oxygen and acetylene gas are used in approximately equal volumes to produce a flame with a temperature of about 3200°C, it is the hottest gas flame obtainable. The oxy-acetylene flame is made up as follows:

(1) The inner cone which consists of unburnt gases, here the oxygen burns with the acetylene to produce the intense heat, it is at the tip of this cone that the flame is at its hottest.
(2) The burning gases form a blue reducing zone of carbon monoxide and hydrogen; this reducing zone of carbon protects the metal being heated from OXIDATION.
(3) The outer burning zone consists of the carbon monoxide burning with the hydrogen, the oxygen to support combustion being supplied by the ambient air (the surrounding air).
(4) The final products of combustion being discharged at the end of the flame, consisting of carbon dioxide and water vapour.

The type of flame chosen can be adjusted to burn with equal amounts of oxygen and acetylene, a neutral flame, or with an excess of acetylene (a carbonising flame) or an excess of oxygen (an oxidising flame) see illustrations. See NEUTRAL, CARBONISING and OXIDISING FLAMES for uses of each flame type.

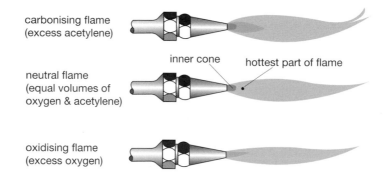

carbonising flame
(excess acetylene)

inner cone   hottest part of flame

neutral flame
(equal volumes of
oxygen & acetylene)

oxidising flame
(excess oxygen)

Oxy-acetylene flame characteristics

OXYGEN

A colourless, odourless, tasteless gas which forms about 21% of the Earth's atmosphere. Oxygen is an important element being vital to all living organisms. Oxygen is a chemically active gas combining readily with most substances and is essential to support combustion of anything burning. The more oxygen present in the vicinity of something alight, the more vigorously it will burn, it is therefore used with many gases to achieve higher flame temperatures. Oxygen is supplied in steel cylinders painted black for easy identification, the thread into which the regulator screws is right handed, see WELDING EQUIPMENT.

OXYGEN DEPLETION DEVICE   See ATMOSPHERE SENSING DEVICE.

P TRAP   See TRAP.

PACKAGED BURNER
Term used to indicate a burner assembly in which all controls are incorporated together as a complete unit which, in turn, is located on an appliance. One example would be a FORCE DRAUGHT BURNER.

PACKING GLAND   See STOPCOCK.

PAD SAW
A tool consisting of a handle into which a small thin saw blade is inserted. It proves handy where a large hand saw or framed hacksaw would be too large for a particular job.

PAN CONNECTOR
The fitting which is used to make the joint between the discharge pipe and a WC pan. See WC SUITE.

PANEL HEATING   See RADIANT HEATING.

PANEL RADIATOR   See RADIATOR.

PARALLEL BRANCH
A branch fitting in which the branch connection looks up and runs parallel in the direction of flow with the main pipe. See BRANCH for example.

PARALLEL GUTTER
A common form of lined gutter found behind a PARAPET wall or between two flat roofs. See GUTTER and TAPERING VALLEY GUTTER.

PARALLEL THREAD
A long thread cut into or onto a fitting or pipe and having a uniform diameter. Generally threads cut onto low carbon steel pipes and fittings are of a tapered thread, so that the pipe slowly tightens as it goes further into the fitting, this does not happen with parallel threads. The parallel thread can have many uses such as on longscrews, ballvalves and pillar taps to assist in their connection and the running down of the backnut along the thread. Nuts and bolts used in general mechanical

fixings also have parallel threads. For example, see longscrew and pillar tap. See also LOW CARBON STEEL PIPES AND FITTINGS.

PARAPET
A low wall found guarding the edge of roofs and balconies, etc. The parapet rises above the lower level of the roof and therefore needs to be weathered by some means, e.g. a DAMP PROOF COURSE. For example of a parapet wall see BUILDING TERMINOLOGY.

PARAPET GUTTER (BOUNDARY WALL GUTTER)
A gutter positioned between a PARAPET wall and a flat or pitched roof. For example, see GUTTER.

PARASITIC CIRCULATION
The unwanted circulation of hot water and its waste of heat in a domestic hot water system. For examples of parasitic circulation see ONE PIPE CIRCULATION.

PARTIAL VACUUM   See VACUUM.

PARTIALLY SEPARATE SYSTEM OF DRAINAGE
A design of below ground drainage which is a compromise between the COMBINED and SEPARATE SYSTEMS OF DRAINAGE. The system consists of two drain/sewer pipes, one for SURFACE WATER and one for FOUL and WASTE WATER plus a limited amount of surface water. This system is as expensive to install as the separate system but the design is usually simpler and also the foul water drain is flushed by some rain during heavy rainfalls. A drawback of this system is that there is a possibility of SURCHARGING during heavy rainfalls. For example, see BELOW GROUND DRAINAGE.

PASCAL (Pa)
A unit of PRESSURE sometimes used. One pascal is equal to one NEWTON per square metre ($1 \, Pa = 1 \, N/m^2$).

PASSIVE STACK VENTILATION
A system of natural ventilation to a room using a pipe duct from a grill in the ceiling to the roof outside. The air movement is created due to the difference between inside and outside temperatures and the effect of wind passing over the roof of the dwelling.

PASS-OVER OFFSET (CRANK OFFSET)
The formation of pipework to permit a pipe to pass over or under an obstruction, for example, see PIPE BENDING TECHNIQUE.

PAT TEST (PORTABLE APPLIANCE TEST)
An electrical inspection and, where applicable, electrical test carried out on all hand-held or movable equipment that is plugged into a 230 or 110 volt supply. These tests are carried out to ensure the appliance is fit for purpose as safe to use. The frequency for PAT testing would vary depending upon usage of the piece of electrical equipment. Records must be maintained for all equipment inspected and tested.

PATHOGENIC BACTERIA
Micro-organisms that are harmful to humans, examples include Salmonella and Cholera. These bacteria are transported and found in faecal matter.

PATINA
The thin film of OXIDE which forms on NON-FERROUS METALS exposed to the atmosphere. The oxide coating which forms is insoluble and quite stable, it therefore protects the metal from further corrosion. Patina often varies in colour from the original metal, for example lead turns dark grey or white and copper turns green or black, in each case according to atmospheric conditions. See also VERDIGRIS.

PEAK DEMAND
The period of time when the maximum amount of water or gas would be required to pass through a mains or distribution pipe to cope with its maximum demand. The period is often between 5 and 7 pm when everyone is doing their various washing and cooking at home.

PEAK FLOW LOAD
The maximum expected discharge of water from a drainage installation.

PEAR-SHAPED MALLET   See BOSSING MALLET.

PEDESTAL WASH BASIN   See WASH BASIN.

PEDESTAL WC PAN
A WC pan which is supported by an integral base which is screwed to the floor. For example, see WASHDOWN WC PAN.

PEDIMENTS
A decorative feature formed over door and window openings. Pediments are usually triangular or semi-circular in shape and constructed out of stonework. It is often the task of the plumber to cover these pediments in some metallic roofing material.

PEGGED VALVE
Various taps such as PILLAR TAPS and BIB TAPS in which the JUMPER has been fixed to the spindle thus ensuring it will be lifted off from the SEATING when the valve is opened. Current Regulations require all valves to have fixed jumpers.

PENETRATION
The depth to which the filler rod passes down to the ROOT of a weld. See WELDING DEFECTS.

PENSTOCK
A gate which rises up and down vertically to control the flow of liquid. The penstock can be designed so that water can discharge over the top of the gate in which case the arrangement is commonly known as a weir penstock.

PERFORMANCE TEST

A test which is carried out to above ground sanitary pipework to ensure that a minimum of 25 mm of water TRAP SEAL is retained in every TRAP when the pipework is subjected to its worst possible working conditions. Each of the following tests should be carried out a minimum of three times and before each test the trap should be recharged. The depth of the water seal should be measured with a dipstick as shown.

Tests for siphonage in branch discharge pipes
The test for SELF-SIPHONAGE is carried out by filling the sanitary appliance to overflowing level and removing the plug. To test for INDUCED SIPHONAGE, the sanitary appliance furthest away from the drain or discharge stack should be discharged, on completion the trap seal remaining in all the traps should be measured.

Tests for siphonage and compression in discharge stacks
A selection of sanitary appliances should be discharged simultaneously. Those selected should be on the highest floors thus giving the worst pressure conditions.

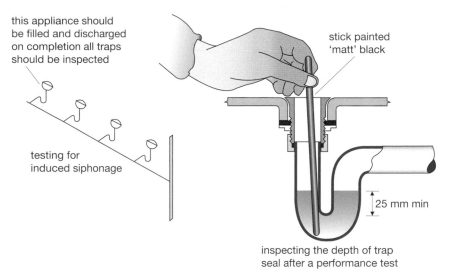

this appliance should be filled and discharged on completion all traps should be inspected

testing for induced siphonage

stick painted 'matt' black

25 mm min

inspecting the depth of trap seal after a performance test

Performance testing

PERMANENT HARD WATER

Water containing the sulphates of calcium or magnesium (limestone salts). It is called permanent hard water because unlike TEMPORARY HARD WATER the hardness cannot be removed from the water by boiling. Permanent hardness in the water is caused when rain water falls upon ground which contains calcium or magnesium sulphates. Due to the natural SOLVENCY of water the sulphates are taken up in suspension within the water. Permanent hardness does not cause appreciable scaling up of pipes and boilers. See also HARD WATER.

PERMEATION

A situation where contamination has occurred to a liquid or gas by another fluid passing through the walls of a container.

PERPENDICULAR
A line at right angles to a given line or surface.

PERSPEX   See ACRYLIC.

PERVIOUS
Any material or rock strata which will permit the passage of water through it. See also IMPERVIOUS.

PET COCK
A small PLUG COCK. For one example of its use see AIR VESSEL. Pet cocks are often found regulating the flow of water into an AUTOMATIC FLUSHING CISTERN. When used for this purpose it is often of a design different from a plug cock in that the flow of water is regulated by turning a screw head at the end of the valve slowly bringing a cone-shaped spindle to rest on the seating.

PETROL INTERCEPTOR
An arrangement of inspection chambers designed to prevent petroleum spirit from entering the public sewer. A petrol interceptor is designed in such a way that any petrol that may be present can rise to the surface of the stored water (petrol being lighter than water) and evaporate safely into the atmosphere. Notice that in the first chamber the exit pipe is shorter than the following two, this is to allow grit and sludge to settle and accumulate without blocking the pipe end. For a single garage a deep GARAGE GULLY may be sufficient.

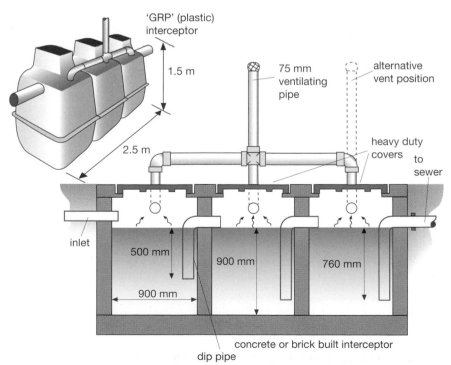

Petrol interceptors

P

pH VALUE

The term pH is the abbreviation for potential of hydrogen which is a measure of the amount of hydroxyl ions (ALKALINE) or hydrogen ions (ACIDIC) in a sample of water or material. The pH value of a substance can be found by using special indicators which are supplied as coloured liquids or in paper strip form, known as litmus papers or universal indicators. The litmus paper has the disadvantage of only showing whether a substance is acidic or alkaline, which is indicated by the paper turning red for acids and blue for alkalis; the universal indicator on the other hand will show how much acidity or alkalinity a substance contains by changing to a range of colours depending upon its pH value. The pH ranges from 0 to 14, 7 being a NEUTRAL WATER.

$$0\ 1\ 2\ 3\ 4\ 5\ 6\ 7\ 8\ 9\ 10\ 11\ 12\ 13\ 14$$

acidic ←   ↑   → alkaline

neutral

PHILLIPS SCREWDRIVER   See SCREWDRIVER.

PHOTOELECTRIC CELL

A device found typically in a pressure jet burner to detect light rays from the flame within the appliance, thus confirming its operation. It consists of a small resistor which, when exposed to light, makes or breaks an electrical circuit. For an example of its application see PRESSURE JET BURNER.

PHYSICAL AND WORKING PROPERTIES

Two terms used to describe the characteristics of various materials. A material may be hard or soft, light or heavy, pervious or impervious, etc. All materials have their own qualities, see the following for a description of some of the physical and working properties that may be present in a certain material ANNEALING, COHESION, DENSITY, DUCTILITY, DURABILITY, ELASTICITY, FATIGUE CRACKING, FUSIBILITY, MALLEABILITY, SOLVENCY, TEMPER, TENACITY, TENSILE STRENGTH, WORK HARDENED.

PHYSICAL CHANGE

A physical change happens when a substance changes its form, for example from a liquid to a solid or a liquid to a gas. Water (liquid) changes to ice (solid) by cooling to 0°C or changes to steam (gas) by heating to 100°C. In either case solid, liquid or gas it was still water (H_2O), and the change made was only temporary due to outside forces such as heat or pressure. As another example lead, if heated to 327°C, would melt and if further heated to 1740°C would change to a lead vapour. See also CHEMICAL CHANGE.

PIEZO-ELECTRIC IGNITION   See IGNITION DEVICES.

PIG IRON

Term used to mean crude cast iron taken directly from the blast furnace. See IRON.

PIGTAIL

The rubber hose which connects between an LPG cylinder and the pressure regulator. All hoses should be manufactured to BS 3212, with the year of manufacture embedded along its edge. With all high pressure applications the hose should be secured to the threaded connection with suitably crimped proprietary hose clamps. Worn hoses must be replaced when necessary.

PILLAR TAP

A valve with a long inlet thread which is fitted vertically through an appliance to which a backnut is fitted clamping the tap to the appliance. The water supply is then made up to this thread. When fitting pillar taps it is essential to ensure that an adequate AIR GAP is allowed for. Pillar taps fitted to appliances which are very thin may require a TOP HAT otherwise there may not be enough thread for the backnut to fully tighten up to the appliance. See also SCREWDOWN VALVES.

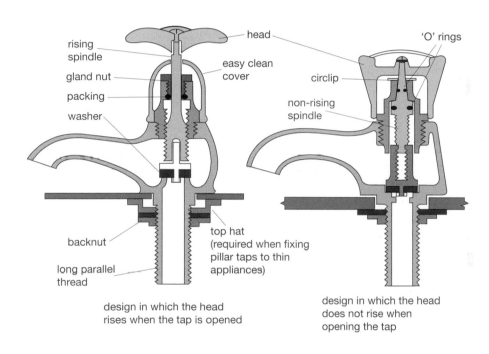

design in which the head rises when the tap is opened

design in which the head does not rise when opening the tap

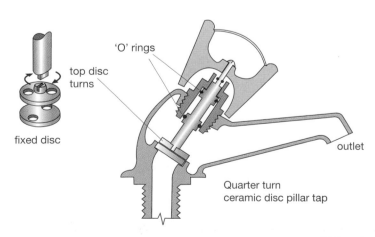

Quarter turn ceramic disc pillar tap

Pillar taps

PILOT FLAME (PILOT LIGHT)

The small gas burner jet which is used to ignite the main burners to a gas installation. In most cases the pilot flame is continuously lit, but, some appliances are designed to have the pilot jet burning only when required. See THERMOELECTRIC VALVE.

PILOT HOLE

A small hole drilled prior to the drilling of a larger hole to assist in ensuring the accurate positioning of the larger hole.

PIPE (TUBE)

A long hollow cylinder used to convey liquids and gases.

PIPE BENDING TECHNIQUES

(1) Heavy gauge tube, using a hydraulic press bender

(Prior to reading about this bending technique, see HYDRAULIC PRESS BENDER.) To form a 90° bend from a fixed point on these machines is very simple. First a line is marked on the pipe at a distance from the fixed point equal to the distance between the fixed point and the centre line of the finished bend (distance X) see sketch. Then

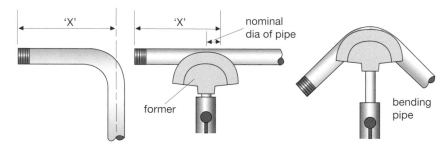

forming a 90° bend

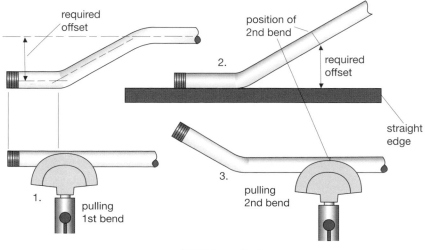

forming an offset

Pipe bending using a hydraulic press bender

from this measurement is deducted the nominal bore of the pipe. The pipe is now placed in the bending machine at this new point and lined up with the centre line of the FORMER. The machine is now pumped to apply pressure and bend the pipe to the required angle. Due to a certain amount of SPRING BACK the bend is over pulled by about 5°.

To make an OFFSET using these machines first measure along the pipe from the fixed point (this time making no deductions) and pull this round to the required angle, as shown. Then using a straight edge or positioning the pipe against two parallel lines the distance of the offset, the second bending point is marked on the tube. The pipe is replaced in the machine and this mark is positioned at the centre line of the former. When pulling this second bend it is essential to ensure that it is pulled running true and on the same plane as the first bend.

(2) Light gauge tube, using a lever type bending machine
(Prior to reading about this bending technique see BENDING MACHINE.)

To form a 90° bend from a fixed point is quite simple. First a line is marked on the pipe at a distance from the end of the tube equal to the distance between the fixed point and the back of the finished bend (distance Y). The pipe is then inserted into the bending machine as shown. Make sure the line marked on the tube is square with the back of the FORMER, this is best achieved using a small engineers square. Then assuming that the roller is adjusted correctly (see BENDING MACHINE) the bend is simply pulled round by pulling at the lever arm. Should a return bend be required the above process is simply repeated, only the bend that has just been pulled now becomes the fixed point.

To form an OFFSET, the first bend is pulled to the required angle. The tube is now pushed further along the bending machine and turned around to make the bend just pulled look upwards away from the former. A straight edge and rule are now used as shown to measure the required offset, when marking out ensure that the straight edge is running parallel with the piece of pipe looking upwards away from the former. Once the correct measurement is obtained the bend can be pulled round until it is the same angle as the first bend and can be seen to run parallel, best checked using a straight edge.

A PASS-OVER OFFSET can be formed using several methods but one of the simplest is as follows. A bend is first pulled, the angle of which depends upon the size of the obstruction, but it should not be too sharp, otherwise difficulty will be experienced pulling the offset bends. A straight edge is positioned over the bend centrally at a distance of the required passover. The tube is marked with two lines as shown; these are the back of the finished offsets. The tube is then inserted into the bending machine and when the first mark is in line with the back of the former the first set is pulled; once completed the pipe is then simply reversed in the machine and the second mark is lined up in the former and pulled to complete the passover.

(3) Using heat
Should a bend need to be pulled without the use of a bending machine two principles have to be understood, firstly the internal diameter of soft or thin walled pipes must be protected from flattening; this is achieved by either inserting a BENDING SPRING, SAND LOADING the tube or in the case of lead restoring its true bore using BOBBINS. Low carbon steel can be bent without support. Secondly to bend accurately the bend must be 'set out' as described below. To form a 90° bend a drawing is first

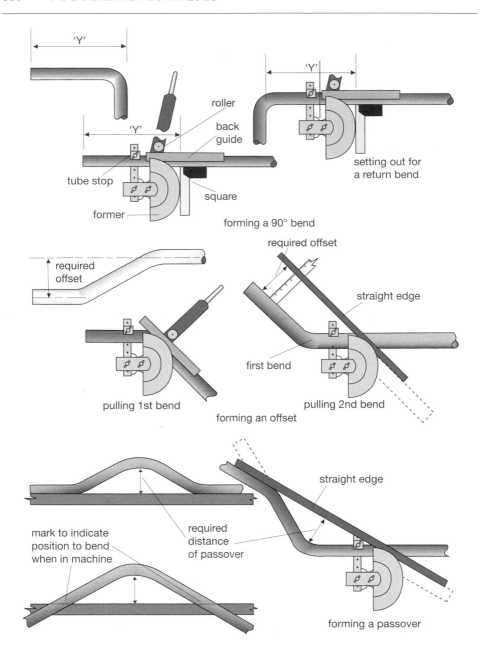

Pipe bending technique for light gauge tube

produced showing the centre line of the pipe. This is achieved by drawing two lines at 90° to each other. The bend is then added, which may be a specified radius or a radius is chosen, which in the absence of a specification is usually four times the nominal diameter of the pipe. To find the position for the point of the compass two lines parallel to the centre line are drawn at a distance equal to the radius; see sketch, where they cross is the compass point centre. These two parallel lines, when extended to the centre line, represent the start and finish of the bend. Upon completion of the centre line drawing the outside pipe walls are added to the drawing by simply running parallel lines, the size of the pipe, either side of the centre line. Distance A is then marked off on the pipe to be bent, followed by distance B (the arc of the circle). It is then between these two lines that heat is applied and the tube bent; as the work is bent it should be offered up to the drawing to check that it is being bent symmetrically.

To form an OFFSET again a drawing is produced showing the centre line of the pipe. This is achieved by first drawing two parallel lines at a distance apart which is equal to the offset required. A line, the chosen angle of the set, is then drawn to join these two lines, see sketch. As with the 90° bend, at a distance equal to the radius required, lines are drawn parallel to the centre line and where these lines cross the compass point is positioned and the two arcs can be drawn in, see illustration. With the 90° bend these lines extended to the centre line which represented the start and finish of the bend. This is not the case with angles other than 90°, therefore these positions must be added and are achieved by running a 90° set square along the centre line until the compass centre is reached, see sketch. The outside pipe lines are added to the drawing and the pipe marked, heated and bent as with the 90° bend.

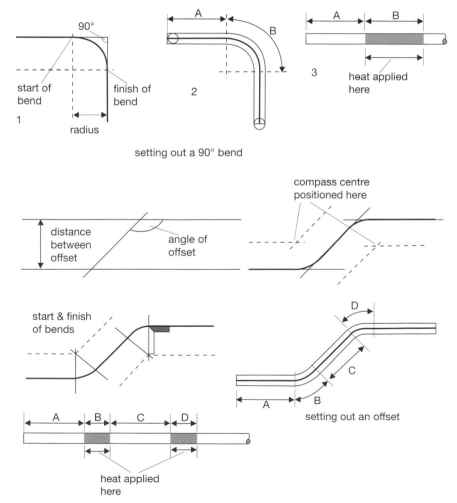

setting out a 90° bend

setting out an offset

Pipework development for heat bending

When bending metallic pipes the pipe should be heated to an ANNEALED temperature. THERMOPLASTIC materials should be heated to a temperature where the material becomes soft and pliable. The softening temperature of plastic is very near the temperature at which it chars, therefore care must be taken not to damage the tube due to heating, which is best carried out using only the hot air from the flame and not actually allowing the flame to touch the plastic material.

PIPE CLIP   See PIPE SUPPORT.

PIPE CUTTERS

A selection of various tools used to cut all types of pipes, there are four basic types, roller, wheel, link and snap action cutters. Roller cutters have one cutting wheel, the smaller sizes are used to cut copper and stainless steel tube whereas the larger roller cutters can cut low carbon steel tube. The purpose of the rollers is to overcome the problems of an external burr being formed. Wheel cutters come with either three or four cutting wheels and prove useful when cutting pipes in confined spaces as the pipe cutter does not need to be fully rotated around the pipe to obtain a cut. When using roller or wheel cutters, care must be taken when starting the cut to avoid TRACKING. Link cutters are used to cut cast iron, clayware and any other large diameter brittle pipes. The cutting wheels are connected to a series of links, which can be added or removed to alter the cutting diameter. With this type of cutter the pipe is actually broken as the cutting wheels are slowly tightened, thus exerting an even pressure all round the pipe. Snap action cutters are similar to link cutters in that they cut the same type of pipes and the cutting wheels exert a pressure at the point of cutting. The difference with snap cutters is that they have a lever arm which is usually pumped using HYDRAULIC POWER or ratcheted to slowly bring the wheels closer together, thus breaking the pipe in a clean cut.

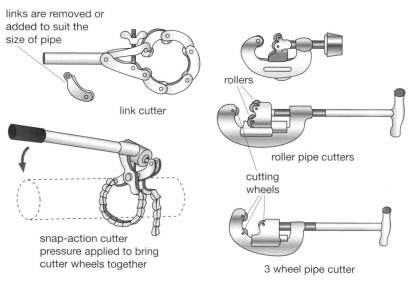

links are removed or added to suit the size of pipe

link cutter

rollers

roller pipe cutters

cutting wheels

snap-action cutter pressure applied to bring cutter wheels together

3 wheel pipe cutter

Pipe cutters

PIPE FREEZING   See FREEZING EQUIPMENT.

PIPE GOUGE   See AUGER.

PIPE HANGER
A pipe bracket designed so that the pipe is suspended from the roof of a structure. The distance between the pipe and the roof can vary and be adjusted depending upon the length of studding (the threaded rod) used. See PIPE SUPPORT.

PIPE HOOK
A pipe clip sometimes used to hold pipes in position especially if they are being secured in a wall chase. See PIPE SUPPORT.

PIPE INTERRUPTER
A BACKFLOW PREVENTION DEVICE without any moving parts. In the event of a vacuum in the pipeline being created, which might be caused by BACK-SIPHONAGE air will be admitted through several holes or slits in its body.

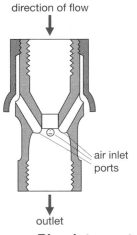

direction of flow

air inlet
ports

outlet
Pipe interrupter

PIPE RING   See PIPE SUPPORT.

PIPE ROLLER AND CHAIR
A special support designed to allow for the THERMAL MOVEMENT of a pipe. See PIPE SUPPORT.

PIPE SIZING
Gas Installations
There are several methods that can be employed to find a suitable size of pipe for a gas installation One such method is to produce a table of calculations, as shown, consisting of twelve columns into which the information can be inserted to confirm, or not, the suitability of a suggested pipe. For a pipe to be adequately sized, a pressure drop of no greater than 1 mbar between the gas meter and the furthest appliance, under maximum load, should be experienced. The illustrated example shown has been completed as follows.

(See illustration over.)

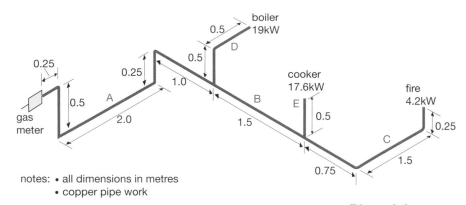

notes: • all dimensions in metres
       • copper pipe work

Pipe sizing

1	2 (m)	3 (m)	4 (m)	5 (kW)	6	7 (m³/h)	8 (mm)	9 (m)	10 (mbar)	11 (mbar)	12
A	4.0	2.5	6.5	40.8	0.094	3.84	22	12	0.54	0.54	✓
B	1.5	0.5	2.0	21.8	0.094	2.05	15	3	0.67	1.21	✗
B	1.5	0.5	2.0	21.8	0.094	2.05	22	30	0.07	0.61	✓
C	2.5	1.0	3.5	4.2	0.094	0.39	10	9	0.39	1.0	✓
D	1.0	0.5	1.5	19	0.094	1.79	15	6	0.25	0.79	✓
E	0.5	0	0.5	17.6	0.094	1.65	15	6	0.08	0.69	✓

Column 1   indicates the section of pipe being sized.
Column 2   indicates the actual pipe length measured.
Column 3   indicates the additional length to be added for fittings used.
           Note allow 0.5 m for each tee fitting or elbow used, including the one that you are entering and 0.3 m where a pulled bend has been made.
Column 4   is the effective length, i.e. the sum total of columns 2 and 3.
Column 5   shows the total kW rating for all the appliances being supplied by the section in question.
Column 6   This is a constant figure, which is used to convert kW to m³/h.
Column 7   This is the required gas flow rate and is found by multiplying column 5 by column 6.
Column 8   indicates a suggested pipe size which, if necessary, is a guess.
Column 9   indicates the maximum pipe length allowed within which it is possible to supply the required volume of gas, thus ensuring that the maximum pressure drop of 1 mbar is not exceeded.
           This information is selected from a table. The table given over is for copper tube, for other materials see BS 6891.
           A straight edge is placed on the suggested pipe size and the column selected is that which will still permit the required gas flow (column 7) in m³/h to pass.

Column 10 is the pressure loss for the section in question and is found by dividing column 4 by column 9.

Column 11 is the progressive pressure loss and is the sum total of all the preceding pipe sections feeding this section. For example only sections A and B precede section E not all sections A–D.

Column 12 This column is ticked if the pipe size selected in column 8 is of sufficient size, i.e. column 11 does not exceed 1 mbar.

Flow discharge (m^3/h) from copper tube with a 1 mbar drop pressure differential between its ends

Pipe dia. (mm)	Maximum length of pipe (m)							
	3	6	9	12	15	20	25	30
10	0.86	0.57	0.5	0.37	0.3	0.22	0.18	0.15
12	1.5	1.0	0.85	0.82	0.69	0.52	0.41	0.34
15	2.9	1.9	1.5	1.3	1.1	0.95	0.92	0.88
22	8.7	5.8	4.6	3.9	3.4	2.9	2.5	2.3
28	18	12	9.4	8.0	7.0	5.9	5.2	4.7

It is possible to confirm that the maximum 1 mbar pressure drop has not been exceeded by undertaking the following test to find the working pressure drop across the system. Position one manometer at the meter and a second at the furthest appliance (prior to any governing device). Zero both gauges and confirm the standing pressure is the same at both points. Ignite all appliances and when they are operating at their maximum heat input read the two manometer gauges, the difference in the two is the total system pressure drop.

PIPE SIZING

Hot and cold water supplies

There are several methods that can be adopted to calculate the pipe size for a particular installation, one such method is as explained here. However, before pipe sizing any hot and cold water supply system an assessment must be made of the probable maximum demand of water, in litres per second, at any given time, because it is very unlikely that all the SANITARY APPLIANCES will be used at once. This assessment depends upon the type of sanitary appliance and the type of building in which they are installed. To perform this assessment a method has been devised based upon the theory of probability in which a unit rating is given to each type of sanitary appliance, see LOADING UNITS, this will give the required flow rate. The diameter of pipe required to produce such a flow rate depends upon the material used and the permissible loss of HEAD per metre run. This is found by adding the actual net length of pipe to the length of pipe due to FRICTIONAL LOSS, then dividing this length into the available head. Frictional loss caused by water passing through a fitting is generally expressed as the loss of an equivalent length of pipe. See example over.

(See illustration over.)

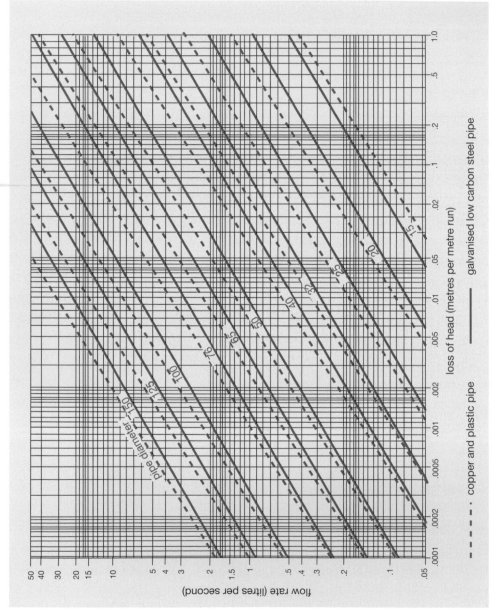

Pipe sizing graph

flow rate (litres per second)

loss of head (metres per metre run)

- - - - copper and plastic pipe   —— galvanised low carbon steel pipe

| | Equivalent pipe lengths in metres | | | | |
| | Copper | | Low carbon steel | | |
Nominal diameter mm	tee	elbow	tee	elbow	bend
13	0.61	0.46	1.14	0.46	0.38
19	1.07	0.76	1.37	1.53	0.46
25	1.52	1.07	1.75	0.69	0.53
32	1.98	1.37	2.29	0.91	0.69
38	2.44	1.68	2.68	1.07	0.84
51	3.51	2.29	3.41	1.37	1.07

Example

Find the diameter of a copper pipe in which 1.3 litres per second would be required. The head available is 4 metres and the length of the pipe run is 12 metres, there will be 6 elbows and 2 tees in the pipe run.

Effective length of pipe (assuming 32 mm copper pipe) is:

$$
\begin{aligned}
6 \text{ elbows at } 1.37 &= 8.22 \text{ m} \\
2 \text{ tees at } 1.98 &= 3.96 \text{ m} \\
\text{plus net length of pipe} &= 12.0 \text{ m} \\
\text{Total} &= 24.18 \text{ m}
\end{aligned}
$$

Permissible loss of head is:

$$\frac{head}{length} = \frac{4}{24.18} = 0.17 \text{ metres per metre run}$$

The final pipe size is found by referring to the graph on the previous page in which a 32 mm copper pipe with a loss of head of 0.17 m/m run would discharge 1.9 l/s. A second calculation is carried out to find the flow for a 25 mm pipe in which the answer proved to be only 1.05 l/s. Therefore the 32 mm diameter tube is chosen.

When pipe sizing to individual appliances it is essential to remember that a head loss is also experienced as the water passes through the draw-off tap or delivery valve. See also BOX'S FORMULA.

PIPE SIZING

Sanitary pipework and drainage

If the number and type of SANITARY APPLIANCES used in a particular building is known the diameter of the pipe is generally found by the discharge unit method in which each appliance is given a discharge unit value. See DISCHARGE UNITS.

PIPE SUPPORT

A bracket which fits around the pipe either tightly to anchor the pipe or loosely so as to support the pipe while allowing for THERMAL MOVEMENT. There are many types of pipe support and the choice depends upon the material of the pipe and the cost allowed for the job. This in turn would then depend upon circumstances, for example, it would be pointless using plastic pipe clips in schools or hospitals, etc. where they could very easily be damaged. Whatever pipe support is chosen, the fixing must be secure to prevent damage and possible development of AIR LOCKS. As a guide the general recommended pipe support spacings are given below, but one clip too many is better than one clip too few and in many cases the plumber has to use their own judgement.

Copper tube pipe support spacings

External diameter of pipe (mm)	Horizontal intervals (m)	Vertical intervals (m)
up to 15	1.2	1.8
22–28	1.8	2.4
35–42	2.4	3.0
54	2.7	3.0
over 54	3.0	3.6

Low carbon steel pipe support spacings

BS thread inches	External diameter (mm)	Nominal bore (mm)	Horizontal intervals (m)	Vertical intervals (m)
$1/2$	21.5	15	1.8	2.0
$3/4$	27	20	2.4	3.0
1	34	25	2.4	3.0
$1^1/4$	42.5	32	2.7	3.0
over $1^1/2$	48	40	3.0	3.6

UPVC plastic pipe support spacings

Diameter of pipe (mm)	Horizontal intervals (m)	Vertical intervals (m)
up to 75	0.75	1.5
over 75	1.0	2.0

Note For certain plastics, such as polythene, a continuous support is to be recommended when they are used for conveying hot water.

(See illustration opposite)

PIPE THERMOSTAT

A thermostat that has been fixed to the side of a pipeline and is generally designed to make or break the electrical switch that allows the flow of water through the system. One example is a frost thermostat, where cold conditions that might freeze the water within the system are identified and the thermostat overrides the heating controls to bring on the boiler and pump. Conversely, the thermostat may be fitted within a pipeline to shut down the heating flow.

PIPE VICE

There are various types of pipe vices used to secure pipes in position whilst cutting or threading, etc. Soft metals should not be held in these vices due to their ease in being flattened and marked, instead these should be held in lead or wooden blocks made specially.

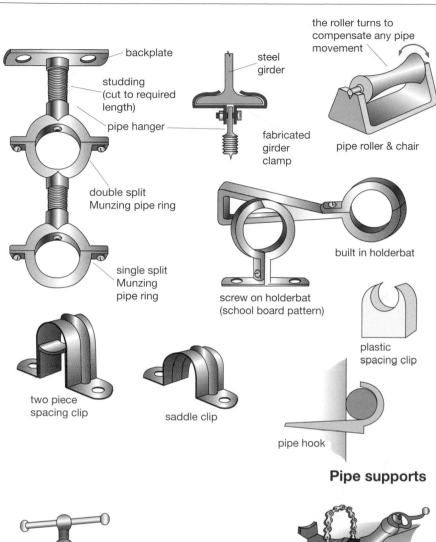

backplate

studding
(cut to required
length)

pipe hanger

double split
Munzing pipe ring

single split
Munzing
pipe ring

steel
girder

fabricated
girder
clamp

the roller turns to
compensate any pipe
movement

pipe roller & chair

built in holderbat

screw on holderbat
(school board pattern)

plastic
spacing clip

two piece
spacing clip

saddle clip

pipe hook

Pipe supports

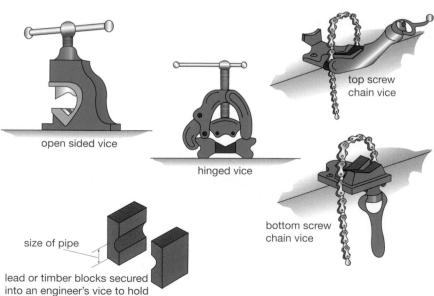

open sided vice

hinged vice

top screw
chain vice

bottom screw
chain vice

size of pipe

lead or timber blocks secured
into an engineer's vice to hold
light gauge tube without
damaging it

Pipe vices

PIPE WRENCH   See WRENCH.

PIPELINE SWITCH
A FLOAT SWITCH fitted in the pipeline of systems such as BOOSTED SYSTEM OF COLD WATER SUPPLY. It is designed to make an electrical contact thus switching a pump on and off.

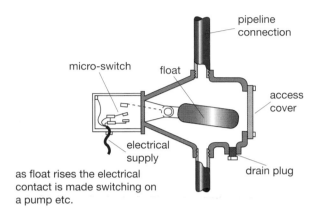

as float rises the electrical contact is made switching on a pump etc.

Pipeline switch

PISTON TYPE FLUSHING CISTERN   See FLUSHING CISTERN.

PITCH
The angle at which a roof lies to the horizontal. See also GRADIENT.

PITCH FIBRE PIPES
Pipes made of coal tar, pitch and impregnated wood fibres. Pitch fibre pipes will be found used on some existing foul and surface water drains. The most common method of joining was made with the use of polypropylene fittings which were pushed onto the pipe wedging an O ring between the pipe and fitting. Connections to this material are made with either a pitch fibre adapter or by a cement mortar joint.

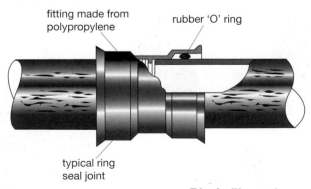

Pitch fibre pipe

PITCHER CROSS
A double branch fitting in which the two branch connections curve through 90° and sweep in to join the main flow pipe. See BRANCH for example.

PITCHER TEE (SWEPT TEE)
A 'tee' piece in which the branch is curved through 90° to join the main pipe.

PLANT ROOM
A room used specifically to house Building Service Equipment such as a boiler, or water softening equipment, etc.

PLAQUE HEATER
A ceramic surface which is heated by and thus radiates heat. These heaters are often found installed high within workshops. See RADIANT PANEL.

PLASTIC
There are many types of plastics each with their own characteristics, but broadly speaking they can be divided into two groups, THERMOPLASTICS and THERMOSETTING-PLASTICS. Plastic materials are very light in weight and simple jointing methods are used but they do have a disadvantage of having a high LINEAR EXPANSION rate and allowance must be made for this in installations. The ultra-violet rays from the sun can also cause this material to become very brittle.

PLASTIC GUTTERS AND FITTINGS   See either EAVES GUTTERS or RAINWATER PIPE.

PLASTIC PIPE AND FITTINGS
Many types of plastics are used in the plumbing industry, being used for mains to drainage supply systems. Four basic methods are used to join plastic pipes COMPRESSION, PUSH FIT, SOLVENT WELD CEMENT and FUSION WELDED JOINTS, although fusion joints are rarely used. The compression joint used on mains supplies should be made of a brass which is DEZINCIFICATION resistant. Whereas compression joints used on waste fittings are usually made from plastic materials. Mains water pipes are colour-coded blue to identify them from the black electric cables, etc. Black-coloured mains supplies are still to be found underground because many were laid before the blue pipe was introduced. When jointing pipes, with solvent weld cement, the correct makers' brand should be used and it should be noted that not all plastics are capable of being joined by this method, see POLYVINYL CHLORIDE.

PLASTIC RANGE
The condition of a metal or some other solids when they are neither liquid nor completely solid. WIPED SOLDER JOINTS on lead are formed using a solder which has a very long plastic range, it remains in a 'plastic' workable condition between the temperatures of about 185–245°C permitting it to be moulded to the desired shape. When the flame is removed, the joint cools and solidifies.

PLASTICISED POLYVINYL CHLORIDE
A form of PVC in which small amounts of rubbery synthetic polymers have been added during the manufacturing process. This type of PVC is more resistant to damage due to impact from a heavy blow. It is much more flexible than PVC and UPVC therefore it is of no use in the manufacture of soil and waste discharge systems with which PVC is associated. Plasticised PVC is put to use in the manufacture of such items as shower curtains and raincoats, etc.

PLASTICITY

Plasticity is a property of a material by which once deformed from its original shape and size it will stay in its new form. One good example is lead bossing in which the metal is worked to form various details. Plasticity is the opposite of ELASTICITY.

PLATE HEAT EXCHANGER

A HEAT EXCHANGER in which thin plates of corrugated metal separate the heating and domestic hot water (dhw) flow of water. The heat from the heating or primary circuit is very quickly transferred to the dhw circuit. This process is fast enough to produce instant hot water. Plate heat exchangers are commonly found in COMBINA-TION BOILERS where no stored hot water is maintained.

PLENUM SYSTEM

A system of central heating in which air is heated and passed through a system of ductwork. See WARM AIR CENTRAL HEATING.

PLIERS

Pliers are basically a tool used to grip various objects, in each case the jaws are brought together due to the action of the handles pivoting like a pair of scissors. There are many variations, some have toothed jaws designed to grip better whereas others are used to cut or form material. The pliers shown are possibly the most common but there are many specialist types, see DIMPLE PLIERS and DOG EARING PLIERS for some examples.

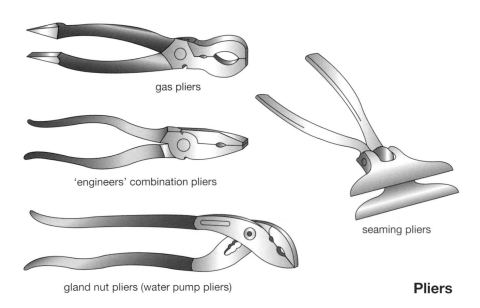

gas pliers

'engineers' combination pliers

seaming pliers

gland nut pliers (water pump pliers)

Pliers

PLUG

A device, fitting or any object used to stop up a hole; for examples, see DRAIN PLUG and WASTE PLUG.

PLUG COCK (QUARTER TURN COCK)

One of the earliest forms of valve which is operated by turning a tapered plug a quarter of a turn to open or close the supply. Plug cocks are used almost exclusively for gas services, they should not be used on high pressure water pipework installations because of the likelihood of WATER HAMMER caused by the rapid closing action of the valve, although they are often used as DRAIN OFF COCKS or ISOLATION VALVES. Plug cocks have a square head into which is cut a slot; if this slot is in line with the pipe, it means the valve is open. Many of the smaller valves used for gas installations are provided with an oval shaped finger plate to assist turning on and off. See also GLAND COCK and DROP FAN COCK.

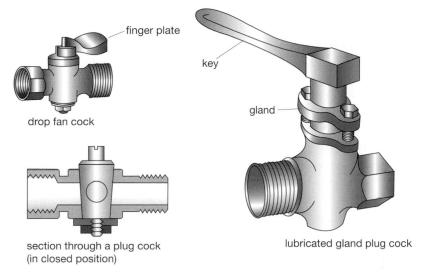

drop fan cock

section through a plug cock
(in closed position)

lubricated gland plug cock

Plug cocks

PLUGGING CHISEL (JOINT RAKER)

A chisel designed to cut out slots in the joints of brickwork. For example, see CHISEL.

PLUMB BOB (PLUMB LINE)

A cone shaped tool usually made of brass to which a string is secured through its centre. The plumb bob is used to find a true VERTICAL line. From the highest point of the proposed line, say on a wall prior to fixing a rainwater pipe, the string is held and the brass plumb bob is lowered. Due to gravitational pull, it will go down directly vertical, to the ground. When the plumb bob stops swinging the line is marked. The string used for a plumb bob is often coated in chalk therefore acting as a CHALK LINE, this assists in the marking of the line. On windy days when trying to achieve a straight line, the plumb bob is sometimes inserted into a pail of water at its lowest end, this tends to steady it from swinging.

PLUMBER

The word plumber is derived from the Latin word for lead, 'plumbum'. The plumber was a person who worked with plumbum (lead). Today, however, the plumber works with many more materials and lead is used less and less by the domestic plumber; much of the lead roof-work is left to specialist firms, the average plumber often

P

carrying out only simple flashings, etc. When the lay-person thinks of plumbing and the plumber, often they only think of the person who installs various SANITARY APPLIANCES; some do not even associate gas installation work or CENTRAL HEATING with plumbing and most do not include the weathering of roofs, etc. But the good skilled plumber carries out all these tasks plus many more from drainage work to THERMAL INSULATION of pipes and electrical wiring. Because today the field of plumbing has become so vast and most people cannot become specialised at everything the word plumbing is generally being regarded as work to do with hot and cold water supplies and sanitation. The multi-skills of heating, drainage, roofwork, etc. now come under the term mechanical services, therefore plumbing works fall within this new category. See also MECHANICAL SERVICES.

PLUMBER'S BLACK (PLUMBER'S TARNISH/SMUDGE)
A substance used by plumbers to paint onto an area of lead prior to soldering, the black applied acts as a tarnish and prevents any solder from sticking to the metal where it is not required. Plumber's black is made from glue size and carbon black, which is the soot deposits produced from incomplete combustion. To best apply plumber's black to the metal, first the surface should be wire brushed to provide a key, then chalk applied, this tends to dry up any remaining oil on the surface, finally the black which has been wetted to make it moist is painted onto the surface.

PLUMBER'S MAIT
A waterproof non-setting putty which is used to assist making joints to waste fittings and metal EAVES GUTTERS among other tasks. Plumber's mait can be used to make spigot and socket joints to WC pans, etc. to allow for thermal movement and the withdrawal of the appliance, although generally a purpose designed plastic fitting, such as a MULTIQUICK, is preferred. Sometimes plumber's mait is used as a bedding material to support sinks and basins, thus avoiding the CERAMIC WARE sitting directly onto hard surfaces possibly causing damage. When using plumber's mait the surfaces onto which it is going to be used must be clean, dry and free from any grease otherwise a watertight seal will not be made.

PLUMBER'S METAL (PLUMBER'S SOLDER)
Name sometimes given to BS grades D or J solder which is used to make wiped solder joints. These two solders have the longest PLASTIC RANGE which enables the joint to be formed and moulded to the required shape. See WIPED SOLDER JOINT.

PLUMBOSOLVENT
The ability of water to dissolve lead. See SOLVENCY.

PLUMING
A situation where the visible clouds of combustion products are seen at the flue terminal. Plume gas is formed as a result of the water vapour that is produced as part of the combustion process being cooled close to the dew point of water (approx. 55°C). Pluming can usually be seen at the outlet of a cold flue upon first lighting a gas appliance or where a condensing boiler is in use.

PLUNGER See SUCTION PLUNGER.

PLUNGER TYPE FLUSHING CISTERN   See FLUSHING CISTERN.

PNEUMATIC CYLINDER   See AUTOMATIC PNEUMATIC CYLINDER.

PNEUMATIC EJECTOR

A device used to pump sewage up a relatively small height. The installation consists of an automatic self-starting air compressor which forces the sewage through the outlet pipe. Pneumatic ejectors are not recommended for large installations requiring the lifting of large volumes of liquid and at a greater speed, for such installations CENTRIFUGAL PUMPS should be considered. When using ejectors at least two should be provided to facilitate repairs.

There are many types of ejectors, one type, known as the lift and force sewage ejector, works by first sucking up the sewage from a low level into an ejector then forcing it out to be discharged into a high level INSPECTION CHAMBER. The lift and force ejector works as follows:

(1) As the sewage level in the intake chamber rises it operates a float switch which automatically switches on the motor, sucking the sewage into the ejector. This is due to a PARTIAL VACUUM being created in the ejector, thus the sewage is in effect pushed up the delivery pipe by the atmospheric pressure pushing down on top of the liquid.

(2) When the ejector is full the suction valve to the ejector closes and the compressed air valve opens forcing the sewage out of the ejector, it cannot travel back down to the intake chamber because of the non-return valve fitted in the delivery pipe. Therefore it is forced along the discharge pipe to the high level drain. See also WC MACERATOR PUMP.

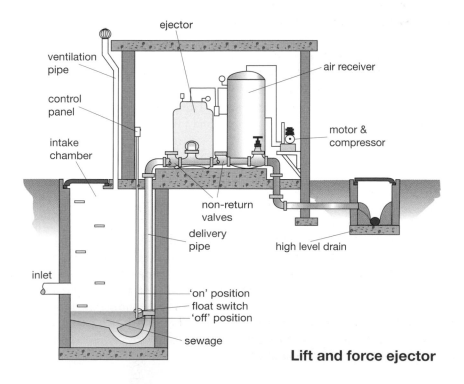

Lift and force ejector

PNEUMATIC TEST   See AIR TEST.

POISONED SOLDER
SOLDER which has become contaminated with other metals, in particular zinc. When solder was frequently melted down, in a melting pot, for wiped solder joints, zinc from brass filings and fittings which were dipped into the pot, for TINNING purposes, often became dissolved into the solder affecting its properties. Poisoned solder can be cleaned by mixing in with the cooling molten solder a quantity of sulphur. When the solder is reheated a scum forms on its surface, consisting of sulphur and zinc, and this can simply be skimmed off leaving purified solder behind. Today because solder is rarely melted down in a pot for joint wiping, poisoned solder is not a problem.

POLY-FUSION
A type of plastic fusion welding. See FUSION WELDED JOINT (PLASTIC).

POLYETHYLENE (PE)
One of the many THERMOPLASTICS materials used within the plumbing industry, its most common use being in yellow gas and blue water main pipes, etc.

POLYPHOSPHONATE CRYSTALS
Crystals which are added to hot water supply pipework in order to reduce SCALE. See WATER CONDITIONERS.

POLYPROPYLENE
A THERMOPLASTIC which has similar characteristics to POLYTHENE.

POLYTETRAFLUOROETHYLENE   See PTFE TAPE.

POLYTHENE
A THERMOPLASTIC used extensively in the manufacture of pipes, which would be classified as high or low DENSITY. High density polythene is much more rigid in its construction than low density. Polythene is highly resistant to chemical attack and is widely used for chemical waste installations. The jointing methods used for poly-thene pipes would be either of an 'O' ring push fit joint or compression type. It is not possible to make joints using a SOLVENT WELD CEMENT. See also PLASTIC PIPES AND FITTINGS.

POLYVINYL CHLORIDE (PVC)
A very common THERMOPLASTIC used extensively for discharge pipes and below ground drainage systems. PVC pipes can be jointed in a variety of ways including 'O' ring PUSH FIT JOINTS, COMPRESSION JOINTS and SOLVENT WELD CEMENT. PVC is sometimes confused with high density POLYTHENE, this causes no real problem except that polythene cannot be solvent welded. A simple test to find out which material is being used is to drop a sample into a pail of water, if it floats it is poly-thene. PVC would sink as it is heavier than water. PVC unlike most plastics will not readily burn upon ignition, it gives off a heavy vapour which excludes the oxygen thus is said to be self-extinguishing. See also PLASTIC, PLASTICISED PVC and UNPLASTICISED PVC.

POP-UP WASTE
A waste fitting with a plug which can be seated or lifted into the outlet by means of a manually operated lever arm device, for example, see BIDET.

PORT-A-LOO
A type of CHEMICAL WC.

PORTSMOUTH BALLVALVE
A very common ballvalve with the piston moving in a horizontal plain. There are several types of Portsmouth ballvalves, some with a body removal nut to assist easy maintenance, others are manufactured as a complete unit, these are cheaper but not so easily repaired. The water level to this type of ballvalve is adjusted by means of bending the lever arm up or down. It is not permissible to install this type of valve into mains fed cisterns without some means of a BACKFLOW PREVENTION DEVICE. For example, see BALLVALVE. See also EQUILIBRIUM BALLVALVES.

POSIDRIV SCREWDRIVER   See SCREWDRIVER.

POSITION HEAD   See POTENTIAL HEAD.

POSITIVE DISPLACEMENT GAS METER
A design of GAS METER that uses diaphragms that are alternately inflated and deflated by the flowing gas causing a 'worm' wheel to turn. This, via a pinion and series of gears, registers the quantity of gas used. The common U6 meter found in domestic premises is of this type.

POST AERATED FLAME   See NON-AERATED FLAME.

POT BOILER
A boiler which has a WATER WAY all around an ANNULUS shaped cylindrical pot. See BOILER and VAPORISING BURNER.

POTABLE WATER
Water which is said to be fit for human consumption.

POTENTIAL DIFFERENCE
The difference, measured in volts, between two conductors, when the current is flowing around an electrical circuit.

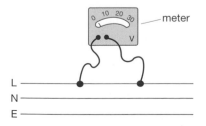
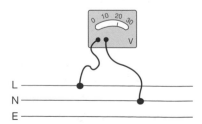

Potential difference

POTENTIAL HEAD (POSITION HEAD)
The vertical height of a column of water which would give X amount of pressure at a given point.

POURING ROPE   See RUNNING ROPE.

PRE-AERATED FLAME
A flame that has had air mixed with the gas prior to the combustion process taking place, such as found in domestic natural gas burning appliances. A good example would be a Bunsen burner in which the primary air is drawn in to premix with the gas prior to combustion and the secondary air is added as the gas burns away from the REACTION ZONE.

PRE-CAST FLUE BLOCKS
Preformed sections of flue that have been designed to be built into a wall structure.

PREFABRICATION
Something that has been made prior to delivery on site, therefore it has only to be assembled. Often sanitary pipework is prefabricated in the workshop, or by specialist manufacturers of prefabricated systems. By having work prefabricated time can be saved on site and work can be carried out before the building is ready for installation.

PRE-MIX BURNER   See FORCE DRAUGHT BURNER.

PRESS FITTING
A pipe fitting made in copper and stainless steel that is fitted to a pipe line using a special machine. Basically the fitting incorporates a butyl rubber seal and the joint, once assembled, is finely crushed on to the pipe, preventing its removal. These fittings cannot be reused once installed.

PRESSED STEEL
A steel sheet which is pressed into various shapes for many uses, it can be painted or GALVANISED. Many plumbing appliances such as baths, wash basins, and sinks are made from pressed steel and once formed are protected with a coating of VITREOUS ENAMEL.

PRESSURE
Pressure can be defined as the force acting upon a given area. Force on the other hand can be defined as any action that tends to hold or alter the position of a body. Since the introduction of the metric system in this country the unit of force is the newton (N).

The newton is the force that gives a mass of 1 kg an acceleration of 1 metre per second per second ($1 \text{ N} = 1 \text{ kg} \times 1 \text{ m/s}^2$)

$$\simeq 1 \text{ newton} = 101 \text{ grams weight } (0.101 \text{ kg})$$

$$\text{The force produced by a 1 kg mass} = 9.81 \text{ N}$$

Therefore for every kilogram force 9.81 newtons is being applied.

A container $1 \text{ m} \times 1 \text{ m} \times 1 \text{ m}$ would contain 1000 litres of water and weigh 1000 kg, therefore the pressure exerted at its base would be $1000 \times 9.81 = 9810 \text{ N/m}^2$ or 9.81 kN/m^2. See WATER PRESSURE. Sometimes other terms are used to identify pressure such as the bar, the pascal or the pound force per square inch. These can be illustrated as follows:

$$1 \text{ bar} \quad = 100000 \text{ N/m}^2 \ (100 \text{ kN/m}^2)$$

$$1 \text{ Pa} \quad = 1 \text{ N/m}^2$$

$$1 \text{ lbf/in}^2 = 6894 \text{ N/m}^2 \ (6.894 \text{ kN/m}^2)$$

PRESSURE GAUGE
A gauge fitted to pipework or on vessels to indicate the internal pressure. For the design of one type of pressure gauge see BOURDON PRESSURE GAUGE.

PRESSURE GOVERNOR   See GOVERNOR.

PRESSURE HEAD
The pressure that would be created at the base of a column of water, or at a point along this distance. This vertical distance or head in metres, multiplied by 9.81, would give the pressure head in kilonewtons per square metre (kN/m^2) see WATER PRESSURE.

PRESSURE JET BURNER
An ATOMISING BURNER used on OIL FIRED BOILERS. The pressure jet burner shown is reasonably quiet in operation, and works as follows:

(1) Oil feeds into the burner and is pumped to the atomiser nozzle, where it is broken up (atomised) and discharged into the boiler as a very fine spray.
(2) The fan blows a certain amount of air through the air turbulator cone to mix with the atomised oil.
(3) Finally the oil/air mixture is ignited by an ignition ELECTRODE situated at the point of atomisation.
(4) When the boiler reaches the required temperature, as determined by the boiler thermostat, the oil pump and fan switch off.

See also OIL STORAGE AND SUPPLY.

(See illustration over.)

PRESSURE LIMIT VALVE   See PRESSURE-REDUCING VALVE.

PRESSURE LOSS
The loss of pressure caused by various factors, e.g. from the PRESSURE HEAD being lost, or a number of appliances being used at the same time. A pressure loss which is often overlooked is that caused by FRICTIONAL RESISTANCE of liquid or gas passing through pipes and fittings, see FRICTIONAL LOSS.

PRESSURE POINT   See NIPPLE.

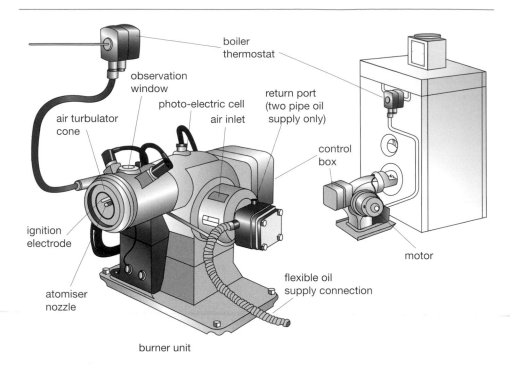

Pressure jet burner

PRESSURE-REDUCING FITTING

A device which is fitted into siphonic WC pans to assist in removing the air from between the two traps of water, thus creating a PARTIAL VACUUM. This allows the atmospheric pressure, pushing down on the surface of the water in the WC pan bowl, to force the liquid and contents from the pan, see SIPHONIC WC PAN. The type shown works by forcing water past an obstruction which reduces the volume of fluid it will allow through, therefore once past the obstruction the water still travelling at the same velocity, pulls air from the void or space, this is caused by the reduction in volume, see sketch. See also VENTURI TUBE for a different type of pressure-reducing fitting.

(See illustration opposite)

PRESSURE-REDUCING VALVE (PRESSURE LIMIT VALVE)

A valve fitted in the pipeline to reduce the water pressure as necessary. A pressure limit valve works on the same principle as a pressure reducing valve, the main difference being that it does not work to such a fine tolerance.

The valve shown works as follows:

(1) The valve, prior to being commissioned, is held closed on the seating. This is due to the pressure of the water and the small spring pushing against the washer.
(2) The pressure-adjusting screw is turned to wind down onto the spring, causing it to exert a pressure onto the diaphragm, this in turn pushes the valve from its seating allowing water through.
(3) If the water pressure, below the diaphragm, increases above the pressure exerted by the screw it will push up on the diaphragm, thus pulling with it the valve closing off the supply. As the pressure below the diaphragm drops, so the valve re-opens.

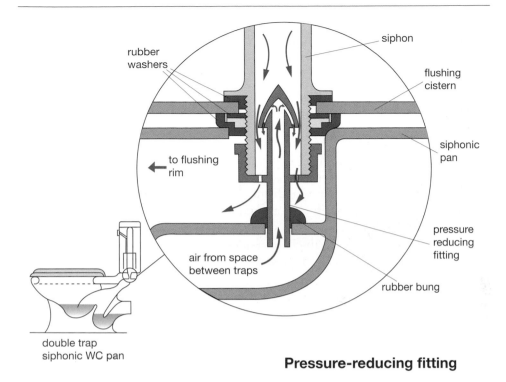

Pressure-reducing fitting

When the pressure-reducing valve is in use it in effect 'beats' or moves continuously back and forth to meet the demand of water. Most pressure-reducing valves are set up during manufacture to the required pressure at the outlet position although this can be done on site as described above. See also GOVERNOR.

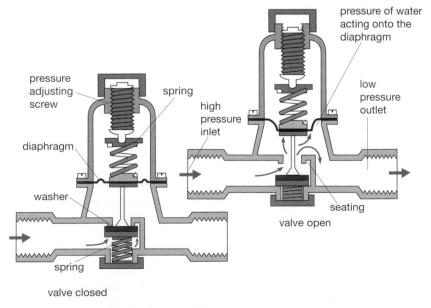

Pressure-reducing valve

P

PRESSURE REGULATOR

A device fitted onto gas cylinders to regulate the pressure to the blowpipe. See REGULATOR.

PRESSURE RELIEF VALVE (EXPANSION VALVE)

A SAFETY VALVE which is designed to open should the pressure inside a hot water system become increased to such a point where it could damage the pipework or vessels within the system. There are two types of pressure relief valves, the spring loaded type, which is by far the most common, and the deadweight type which can only be fitted in the vertical position. When installing pressure relief valves ensure that the hot water discharging from them is at a position where it is visible but will not cause any danger to persons in or around the building. Often the type which has a connection for a discharge pipe is specified. Should a discharge pipe be used it should be the same size as the discharge outlet and discharge via a TUNDISH which gives an AIR GAP. See also TEMPERATURE RELIEF VALVE.

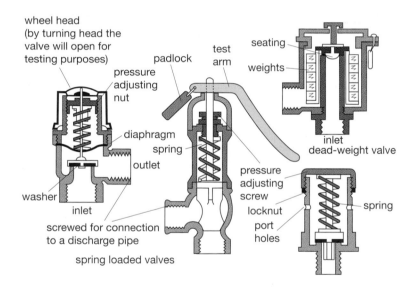

Pressure relief valves

PRESSURE SWITCH

A device installed in systems/appliances such as boosted cold water supplies, forced draught burners, fan assisted boilers and air handling equipment, etc. It is designed to bring on or close the operational control mechanism. For example, the pressure switch found inside a fan assisted boiler gives an indication that the fan is operating and therefore allows gas into the combustion chamber and brings on ignition, 'knowing' that the combustion products will be expelled by the operating fan. Conversely, with a boosted system of cold water supply, the pumps are either switched on or off, depending upon the pressure within the pipework. Pressure switches use either positive or negative forces applied to one side of a diaphragm; the flexing diaphragm makes or breaks an electrical micro switch. Some, however, use transducers, which effectively are electrical resistors, in an electrical circuit, again opening or closing the electrical current flow through a switch.

PRESSURE TEST

A test to determine if a system is sound or not. See TIGHTNESS TESTING and HYDRAULIC PRESSURE TEST.

PRESSURE TEST PUMP

A FORCE PUMP used to test completed hot and cold water pipework, see HYDRAULIC PRESSURE TEST.

PRESSURISED HOT WATER SYSTEM   See UNVENTED DOMESTIC HOT WATER SUPPLY.

PRESSURISED SYSTEM OF CENTRAL HEATING

A hot water heating system in which the water is filled into the system and sealed from the influence of atmospheric pressure, therefore as the water heats up a pressure is created within the system. See SEALED SYSTEM.

PREVAILING WIND

The direction in which the wind usually blows.

PRIMARY CIRCUIT

The pipework going to and from the boiler and hot storage vessel, consisting of a primary flow and return. See CENTRALISED HOT WATER SUPPLY.

PRIMARY CIRCULATION

The water which flows around the PRIMARY CIRCUIT. The circulation of water is due to either the operation of a pump or a CIRCULATING PRESSURE caused by CONVECTION CURRENTS.

PRIMARY FLOW AND RETURN

The pipes which make up the PRIMARY CIRCUIT. See CENTRALISED HOT WATER SUPPLY.

PRIMARY FLUE

The short length of flue pipe which leaves the boiler, prior to entering the chimney or DOWN DRAUGHT DIVERTER. For example, see OPEN FLUE.

PRIMARY GAS METER   See GAS METER.

PRIMARY VENTILATED STACK SYSTEM (SINGLE STACK SYSTEM)

A design of above ground sanitary pipework. Most buildings should be designed to meet its criteria thereby reducing the cost of installation. The system is designed so that no separate ventilating pipes will be required to prevent TRAP SEAL LOSS. This can only be achieved by observing the following guidelines.

The vertical discharge stack must be as straight as possible with a large radius bend (200 mm minimum radius) fitted at the base of the stack.

All sanitary appliances must be closely grouped to the discharge stack, within the limits shown in the diagram. See also FLOAT for examples of branch discharge pipes with more than one appliance connected to them.

P

The lowest connection to the discharge stack must be a minimum distance of 450 mm above the INVERT of the drain. Should the building be over three storeys high, this distance should be increased to 750 mm. For buildings over five storeys, all ground floor appliances should not connect into the stack. For buildings over twenty storeys, all first floor appliances, as well as ground floor ones, should not be connected to the main discharge stack.

All appliances, as far as possible, should be fitted with a P TRAP and bends in branch pipes should be avoided.

For pipes less than 75 mm in diameter, the branch connections into the vertical discharge stack must be swept into the stack with a minimum radius of 25 mm, or at an angle of 45°.

For pipes over 75 mm in diameter, the branch connections into the vertical discharge stack must be swept into the stack with a radius of 50 mm, or at an angle of 45°.

The branch discharge pipes that exceed the criteria above should be vented. In that case the system is generally referred to as a ventilated discharge branch system. See also SANITARY PIPEWORK.

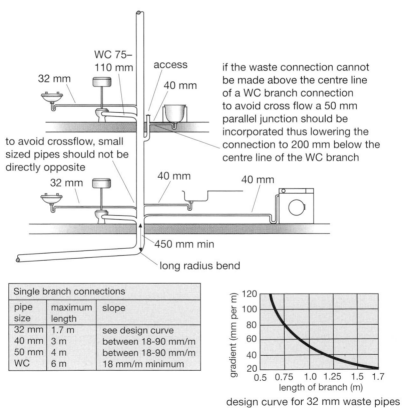

Primary ventilated stack system

PRIMATIC CYLINDER
A SINGLE FEED CYLINDER, see SINGLE FEED HOT WATER SUPPLY.

PRINTED CIRCUIT BOARD (PCB)

A thin insulated board onto which electronic components have been soldered. These are found typically within the control panel of a modern boiler.

PRIVATE SEWER   See SEWER.

PROCESS DRAIN

A drain designed and intended to carry TRADE EFFLUENT only.

PRODUCTS OF COMBUSTION

Should fuel be completely consumed and burnt the products of combustion would consist mainly of carbon dioxide and water vapour. If the fuel was burnt using the ambient air (that surrounding the fuel), nitrogen would also be present. Nitrogen plays no part in combustion it is just in the air used. Therefore the products of complete combustion should cause no harm to the occupants of a room assuming an adequate fresh supply of air is available to replace the oxygen used up. Carbon dioxide, nitrogen and water vapour are all non-toxic gases. Should incomplete combustion result due to lack of oxygen being available carbon monoxide gas would be produced; this is a highly toxic gas and can prove fatal to anyone breathing it in. The water vapour present in flue gases can sometimes give rise to condensate water running back down the flue pipe.

PROGRAMMER

A device which is used to control a CENTRAL HEATING system. A programmer generally consists of a time clock to automatically switch on and off the boiler and pump, etc, and other controls to enable the user to overide the time clock settings. The programmer also allows the user to control the system to only heat the water they require, for example, to heat up the domestic hot water only, as may be required during the summer months.

PROJECTION DRAWING

A method of drawing three-dimensional objects so that their appearance can be easily understood. There are many different methods used but the three most common tend to be the isometric, the oblique, and the orthographic projections.

With isometric drawing the basic principle is that all VERTICAL lines of an object stay vertical on the drawing and all HORIZONTAL lines are drawn at an angle of 30°. The oblique projection differs in that first a normal front elevation is drawn of the object then the depth of it is portrayed by drawing lines usually at an angle of 45° from the elevation sketch. With this method of projection it is usual to reduce the depth lines to $^1/_2$ or $^3/_4$ their true length, this prevents an exaggerated impression of depth being given.

Finally the orthographic projection which usually consists of three drawings, a front, a side and a plan elevation. The front elevation is first drawn, the lines are then transferred across and the side elevation is completed next to it, then the lines from the front and side elevations can be transferred to a position below the front elevation where a plan elevation is drawn. See sketch.

(See illustration over.)

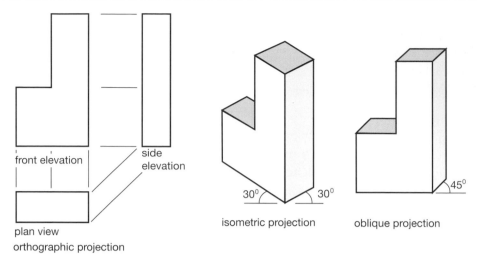

front elevation

side elevation

plan view

orthographic projection

isometric projection

30°   30°

oblique projection

45°

Projection drawing

PROPANE GAS
A gas which is extensively used by plumbers to fuel their blowlamps. It is supplied in steel containers usually coloured red. See LIQUEFIED PETROLEUM GAS.

PROPERTIES   See PHYSICAL AND WORKING PROPERTIES.

PROTECTED SHAFT
A shaft or enclosure made of fire-resistant materials through which people, objects or air may pass from one room through to another.

PTFE TAPE
Abbreviation for polytetrafluoroethylene. It is a white coloured jointing tape used when making joints on gas, water and steam pipes. PTFE tape is highly resistant to chemical attack and has a working temperature ranging from –80°C to 250°C, thus it covers a wide range of uses. Care should be taken when using PTFE tape on joints which do not fit too well together or where the joint is slightly distorted, such as the welded connections at the end of steel radiators.

PUBLIC SEWER   See SEWER.

PUDDLE FLANGE (PUDDLING FLANGE)
A FLANGE which is formed on or fixed onto a pipe which passes through a water retaining surface. The puddle flange is used to assist in preventing any leakage of water along the exterior of the pipe and through the hole. Where the flange butts up to the structure a jointing medium or washer is used.

PUFF PIPE
A short VENTILATING PIPE which was sometimes used to ventilate WASTE APPLIANCES on the old TWO PIPE SYSTEMS OF ABOVE GROUND DRAINAGE. The puff pipe, connected to the CROWN of the TRAP, went out through the wall to the external air where it terminated as far away as practicable from any openable windows and

above the spill-over level of the appliance. Puff pipes are no longer permitted in modern sanitary pipework installations, but they are occasionally seen in old buildings, generally recognisable by the bead formed on the end of the pipe and crossed wires soldered across the opening.

PUMP

A mechanical device used to provide a pressure to move a body of water. A pump could be used to move a quantity of water from A to B or to circulate water around a system. For several designs/types of pump see sketch below. See also LIFT, CENTRIFUGAL, DIAPHRAGM or FORCE PUMP and HYDRAULIC RAM.

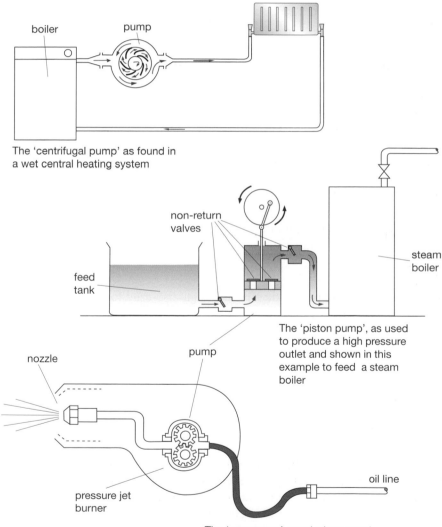

The 'centrifugal pump' as found in a wet central heating system

The 'piston pump', as used to produce a high pressure outlet and shown in this example to feed a steam boiler

The 'gear pump', used where good suction & pressure are needed, seen here in a pressure jet burner feeding an oil boiler

Pumps

PUMP OVERRUN

A situation where the central heating pump from a boiler has been designed to continue running for a period until the boiler has sufficiently cooled down.

PUMPING MAIN

Any mains water or sewerage pipe in which the fluid is pumped from one point to another.

PURGE STACK

An assembly of pipe work with meters, etc. fitted. It is used to ensure that the correct volume of gas is discharged, at the correct velocity from a 'large volume' gas installation when it is initially commissioned into operation with gas or, conversely, when de-commissioning it.

PURGING

Term used to mean the expelling of all air from within a gas installation. Purging is effected by passing a volume of not less than 0.01 m^3 (metric index) or 0.35 ft^3 (imperial index) of gas through the meter and ensuring gas is smelt at the furthest appliance and each individual leg of pipe run. During the purging procedure care must be observed with the discharge of gas into the building.

PURIFICATION OF WATER

Water taken from the ground via rivers, lakes, reservoirs, etc. must be treated to make it fit for human consumption. This is one of the duties of the various WATER AUTHORITIES. Most of the purification is carried out by passing the water through SAND FILTERS, these destroy about 95% of any pathogenic bacteria (diseased bacteria) and the remainder is generally destroyed by CHLORINATION.

PUSH FIT JOINT

A joint commonly used on plastic pipe although also used for other materials. The joint is made water tight by pushing home the pipe spigot into the socket which houses an 'O' ring or some other rubber sleeve. If a push fit joint is to be used as an expansion joint the pipe once pushed fully into the socket must then be pulled out about 10 mm, this allows it to expand. For one of many examples, see EXPANSION JOINT.

PVC   See POLYVINYL CHLORIDE.

PYROMETER

A device that is used to measure very high temperatures, unlike thermometers, which are used to measure the lower temperatures. The principal difference between pyrometers and thermometers is that a thermometer often uses direct contact and is based on expanding liquids and/or gases. The pyrometer, however, uses an indirect method to record the temperature rise, for example by employing a THERMOCOUPLE device to generate a current flow as the tip gets hot. The hotter the tip gets, the greater will be the electrical current flow. This current flow is measured and indicated as a temperature rise in degrees.

Q

Q TRAP   See TRAP.

QUANTITY OF HEAT (AMOUNT OF HEAT)

The quantity of heat means the amount of heat contained in a substance. The amount of heat contained in a substance and its temperature are two completely different things. A burning match has a much higher temperature than an iceberg, but the total heat energy in the iceberg is much greater than the match. To raise 500 litres of water to 65°C requires a greater amount of heat energy than to raise 5 litres of water to 65°C, the temperatures may be the same but the 500 litres of water contain much more heat and equally give off much more heat when cooling. The quantity of heat a substance contains is measured in joules (J).

1 joule = the power supplied by 1 watt in 1 second

To heat 1 gram of water 1°C 4.186 J would be required. Therefore to heat 1 litre or 1 kg (1000 g) of water 1°C 1000 × 4.186 J = 4186 joules or 4.186 kJ would be required.

Since 1 litre is the unit of liquid capacity in the metric system it can be stated that 4.186 is the specific heat (SH) of water. The quantity of heat energy required to heat up a given volume can very easily be found using the following simple formula:

specific heat × weight in kg × temperature rise

Example 1
Find the amount of kilojoules required to heat 200 litres of water from 8°C to 60°C (note 1 litre = 1 kg)

SH × kg × temperature rise

$$4.186 \times 200 \times (60 - 8) = 43534.4 \text{ kJ } (43534400 \text{ J})$$

As previously stated 1 joule = 1 watt per second therefore in the above example 43534.4 kJ would be required to heat up the water in one second. However in reality the heating up period allowed for would be much longer than one second thus the amount of joules is divided by the amount of seconds available to heat the water.

Find the power in kW to heat up 43534.4 kJ in two hours.

$$\frac{\text{quantity of heat in kJ}}{\text{seconds in 2 hours}} = \frac{43534.4}{7200} = 6.046 \text{ kW}$$

Example 2
Find the power required to heat 136 litres of water in one hour from 6°C to 60°C. The water is held in a copper vessel which weighs 5.5 kg. (SH of copper = 0.385 kJ/kg°C).

$$\text{power} = \frac{\text{QH for water} + \text{QH for copper}}{\text{seconds in one hour}}$$

$$\text{water} = \text{SH} \times \text{kg} \times \text{temp. rise}$$

$$= 4.186 \times 136 \times 54 = 30741.984$$

$$\text{container} = \text{SH} \times \text{kg} \times \text{temp. rise}$$

$$= 0.385 \times 5.5 \times 54 = \quad 114.345$$

$$\text{total heat energy required} \quad = \overline{30856.329 \text{ kJ}}$$

$$\therefore \frac{30856.329}{3600} = 8.57 \text{ kW}$$

See also CALORIFIC VALUE, LATENT HEAT, SPECIFIC HEAT and RECOVERY TIME.

QUARTER BEND
A bend which turns through 90°, see BEND.

QUARTER TURN COCK   See PLUG COCK.

QUARTER TURN TAP
A tap which is either fully open or closed by just a quarter turn of its turning head or handle, see ELBOW and WRIST ACTION TAPS.

QUARTZOID BULB SPRINKLER   See SPRINKLER HEAD.

QUENCHING
A term used to mean the rapid cooling of a metal by immersion in oil or water. This is usually carried out to maintain various properties associated with a crystalline structure that would be lost by slow cooling. See TEMPERING.

R

RADIAL DUCT SYSTEM   See WARM AIR CENTRAL HEATING.

RADIANT
A component designed to glow when heated by a flame. It therefore radiates heat.

RADIANT HEAT   See RADIATION (THERMAL).

RADIANT HEATING
A method used to raise the temperature of a room space by the admission of infrared energy, which is basically thermal RADIATION. Radiant heat passes directly through air and will only heat the more solid surfaces upon which it falls. Radiant heaters are usually mounted in the floor, walls or ceiling of a room. The panels are heated electrically or by circulating hot water, steam or hot air through them. Unlike other forms of CENTRAL HEATING, the effectiveness of radiant heating does not rely on the efficient circulation of air due to direct contact with the heat source. For an equal state of comfort in a room, systems relying mainly on CONVECTION CURRENTS must provide a higher air temperature within the room, this is because the cold surfaces of walls and windows etc. are removing the human body heat, that can only be replaced by the surrounding/ambient air. Radiant heat on the other hand warms up the floors, walls, windows, etc. and thus reduces the heat lost from the human body. Therefore lower air temperatures within the room are required which also gives a greater feeling of freshness. With the reduction of convection currents in the room cold draughts and dust problems are reduced to a minimum. Radiant heating can save fuel, unless the heating of the room is intermittent such as a building which lowers its air temperature at night and requires it to be rapidly raised in the morning.

The design of radiant heating systems is quite straightforward as shown in the sketches, requiring only a coil of embedded pipe running through the surface of a floor, wall or ceiling; the only important requirement is that the surfaces of the radiant heaters should not be metallic. Fitted behind the heater should be a means of thermal insulation to give heat emission only into the room. With this system one major disadvantage is the danger of a pipe leakage which can prove difficult to find and expensive to repair. The following table gives the recommended surface temperatures of walls, etc. fitted with embedded pipe panels. Notice how panels located in the ceiling can be used to give off a higher radiant heat emission, therefore they give a quicker room heating period and are generally chosen.

(See illustration over.)

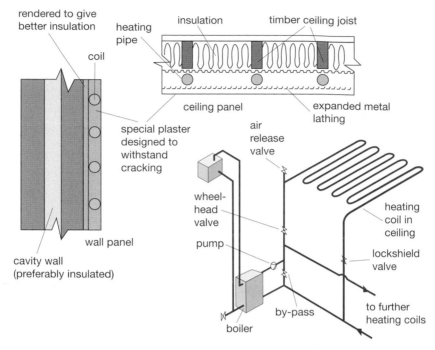

Radiant heating

embedded heating panel system

Panel location	Surface temperature	Heat given off
Ceiling	40°C	65%
Floors/walls	24°C	50%

The water temperature flowing through the pipes should be between 40 and 55°C. To achieve the higher surface temperatures used in a ceiling the heating pipes are spaced closer together.

RADIANT PANEL (HEATING PANEL or RADIANT HEATER)
A manufactured or custom-built heating panel used in a RADIANT HEATING system.

RADIANT STRIP
A form of RADIANT HEATING consisting of a continuous horizontal heating pipe extending the length of the room to be heated. It is supported from the roof structure well above head level and sometimes inclined at an angle towards the floor. The back of the heating pipe is fixed with an insulated metal casing to direct the heat in one direction only. Radiant strips have an advantage over other forms of radiant heating as maintenance is easier.

(See illustration opposite)

RADIATION (THERMAL)
The transference of heat. Radiation is heat energy in the form of infrared rays (invisible electromagnetic energy) emitted by a heated surface and travels directly to its point of absorption at the speed of light. It is the direct heat felt, for example, if

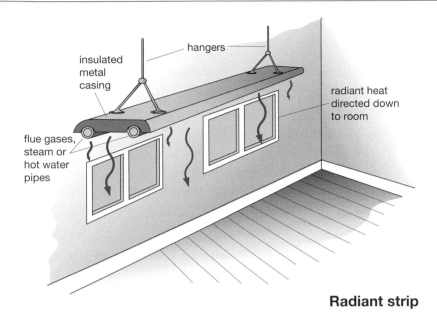

Radiant strip

standing in front of a fire, or the warmth felt when lying in the sun on a hot day. Radiant heat will pass through the air or space until it meets something which will absorb it. The rate at which a surface absorbs heat depends upon its colour. A blackened surface is an excellent absorber of heat as well as an excellent emitter. Objects that are good absorbers of heat are also good emitters, a surface that is painted silver will not absorb or emit heat readily. An ideal colour to paint a radiator would be matt black but for reasons of appearance this is not done.

RADIATOR

A HEAT EMITTER, designed to expose its hot surface to the air in a room. Despite their name radiators do not heat up a room by RADIATION, but mainly by CONVECTION CURRENTS. Originally all radiators were made of cast iron but due to their size,

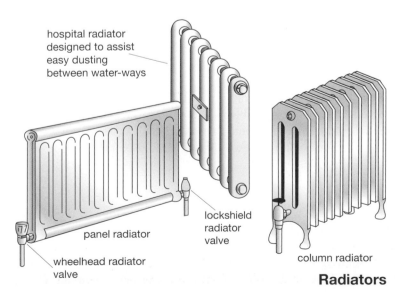

Radiators

cost and weight most modern ones are made from PRESSED STEEL. There are three basic designs of radiators, column, hospital and panel type radiators. Hospital radiators are designed to assist in the easy cleaning between the WATER WAYS and behind it, thus preventing the build up of dust, etc. See also CONVECTOR HEATERS.

RADIATOR KEY   See AIR COCK KEY.

RADIATOR VALVES

Two valves fitted to radiators and CONVECTOR HEATERS. In general one valve is fitted to each end. The valves are identical except that one is fitted with a lock-shield head, this prevents unauthorized people tampering with the regulated flow of water. The lock-shield valve is used only when BALANCING out the heating system, ensuring an equal distribution of hot water, or if a radiator has to be removed from the wall the valve could be shut down. The other radiator valve has a wheel head which can be used to turn the heater on and off. See also THERMOSTATIC RADIATOR VALVES.

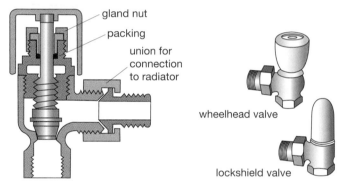

Radiator valves

RADIUS

The distance between the centre of a circle and any point on the circumference of that circle measured in a straight line. The radius is half the diameter of a circle.

RADON GAS

A radioactive gas given off by uranium in the ground. Houses built over granite areas experience the biggest problems. Sometimes radon gas extraction systems need to be incorporated to overcome particularly bad cases. Note: where radon gas is identified, no ventilation into the building should be taken from below ground floor level.

RAFTER

A sloping timber beam which extends down from the ridge to the eaves of a roof. See BUILDING TERMINOLOGY.

RAFTER BRACKET

An EAVES GUTTER bracket designed to screw into the side of roof RAFTERS.

RAG BOLT

A bolt that has been set into a wall or floor using cement mortar or placed into the setting concrete.

RAINWATER CHUTE   See CHUTE.

RAINWATER CONNECTIONS TO SOIL DISCHARGE STACKS

In some areas with a COMBINED SYSTEM OF DRAINAGE the local authority will permit the foul and waste water discharge stacks to receive rainwater from roofs. Designing a system in this way can save on the cost of installing separate rainwater pipes. To avoid excessive air pressure fluctuations within the discharge stack this method of rainwater disposal is not to be recommended for buildings over ten storeys in height or for the removal of rainwater from roof areas exceeding 40 m^2 per stack. The main disadvantage of designing above ground drainage in this way is the problem of flooding if a blockage occurs in the discharge stack. The rainwater pipe should be connected to the stack via a branch connection; this ensures a free unrestricted flow of air through the ventilating pipe, see sketch.

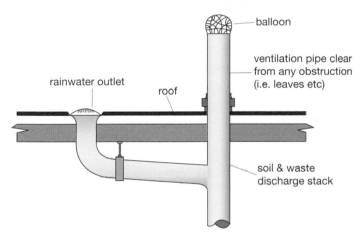

Rainwater connections to soil discharge stacks

RAINWATER HEAD   See HOPPER HEAD.

RAINWATER PIPE (FALL PIPE)

A pipe used to convey the surface water collected from the roof to the drain at ground level. In most cases the rainwater pipe is run externally down the outside of the building in which case, normally, no jointing medium is used but the spigot simply enters the socket. As with EAVES GUTTERING it is essential to allow for the expansion of the material otherwise buckling or cracking of the pipes will result. For appearance sake it is essential to ensure that the rainwater pipes are fixed perfectly vertical and to assist in this operation it is best to use a PLUMB BOB. The termination of the rainwater pipe at the lower level could be either by a rainwater shoe discharging into a GULLY or preferably the pipe is run directly into a back inlet gully. When connecting rainwater pipes into a surface water drain it is not essential to trap the bottom of the pipe at ground level. If the rainwater pipe is run internally, the joints must be made watertight.

RAINWATER SHOE

A rainwater fitting fixed at the lower end of a rainwater pipe to discharge the water clear of the building. The term is also used to represent a drainage fitting fixed at the base of a rainwater pipe into which an inspection opening is fixed.

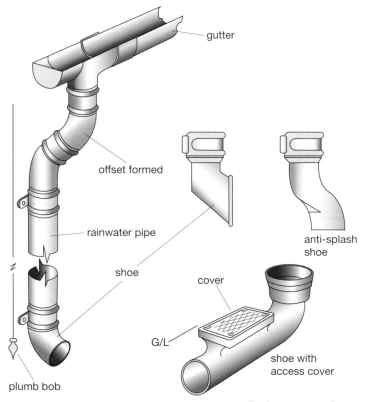

Rainwater shoes

RAISING PIECE (GULLY RISER)

A drainage fitting used to extend the height of a gully. The raising piece may or may not have branch inlets. See GULLY for example.

RAKING   See STEP AND COVER FLASHING.

RAMP

An open channel formed in the BENCHING of an INSPECTION CHAMBER to join two drains run at different levels. It would be used on occasions where it would be difficult to achieve a constant self-cleansing gradient in the pipeline. The ramp should have a maximum drop in height of 680 mm, for invert levels between drains in excess of this a BACK-DROP CONNECTION should be used.

(See illustration opposite)

RANGE   See FLOAT and COMMON DISCHARGE PIPE.

RANGE

The name given to a heavy-duty cast iron cooker.

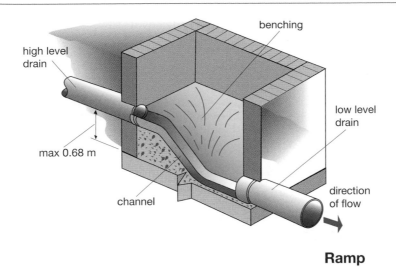

Ramp

RANGE BOILER   See BACK BOILER.

RANGE PIPE
A branch pipe connected to the main distribution pipework in a sprinkler system of fire prevention. For example, see SPRINKLER SYSTEM.

RANGE RATED
An appliance that operates between various settings, i.e. minimum and maximum. See OPERATING PRESSURE.

RANGES OF APPLIANCES
A series of similar SANITARY APPLIANCES fitted in a row. The appliances generally discharge their waste water into a COMMON DISCHARGE PIPE often called a range or float.

RASP
A tool very similar to a file and used in the same manner, the only difference being that it has coarser cutting teeth. It is generally used to remove the excess surface from materials which would soon block up the teeth of a file, such as wood, plastic or lead. See FILE.

RCD   See RESIDUAL CURRENT DEVICE.

REACTION ZONE
That part of a PRE-AERATED FLAME in which combustion occurs, resulting from the primary air intake to the burner. During this stage the gas is not completely burnt. At the end furthest from the burner head, just beyond the tip of the reaction zone will be the hottest part of the flame.

REAMER
A tool used to enlarge the hole size of an existing hole drilled without the drill snatching, which may happen if using a twist drill bit to enlarge its size. The reamer

is designed to slowly make the hole larger and consists of a series of blades set in a circle which taper down at one end. A reamer designed with its blades tapering at a much larger angle is known as a de-burring reamer and is used to remove the burr from around the inside of a pipe which has been cut with pipe cutters.

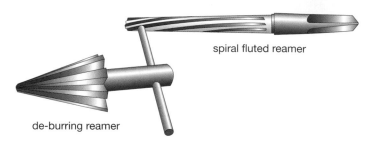

spiral fluted reamer

de-burring reamer

Reamers

REARRANGING FORMULAE   See TRANSPOSITION OF FORMULAE.

RECEDER DIES   See STOCKS AND DIES.

RECEIVER   See DILUTING RECEIVER.

RECIPROCATING PUMP
A type of pump in which the water is alternately drawn into the pump body through an inlet pipe fitted with a non-return valve, and expelled out through another pipe. There are several types of reciprocating pumps such as FORCE PUMPS, LIFT PUMPS and the HYDRAULIC RAM.

RECOVERY TIME (RECOVERY PERIOD)
The amount of time required to heat up a quantity of water. The recovery time to heat up a quantity of water varies, depending upon the power rating of the heat source used, but basically the higher the power rating the faster will be the heating up period of the water, see QUANTITY OF HEAT. Sometimes a recovery chart is used to give an approximate time and proves a useful guide when sizing up small water heaters or boilers, but one should bear in mind that no allowance has been given for the continuous heat loss from the system to the surrounding air or for the heating up of the hot water pipework and vessels.

Recovery chart: to heat water through 55°C

Litres	Power rating (kW)									
	1	2	3	4	6	8	10	12	15	18
5	19	9.5	6	5	3	2	2	1.5	1	1
10	38	19	13	9.5	6	5	4	3	2.5	2
15	57.5	29	19	14	9.5	7	6	5	4	3
25	96	48	32	24	16	12	9.5	8	6	5
50	192	96	64	48	32	24	19	16	13	10.5
100	384	192	128	96	64	48	38	32	25.5	21
150	576	288	192	144	96	72	57.5	48	38	32
200	767	384	256	192	128	96	77	64	51	43

Time in minutes

Example

To heat 150 litres of water from 5°C to 60°C (i.e. through 55°C) using a 3 kW immersion heater would take 192 minutes.

See also QUANTITY OF HEAT.

RECTIFICATION   See FLAME RECTIFICATION.

RECYCLED WATER

Waste water, such as grey water from baths and basins, etc. that has been treated and re-used for applications such as watering the garden or flushing the toilet.

RED LEAD (TRILEAD TETROXIDE)

A jointing material no longer used in the plumbing industry, but can still be found as a result of old plumbing works. Red lead is basically a form of lead OXIDE produced by heating lead in air. After being ground to a powder the red lead was mixed with linseed oil or putty to make a jointing compound used to make waste connections to sanitary appliances. Red lead is rarely used today except in the production of some paints to give them excellent protection qualities.

RED LEAD PUTTY   See RED LEAD.

REDUCED PRESSURE ZONE VALVE (RPZ VALVE)

A special valve that consists of two CHECK VALVES and is designed to overcome the problems associated with water 'back flowing' into the supply main. Where these valves have been fitted it is a WATER SUPPLY REGULATION requirement that they are checked for correct operation at least once a year.

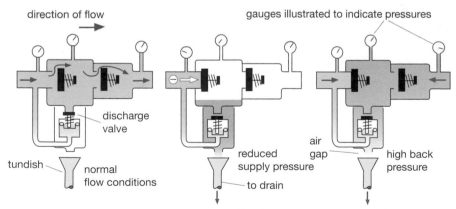

Reduced pressure zone valve

REDUCER (DIMINISHING COUPLING)

A pipe fitting used to join two pipes of different diameters. There are two types of reducers, the concentric and the eccentric. The concentric type reduces the diameter of the pipe by connecting the larger pipe to the smaller one with the pipes having their axes in the same line. The eccentric type is used where it is essential that the change in pipe diameter does not cause a trap in the pipeline for air in water filled systems or condensate in steam mains. See also TAPER.

(See illustration over.)

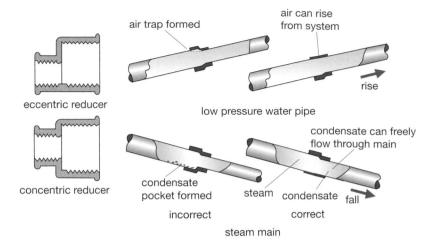

eccentric reducer

air trap formed

air can rise
from system

rise

low pressure water pipe

concentric reducer

condensate
pocket formed

steam

condensate can freely
flow through main

condensate   fall

incorrect

correct

steam main

Reducers

REDUCING FLAME
The part of an OXY-ACETYLENE flame which burns with an outer reducing zone to prevent the metal OXIDISING. See DEOXIDATION.

REDUCING JOINT
A joint which is formed by belling out the end of a piece of copper tube with a FLARING TOOL to the size of a larger pipe. The bell joint is then offered up to the larger pipe and bronze welded. A reducing joint is therefore a means of reducing the diameter of a pipe. For example, see BRONZE WELDING. See also REDUCER.

REDUCTION   See DEOXIDATION.

REFLECTION TEST (MIRROR TEST)
The reflection test is carried out to inspect the internal surfaces of a drainage pipe. It is carried out by using two reflectors or mirrors, which are positioned into the pipe at two adjoining inspection chambers in a straight line. The light from the first chamber is reflected along the drain and by looking into the second chamber a picture can be seen of the internal bore. For example, see ALIGNMENT TEST.

REFLUX VALVE   See NON-RETURN VALVE.

REFRIGERANT
A liquid that evaporates very easily at low temperatures. Such liquids are so volatile that if spilled in a room at a normal temperature they would very quickly evaporate away to form a gas. See also CHLOROFLUOROCARBONS (CFCS).

REGISTER
An opening from or into a system of duct work. The register is fitted with a DAMPER or register plate which can be adjusted to regulate the volume of air passing through it. A register is sometimes found fitted to chimneys to vary the chimney draught or in a warm air central heating system to blow warm air into the room.

REGISTER PLATE

A fire resistant (usually metal) plate that seals the opening at the base of a chimney where a vertical flue pipe or liner enters. Should a pipe enter the chimney, an access door would also be required for inspection purposes.

REGULAR BOILER

The name given to a boiler that is used in conjunction with a separate cylinder for the storage of the domestic hot water. Such a system can be seen by referral to centralised hot water supply. A COMBINATION BOILER or a CPSU are not regular boilers because they do not use a separate hot storage vessel.

REGULATOR

As used for LPG installations

A device to reduce the high pressures from within a BULK TANK or cylinders to a usable pressure of 37 mbar for propane and 28 mbar for butane. These regulators work on a similar principle to those identified in the following regulator entry in which the adjustable screw acting upon the diaphragm forces the valve open. As the pressure below the diaphragm equals or exceeds that of the spring due to the gas supply, the valve closes. The regulators found in LPG installations vary depending upon the gas and appliances to be used but in general either two single stage regulators will be found reducing the pressure in two stages (i.e. first and second stage) or a two stage regulator will be used to reduce the pressure through one control only, such as in the automatic CHANGE-OVER VALVE which automatically opens the back-up cylinders when the supply cylinders have run out. See also GOVERNOR.

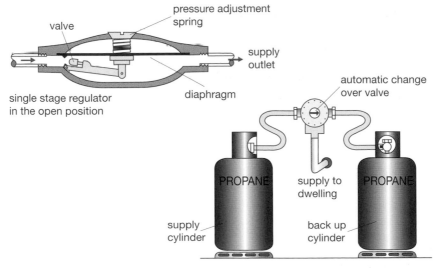

LPG regulators

REGULATOR

As used for oxy-acetylene equipment

Basically a PRESSURE REDUCING VALVE or GOVERNOR designed to reduce the high pressures of gas contained within the cylinders to a lower working pressure. Generally the larger the blowpipe tip or nozzle the higher should the gas pressure be. Most regulators are fitted with two pressure gauges, one to indicate the gas pressure within the cylinder, and one to show the pressure which is available at the blowpipe. To adjust the pressure as required the regulator is fitted with a regulating screw which must be turned clockwise to increase the pressure to the blowpipe and anti-clockwise to decrease it. There are two basic types of regulator either the single or multi-stage type. The single-stage regulator is mostly confined for use with gas cutting and GAS SHIELDED ARC WELDING processes, it would prove difficult to use when welding due to the very precise working pressures required. Shown in the illustration above is a two-stage gas pressure regulator which operates as follows:

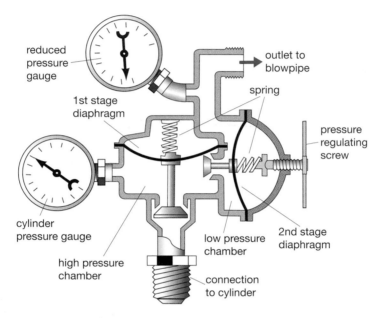

Two-stage gas regulator used for oxy-acetylene

(1) Gas flows from the cylinder into the high pressure chamber where, should the pressure be too high, it compresses onto the diaphragm and spring which, in turn, closes off the supply. The spring in the first chamber is not adjustable and is designed to reduce the pressure which may be as high as 137 bar (2000 psi) to 4 bar (60 psi).

(2) When the regulating screw is turned clockwise, it exerts a pressure on top of the second stage diaphragm. This in turn pushes the valve from its seating to allow gas through to the blowpipe.

(3) If the pressure builds up in the low pressure chamber, it will exert a force on the underside of the diaphragm greater than that being exerted by the regulating screw and therefore will close.

(4) When the blowpipe is in use gas has to be continuously allowed to flow to the nozzle thus in effect the diaphragm moves back and forth very quickly to meet the demand of gas supply.

The single-stage regulator differs only in that there is no first chamber with its pre-set spring giving the gas an initial reduction.

Before connecting the regulators to the oxygen and acetylene cylinders it is essential that any dust, etc. is first blown from the cylinder opening, by purging the gas, this is simply done by quickly opening and closing the cylinder valve. Once the regulator has been fitted and the equipment is assembled ready for use, the gas should be turned on slowly from the cylinder to prevent damaging the regulator. When the equipment is not in use, the cylinder valves should be shut off and any pressure in the hoses released by opening the blowpipe valves. Also the pressure regulating screw should be turned anti-clockwise to release the pressure from on top of the diaphragm. Releasing the pressure from the regulator, in this way, ensures that the diaphragm does not become distorted over a period of time thus resulting in a faulty reading at the pressure gauge.

RELATIVE DENSITY   See SPECIFIC GRAVITY.

RELATIVE HUMIDITY   See HUMIDIFICATION.

RELAY
An electrical device generally used to control another switch remotely. One example would be where a low voltage supply is used to switch on a mains-voltage supply. The relay consists of an electromagnet and moving armature. When current is passed through the coil of wire (winding) it creates a magnetic field, which in turn draws the iron armature into the windings, thus either making or breaking the switch line it serves.

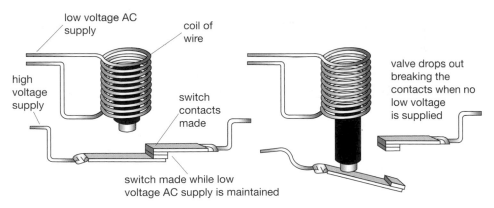

Electrical relay valve

RELAY POINT
Due to the recommended fall or gradient within a STEAM MAIN often the mains has inclined down to its lowest possible level before reaching its destination. Therefore it is necessary to return the steam up to its original height; this is achieved by creating a point in the system which can be suitably drained via a STEAM TRAP. This raising of a steam main is known as a relay point. A relay point is also necessary to raise the steam main up and over an obstruction such as a door opening. See also STEAM HEATING.

(See illustration over.)

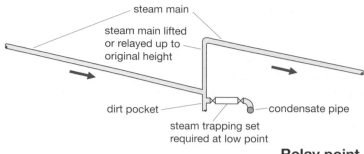

Relay point

RELAY VALVE

A gas valve, not too commonly used nowadays, which is designed to automatically open and close the gas pipeline to the main burner of a gas boiler or CIRCULATOR. These valves operate by the pressure of the gas only and no electrical supply is needed, therefore they can have an advantage for systems which need to be designed with no electrical components (i.e. it will work during power cuts). The relay valve works as follows:

(1) Gas entering the valve cannot pass beyond the closed valve but can escape around the by-pass into the weep pipe.
(2) If the gas thermostat valve is open (calling for heat) and any other valves fitted on the weep pipe are open, gas will pass through this pipe and discharge into the combustion chamber and be ignited.
(3) Gas passing through the weep pipe reduces the pressure above the diaphragm and thus the valve is pushed up by the greater pressure below the diaphragm, allowing gas to the main burners. This condition will continue as long as the weep pipe is open.
(4) Should the thermostat or any other valve on the weep pipe be closed, the pressure above the diaphragm will equalise and the valve will fall under its own weight closing off the supply.

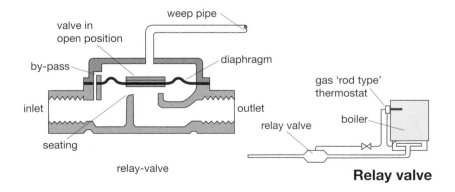

Relay valve

RELIEF VENT

An extended VENTILATING PIPE connected to a DISCHARGE STACK where excessive pressure is likely to occur. A good example of a position where a relief vent could be found is at the base of a discharge stack to prevent COMPRESSION.

RESEALING TRAP (ANTISIPHON TRAP)

A TRAP designed to maintain its water seal should a partial vacuum be created in the waste discharge pipe. There are three basic designs of a resealing trap, the most common one used today incorporates an ANTIVACUUM VALVE; should the pressure drop inside the discharge pipe this valve opens under atmospheric pressure giving a state of equilibrium inside the pipe. Unfortunately these traps often tend to leak through this valve. The other two types of resealing traps work on the principle of retaining the water in a reserve chamber should the conditions be right for SIPHONIC ACTION to take place.

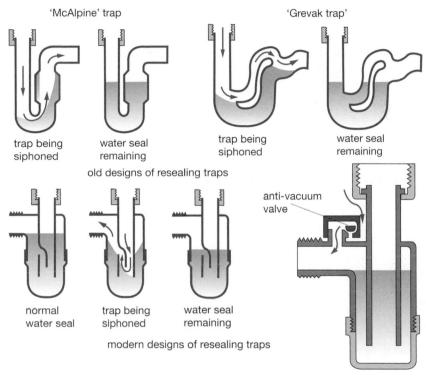

'McAlpine' trap

'Grevak trap'

trap being
siphoned

water seal
remaining

trap being
siphoned

water seal
remaining

old designs of resealing traps

anti-vacuum
valve

normal
water seal

trap being
siphoned

water seal
remaining

modern designs of resealing traps

valve would open should the
pressure drop inside the
discharge pipe
note: with this design of trap
no depth of trap seal is lost

Resealing traps

RESEATING TOOL

After a period of time the brass seatings in valves often become worn or suffer pitting due to corrosion or CAVITATION, resulting in their not fully shutting off the water supply. A reseating tool is specially designed to recut the seating making it smooth and therefore more acceptable to receive the washer. There are two basic types of reseating tool, those designed to recut the seatings to BALLVALVES and those designed to recut the seatings in SCREWDOWN VALVES, such as stopcocks and bib taps etc. The main difference is that the type used for ballvalves does not have a body cone which can screw into the body of the valve, just a plain cone exists which has to be pushed hard against the end of the valve when using the tool. Tap reseating tools are first secured to the tap body then the reseater is slowly turned down. Note the body cones for tap reseaters are available in different size threads or tapered with both left and right handed threads.

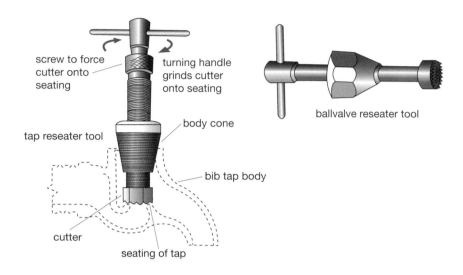

screw to force cutter onto seating

turning handle grinds cutter onto seating

body cone

tap reseater tool

ballvalve reseater tool

bib tap body

cutter

seating of tap

Reseating tools

RESERVOIR

Reservoirs are places where large quantities of water are stored. There are basically two types of reservoir: the storage or impounding type and the service type. The storage type is generally a purpose made lake designed and sited to collect and store water for use as and when required. Reservoirs are often formed by the construction of a dam across rivers impounding the water in a valley. But sometimes reservoirs are provided by diverting canals or pipelines from rivers to natural or artificial depressions in the Earth's surface. Service reservoirs are different in that filtered water is held in these 'closed in' structures, which have a much smaller capacity than the storage reservoirs. Service reservoirs are generally sited close to the area which they serve and usually provide a means of regulating the water pressure. See also WATER TOWER.

RESIDUAL CURRENT DEVICE (RCD)

An electrical device designed to break the circuit should there be a leakage to earth. An RCD plugged into a socket prior to a 230 V portable electric tool will cause the trip to operate in the event of a fault scenario. The RCD works by detecting a current flow that is out of balance. If there were a fault going to earth in a cable feeding the load, the current flow through the phase winding would not equal that flowing through the neutral winding and as a result a magnetic flux would flow round the iron core. Normally this is prevented because the two opposite electron flows cancel each other out. The magnetic flux would induce an electron flow in the search coil, thus identifying the residual current and as a result cause the trip solenoid coil to operate, thus breaking the double pole switch.

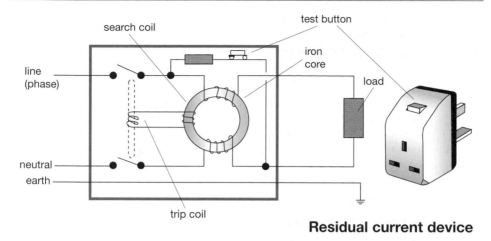

Residual current device

RESIN   See ROSIN.

RESISTANCE
In electrical terms it refers to the restriction or inability of free electrons to pass along or through a material. Resistance is measured in ohms (Ω) and the resistance that a material offers depends upon several factors, including length, cross-sectional area and temperature.

REST BEND   (DUCKS FOOT BEND)
A bend provided with a foot formed on its base, it is so designed to give support at the base of a cast iron DISCHARGE STACK. For example, see BEND.

RESTRICTOR ELBOW
A chrome-plated isolation fitting that is sometimes used on the supply pipework serving a gas fire located within a chimney.

RETENTION FLAME
The small flame used on some designs of gas burners to ensure that the flame does not LIFT-OFF from the burner jets, for example, see GAS BURNER.

RETENTION PORTS
Small holes located in a gas burner, next to the larger burner ports and designed to restrict the flow of gas in order to allow a small RETENTION FLAME to become established, preventing the main burner flames lifting from the burner.

RETURN BEND   (U BEND)
A bend designed to return the run of the pipeline to the direction from which it has just come. The return bend can have many uses, one example being to assist making a DRYING COIL. See BEND.

RETURN PIPE
A pipe in which water flows returning cooler water back to the heat source. The return pipe could be either a central heating return pipe, a PRIMARY RETURN pipe or a SECONDARY RETURN pipe.

REVERBERATION   See WATER HAMMER.

REVERSE ACTION INTERCEPTOR
An INTERCEPTING TRAP connected to the inlet of inspection chambers with the rodding access facing back up the incoming pipe. They are primarily used to connect untrapped SURFACE WATER drains to FOUL WATER drains or sewers. It does away with the need to trap individual surface water connections to COMBINED SYSTEMS OF DRAINAGE.

REVERSED CIRCULATION
Reversed circulation is the flow of water around the PRIMARY CIRCUIT, going in the opposite direction to the intended direction of flow. It is a situation that can arise in domestic hot water systems which do not have pumped primary flow and return pipes and no temperature controls over the water being heated. Water in the hot storage vessel slowly reaches the same or nearly the same temperature as that in the boiler. For example; if a solid fuel back boiler has been alight all day and not much water is drawn off from the hot storage vessel the temperature of the water will continue to rise until at the end of the day the return pipe temperature is very nearly the same as the flow pipe temperature and as a result the circulation of water, due to CONVECTION CURRENTS, tends to cease. During the night as the fire dies down and goes out should the heat loss in the flow pipe be greater than in the return pipe the water will cool quicker and become less dense and as a result fall and push the hotter, lighter water back up the return pipe. This reversal of the circulation will continue until the water completely cools. If during this period of reversed circulation the fire is relit the hot water will rise up the flow pipe from the boiler and meet the descending cooler water resulting in very loud banging and knocking noises within the system, which will continue until normal circulation is resumed. As a preventative measure against reversed circulation, thermal insulation material is placed around the primary flow pipe and the return pipe is left uninsulated, this should ensure that the flow pipe is always warmer than the return.

REVERSED CIRCULATION
A situation in which the heating circuit gets hot when not required. Sometimes experienced with a badly designed FULLY PUMPED CH SYSTEM. It is the result of connecting a central heating return pipe into the domestic hot water circuit return pipe, which may be under a positive pump pressure. The problem is easily avoided by ensuring that all central heating returns are rejoined together prior to rejoining the domestic hot water circuit, see sketch.

(See illustration opposite)

REVERSED RETURN   See TWO PIPE REVERSED RETURN.

RIDDOR
Acronym for 'Reporting of Injuries, Diseases and Dangerous Occurrences Regulations'. This is a mandatory Regulation that must be observed. Under it, it is essential that the Health & Safety Executive (HSE) should be informed as to the specific requirements of a particular incident. For example, where a major accident has occurred or where, without appropriate action, an accident may occur. Reporting

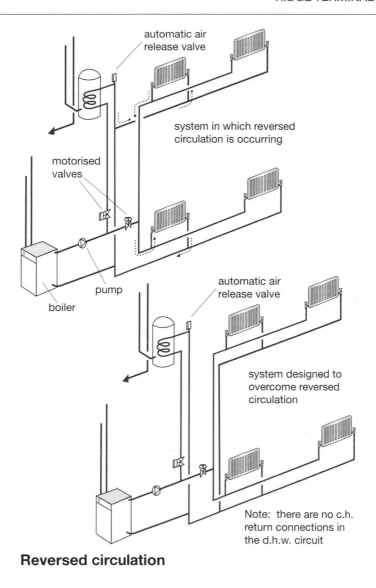

Reversed circulation

under RIDDOR allows the HSE to act as necessary or to prosecute the offending individual(s).

RIDGE
The highest point of a pitched roof, see RIDGE PIECE and BUILDING TERMINOLOGY.

RIDGE PIECE (SADDLE PIECE)
The sheet weathering which is formed to cover the joint where a pitched roof meets an ABUTMENT.

(See illustration over.)

RIDGE TERMINAL
A flue terminal which terminates at the ridge of a roof. See FLUE TERMINAL.

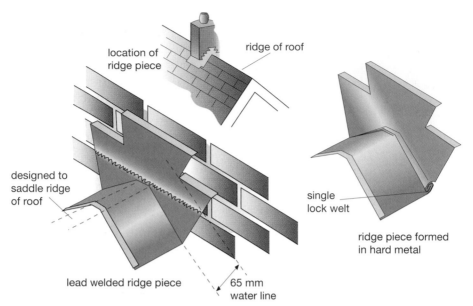

location of
ridge piece

ridge of roof

designed to
saddle ridge
of roof

single
lock welt

ridge piece formed
in hard metal

lead welded ridge piece

65 mm
water line

Ridge piece

RIGHTWARD WELDING TECHNIQUE (BACKHAND WELDING)
A welding technique in which the welding is started at the left side of the joint to be welded and the welding continues along the weld from left to right. With this technique the blowpipe tip is held down at a much lower angle to the weld than the leftward technique and the blowpipe is moved progressively along the joint to be welded in front of the molten pool of metal, the filler rod follows along the weld being used as required. See LOW CARBON STEEL WELDING. See also LEFTWARD WELDING TECHNIQUE.

RING SEAL JOINT
One of many types of push fit joints in which a rubber ring is used to assist making the joint between the spigot and socket of a pipe and fitting. For examples, see PITCH FIBRE PIPE and CLAYWARE PIPE. See also SHRUNK RUBBER JOINT.

RING SPANNER   See SPANNER.

RING TYPE WC SEAT   See WC SEAT.

RIPPLING
Ripples formed on the bend of a pipe. They are caused by a bend being pulled while the pressure is not adjusted correctly at the FORMER, e.g. it is too loose. See BENDING MACHINE. See also THROATING.

RISER
The name given to a vertical pipe rising up through the building for the conveyance of water for fire fighting purposes; see DRY RISER and WET RISER. The term riser is sometimes also given to the mains water supply pipe raising water up to the higher floors and to any storage cisterns which may be fitted. See also GAS RISER.

RISK ASSESSMENT

A pre-assessment of a work situation where possible hazards to individuals, both undertaking the work and passers-by, are identified. This assessment is a requirement by law and all activities need to be evaluated prior to beginning the work, listing control measures and actions to reduce the risk to a safe level.

ROD TYPE THERMOSTAT   See THERMOSTAT.

RODDING EYE (RODDING ACCESS)

A capped extension on a pipe where access can be gained to a drain or any discharge pipe for the purpose of cleaning with rods or inspection. See also RODDING POINT SYSTEM.

RODDING POINT SYSTEM

A 'closed' system of drainage designed to reduce the number of inspection chambers, found in the more traditional systems. The rodding point system is generally run in PVC pipe and, with the omission of inspection chambers, reduces the cost of installation considerably.

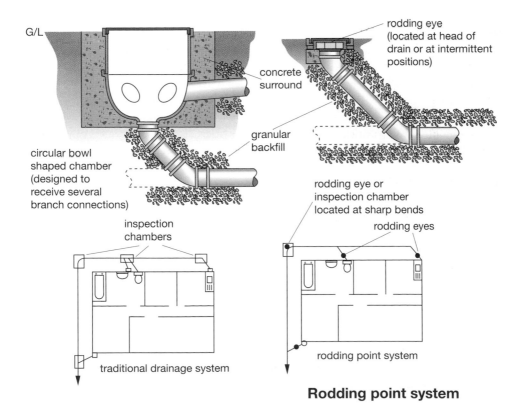

Rodding point system

ROLL

A method of allowing for expansion on metallic roof coverings. It can be used for non-metallic coverings (NURALITE) although if so it is usually for reasons of appearance only. A roll is an expansion joint which runs with the fall of the roof, i.e., in the same direction as the flow of water. There are several types of rolls depending upon the material used and specifications. For examples see LEAD, COPPER and ZINC SHEET AND ROOF COVERINGS.

ROLL CAP ROOFING
Roll cap roofing is a design of DRIP and ROLL developed for zinc weathering materials due to the fact that, compared with lead, copper and aluminium, it is very stiff and has a greater amount of SPRING BACK. Stainless steel and NURALITE, because they also do not possess the WORKING PROPERTIES of lead, copper and aluminium, are also used with the roll cap system. Using Nuralite with this design of roofing is not a common practice and it is generally only carried out for reasons of appearance. For examples and details, see ZINC SHEET AND ROOF COVERINGS.

ROLL RING JOINT
A flexible joint used to join cast iron pipe. The roll ring joint consists of using the standard cast iron pipe, which is normally used for CAULKED JOINTS and a specially designed rubber. As the spigot is being pushed into the socket the roll ring rubber is inserted with the spigot end and the rubber rolls into place. For example, see CAST IRON PIPES AND FITTINGS.

ROLLED JOINT
A method of making an underhand wiped solder joint. Jointing in this fashion is no longer carried out except for special cases, but proved useful in the past to join short lengths of lead to brass fittings, such as a CAP AND LINING. The rolled joint was formed by preparing the joint as for an underhand joint and applying the solder as necessary, but instead of moving a wiping cloth around the pipe to form a uniform shape the pipe was positioned on two wooden blocks and rolled, the wiping cloth was held still under the molten solder. See WIPED SOLDER JOINT.

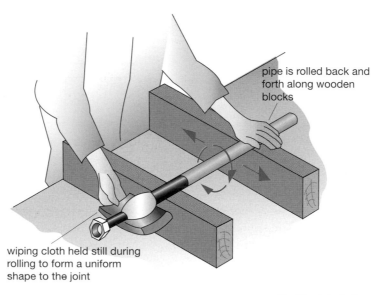

pipe is rolled back and forth along wooden blocks

wiping cloth held still during rolling to form a uniform shape to the joint

Rolled joint

ROLLER PIPE CUTTER   See PIPE CUTTERS.

ROOF WEATHERINGS
The material used to cover a building thus keeping the weather out. Roof weatherings can consist of a great amount of materials, for examples see NON-METALLIC ROOF WEATHERINGS and METALLIC ROOF COVERINGS.

ROOM SEALED
A gas or oil burning appliance in which the air supply is taken from and the flue gas products are discharged outside the room in which the appliance is situated. Examples of room sealed appliances include BALANCED FLUE appliances, VERTEX FLUE systems or appliances connected to SE or U DUCT systems.

ROOM THERMOSTAT
A thermostat designed specifically to control the temperature within a building. When the desired room temperature is achieved the thermostat breaks the switch contact turning off a pump, boiler, or some other control therefore preventing the flow of heat to the room.

A room thermostat should not be located where it would be affected by extremes of temperature, for example, cold draughts or near any heat source. A good location would be in the living room, or lounge, at a height of about 1.6 metres. Halls should be avoided for positioning due to their temperature not being critical and positioned here they are often subjected to draughts.

ROOT
Term used when welding. The root is the position in any welded joint where the two adjoining surfaces to be welded come nearest together. For example, see WELDING DEFECTS.

ROSE
The shower head or outlet from a SHOWER MIXING VALVE.

ROSIN (RESIN)
An INACTIVE FLUX sometimes used in the process of soldering. Rosin consists of the residue obtained by distilling the natural gum-like fluid from pine and conifer trees. See FLUX.

ROTARY BENDING MACHINE   See BENDING MACHINE.

ROTARY VAPORISING BURNER (WALLFLAME BOILER)
A design of OIL-FIRED BOILER which burns its fuel in a circular 'wall of flame' around the internal surfaces of the boiler waterway. Although this type of boiler is classified as a vaporising burner, it should be noted that the oil is partly atomised as it leaves the oil outlet point. There are several designs of rotary vaporising boilers but all basically work as follows:

(1) Oil from a CONSTANT OIL LEVEL CONTROL is fed into a pipeline which leads to an oil well located at the centre of the base of the boiler. Into the oil well dip

two oil distribution tubes which are attached to a rotor arm allowing them to turn. The oil distribution tubes project up through the base of the boiler and are inclined outwards, above these tubes is located an air fan which draws its air through the hollow casing of the motor.

(2) When the boiler requires heat an electrical element rim heater turns on for a short period to assist the ignition of the fuel (some boilers do not have this function).

(3) When the rim heater switches off, the high tension ignition electrode switches on to cause a spark to jump across to earth at the rim of the boiler; some models of boiler such as the 'firefly dynaflame' have an igniter which glows white hot.

(4) The motor starts up which spins the fan/rotor assembly to cause oil to be lifted and spun out by centrifugal force to the rim of the burner in a fine spray.

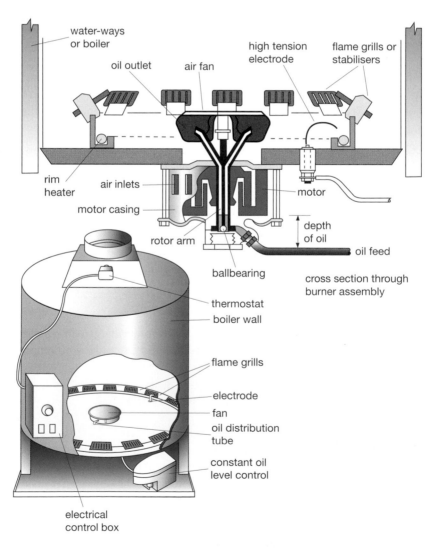

Rotary vaporising burner

(5) As the droplets of oil fall onto the heated rim (caused due to the previously heated rim heater or the high tension spark) they quickly vaporise and when mixed with the oxygen in the air the fuel is ignited by the spark.

(6) A flame quickly becomes established around the circumference of the combustion chamber and the grills or flame stabilisers soon glow red hot and radiate heat to cause the oil to vaporise almost instantly; the grills also direct the flame onto the waterway of the boiler.

(7) When the boiler is shut down a minimum of 7–10 minutes has to elapse before the boiler will relight while the electrical control box resets itself.

See also VAPORISING BURNER.

RPZ VALVE   See REDUCED PRESSURE ZONE VALVE.

RUBBLE DRAIN
A trench which has been partly filled with brick rubble or broken stones. If possible the trench is normally dug with a fall terminating at a suitable water course. The purpose of a rubble drain is to drain away surplus surface and ground water. See also LAND DRAIN.

RUNNING NIPPLE   See BARREL NIPPLE.

RUNNING ROPE (MONKEY'S TAIL/SQUIRREL'S TAIL)
Tool used to retain the molten lead in the socket of a cast iron pipe when forming a caulked joint in the horizontal position. See CAULKED JOINTS. See also CAULKING CLIP.

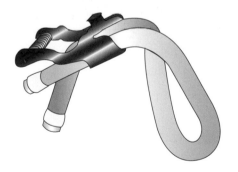

Running rope

RUNNING TRAP   See TRAP.

RUN-OFF
The discharge of water from a surface.

RUST POCKET (RUST TRAP)
A drainage fitting fitted at the base of a cast iron VENTILATING PIPE. It is designed to retain any rust deposits which may fall and collect at the base of the pipe. Periodically

the debris must be removed. Should the ventilating pipe be connected to a discharge pipe the rust pocket is not required as the water will wash away any fallen rust. See also WET VENTING.

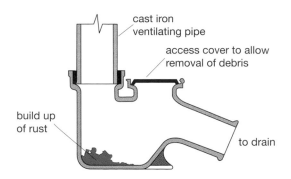

cast iron
ventilating pipe

access cover to allow
removal of debris

build up
of rust

to drain

Rust pocket

RUST TRAP   See RUST POCKET.

S TRAP   See TRAP.

SACRIFICIAL PROTECTION   See CATHODIC PROTECTION.

SADDLE CLIP   See PIPE SUPPORT.

SADDLE JUNCTION
A drainage fitting used to make the connection of a branch drain into an existing drain run which is over 225 mm in diameter. A hole is cut in the top half of the pipe into which the saddle connection is made. For example, see CONNECTIONS TO EXISTING DRAINS.

SADDLE PIECE   See RIDGE PIECE.

SAFE
A tray fitted under a storage cistern or shower tray to catch any water which may leak or spill over the sides. Its purpose is to prevent damage to the structure of the building. The safe, if used in conjunction with a storage cistern, usually has a warning pipe fitted to discharge at a position where it can readily be seen, thus indicating a problem. For example, see STORAGE CISTERN.

SAFE EDGE
A small closed welt or U fold on the edge of the metallic roof covering formed to ensure that no one can cut themselves on the sharp edges of the material. The safe edge also has the advantage of being a STIFFENER and improves the appearance.

SAFE FLUX
Also called an inactive flux. See FLUX.

SAFETY INTERLOCK   See INTERLOCK DEVICE.

SAFETY PIPE
The VENT PIPE from a boiler.

SAFETY SHUT OFF VALVE
One of several designs of valve, such as an electro-hydraulic or electro-mechanical valve, that allows for a relatively slow opening and rapid closure of the gas supply to workshops or large burner control TRAIN. One type, the electro-hydraulic, works by

allowing a supply of oil to be pumped from one chamber to another. This in turn acts upon a diaphragm forcing the valve open. The diaphragm when fully extended breaks the contact of a limit switch to shut off the pump. Should the electrical supply be isolated a powerful compression spring forces the valve closed, allowing the oil to flow quickly back through the relief valve. Another design, the electro-mechanical, uses an electric motor to turn a cam that slowly pushes the valve against the spring to the open position. As with the previous valve, a limit switch controls the amount of valve travel. Should the supply be isolated, the solenoid would be de-energised, allowing the latch to move away. Thus the valve would rapidly close in the small turn of the cam. These valves, once fully opened, conserve energy in that no major power consumption is needed.

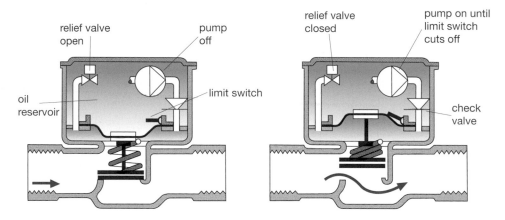

Electro-hydraulic safety shut off valve

SAFETY VALVE

A special valve designed to open only if an increased pressure or temperature builds up in the hot water system due to a fault. There are various designs of safety valves but basically there are two types, those that open should the pressure build up within the system (see PRESSURE RELIEF VALVE) and those that open should the temperature rise above its normal working temperatures (see TEMPERATURE RELIEF VALVE). It is also possible to purchase a combined pressure and temperature relief valve.

SAL-AMMONIAC (AMMONIUM CHLORIDE)

A colourless, crystalline substance that is soluble in water and forms a slightly ACIDIC solution which is sometimes used as a flux in soldering materials such as copper, brass and zinc although when used for zinc it is generally mixed with ZINC CHLORIDE. See FLUX.

SALT GLAZED PIPE

Salt glazed earthenware was the drainage material used for many years. Salt glazed pipes were produced by adding salt to the fire during the firing process; the salt fused onto the clayware, coating its surfaces and giving it a waterproof glaze. Today, however, this process is rarely carried out due to more modern firing methods and the production of materials such as vitrified clay, a non-porous mixture of clays, therefore the term is used less. See CLAYWARE PIPES.

SAND CAST IRON PIPES

CAST IRON PIPES produced by molten metal being poured into a sand cast and allowed to solidify. Pipes are usually cast upright with the socket end at the bottom; this assures a greater density of metal at this point, thus giving greater strength for CAULKING. See also SPUN CAST IRON PIPES.

SAND CASTING

A process occasionally carried out to produce various ornamental finishes to lead work. Generally small moulded designs which have been cast are soldered or lead welded onto the surface of the finished work, but sometimes large projects are undertaken such as ornamental CANOPIES or HOPPER HEADS. Simple sand casting is carried out using sand of a fine loamy nature which is slightly moistened to allow it to just bind or stick together. Ordinary building sand is of no use in casting and generally a special casting sand, such as 'Mansfield red', is purchased. To produce a simple one-sided cast figure, first the figure to be cast is placed on a flat piece of wood. On this is positioned an open-ended box or tube, with the cast remaining somewhere in the centre. Parting powder, or finely ground chalk, is then gently sieved over the whole cast. This is done to prevent the sand sticking to the proposed mould. Now a layer of casting sand is carefully sieved on top of the cast, followed by more sand (although there is no further need to sieve), slowly compacting it down hard over the cast figure until the top of the open box is reached. The sand is now levelled off and another piece of wood is placed on top so that the whole lot can be picked up and turned over. When the piece of wood, now on the top, is removed, the flat side of the lead cast is exposed. The cast is gently tapped a few times to loosen it and then it is carefully lifted out of the sand. Finally, the molten lead is poured into the sand impression.

SAND COMPRESSOR   See SAND LOADING.

SAND FILTER

A means of straining or filtering water to assist in its purification. There are two types of sand filters, the slow sand filter and the pressure filter. The pressure filter tends to be preferred as it is much quicker in operation. Pressure filters also take up much less room and it is easier to clean the sand as required. Freshly cleaned sand filters work by at first only straining the water to remove any suspended matter, then about two days after being filled a gelatin film forms over the sand which forms a barrier to any bacterial growth. Occasionally the sand has to be cleaned, this is usually carried out by back washing the sand filter with previously treated water. After cleaning, compressed air is often blown through the sand to stir up the surface and prevent any mud pockets forming. After the water has been passed through sand filters it is then sterilised with chlorine and ammonia, see CHLORINATION.

(See illustration over.)

SAND LOADING

A method used to support the internal bore of a tube when using heat to bend the pipe; generally a BENDING SPRING is used in preference to this method which can be somewhat slow. The sand used must be a fine dry sand which is free from any small stones or grit. To carry out a bend using this method one end of the tube must first

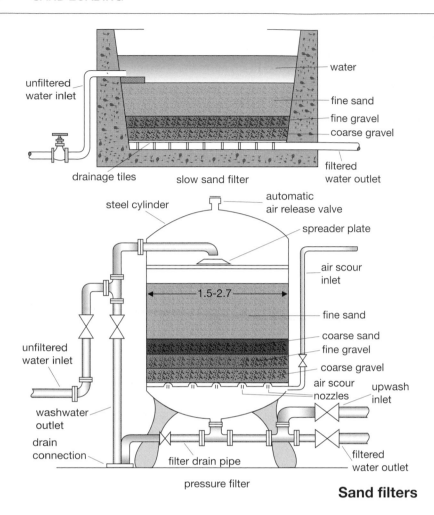

Sand filters

be blanked off, this can be done by inserting a wooden plug or by using a special sand compressor. Sand is slowly poured into the tube and is tightly packed down with a rod, tapping the side of the tube whilst filling the pipe helps this consolidation. When the pipe is full, the other end is also plugged off and bending the tube can proceed. See PIPE BENDING TECHNIQUE (USING HEAT).

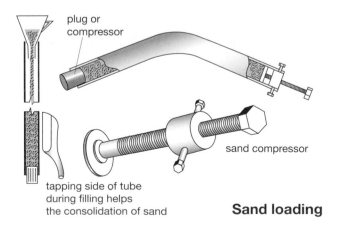

tapping side of tube
during filling helps
the consolidation of sand

Sand loading

SANIFLO

Saniflo, Sanitop and Saniplus are trade names of small bore sanitary waste discharge pipe systems. See WC MACERATOR PUMP.

SANITARY ACCOMMODATION

Sanitary accommodation is a term sometimes used to identify a room within a building which contains a WC pan or a urinal.

SANITARY APPLIANCES (SANITARY WARE)

Fittings used for cleansing and disposing of waste products, most sanitary appliances fall into one of two groups, WASTE APPLIANCES or SOIL APPLIANCES. All sanitary appliances are made of non-absorbent, non-corroding, smooth and easily cleaned material and usually made from ceramic ware, vitreous enamelled cast iron, vitreous enamelled pressed steel, stainless steel or plastics (thermosetting and thermoplastic).

SANITARY BRANCH (WC BRANCH/WC JUNCTION)

A branch fitting designed with an extra long branch inlet which can be used to connect up to a WC pan the other side of a wall. The length of the branch connection varies depending upon the building wall thickness. For example, see BRANCH.

SANITARY CONNECTOR (WC CONNECTING PIPE)

A short length of discharge pipe with a socket on one end to receive the outlet from a WC pan.

SANITARY PIPEWORK

A system of above ground drainage designed to remove all the foul and waste water to the below ground drainage system. Originally the foul water from SOIL APPLI-ANCES was kept separate from the water from WASTE APPLIANCES and two separate discharge stacks were required, the water only joined at ground level in the below ground drainage system. This type of system was known as the TWO PIPE SYSTEM. Such a system installed today would be rare, the only exception might be where the soil appliances and waste appliances are not too closely grouped. All sanitary pipework today is based on the ONE PIPE SYSTEM in which one discharge stack is used to convey both foul and waste waters. There are four basic systems in use: the primary and secondary ventilated stack systems and the ventilated or unventilated branch systems.

With the ventilated discharge branch system, a ventilating pipe is extended to connect to each of the individual branch pipes throughout the system, so designed to safeguard against TRAP SEAL loss. This system is generally adopted in situations where it is not possible to have close groupings of sanitary appliances and long branch discharge pipes can be expected. In the secondary ventilated stack system, only the main discharge stack is ventilated, to overcome pressure fluctuations. With this system the branch discharge pipes connect directly into the main stack without the need for a branch ventilating pipe, therefore this system is only suitable for buildings in which the sanitary appliances are closely grouped to the main stack.

The primary ventilated stack system is used in similar situations to the secondary ventilated stack system, the difference being that the stack ventilating pipe can be omitted if the discharge stack is large enough to limit pressure fluctuations. Finally,

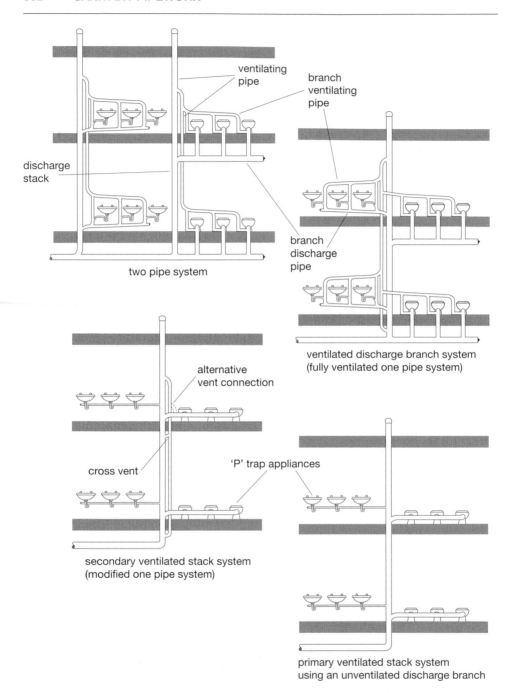

two pipe system

ventilating pipe

branch ventilating pipe

discharge stack

branch discharge pipe

ventilated discharge branch system
(fully ventilated one pipe system)

alternative vent connection

cross vent

'P' trap appliances

secondary ventilated stack system
(modified one pipe system)

primary ventilated stack system
using an unventilated discharge branch

Sanitary pipework

the unventilated discharge branch system allows for a system in which no ventilation pipes are needed. This can only be achieved where branch pipes are suitably sized. See also BRANCH DISCHARGE PIPE, BRANCH VENTILATING PIPE, PIPE SIZING (SANITARY PIPEWORK AND DRAINAGE) and VENTILATION OF DRAINS AND DISCHARGE PIPES.

SANITARY WARE   See SANITARY APPLIANCES.

SAP RATING
Acronym for the Government's 'Standard Assessment Procedure for energy rating of dwellings'. It was designed to enable a reliable means of calculation for the energy efficiency performance of a dwelling. The SAP uses a scale of 1–100, the higher the rating the better the efficiency. Figures above 80 can be regarded as highly efficient, whereas the national average for existing homes is 40. SAP rating takes into account a number of features that include: insulation, design, location, ventilation, and the type of heating and hot water control.

SARKING FELT   (UNDERSLATING)
The bitumen roofing felt used on pitched roofs under slates or tiles.

SATURATED AIR
An environment in which the air has as much water vapour as it can hold at the ambient temperature. Should the temperature of the air be raised, then it could hold more water vapour. See also HUMIDIFICATION.

SCALE
Deposits of calcium or magnesium salts found in boilers and pipework, caused mainly by TEMPORARY HARD WATER. The scale tends to build up over a period of time and eventually can obstruct the pipework completely.

SCALE REDUCER   See WATER CONDITIONERS.

SCALLOPE   See AUGER.

SCENT TEST   (CHEMICAL TEST)
A test which is carried out on below ground drainage systems to determine whether it is the drain which is causing unpleasant odours inside premises. A strong smelling scent or chemical such as calcium carbide or oil of peppermint is flushed into the drainage system by means of a special sealed container which opens on contact with water. See also TRACING EXISTING DRAINS.

SCHOOL BOARD BRACKET   See PIPE SUPPORT.

SCREWDOWN FERRULE   (SERVICE UNION)
A bent FERRULE which incorporates a SCREWDOWN VALVE to enable the valve to be opened and closed as required with a ferrule key.

(See illustration over.)

SCREWDOWN VALVE
Valves made in a variety of patterns. Screwdown valves are characterised by a plate or disc, which shuts against the incoming water pressure at right angles to the valve SEATING aperture; the plate is moved by the rotation of a spindle. Screwdown valves are designed to shut off the water supply slowly and by doing so prevent WATER HAMMER. The headgear to screwdown valves are all basically the same, incorporating

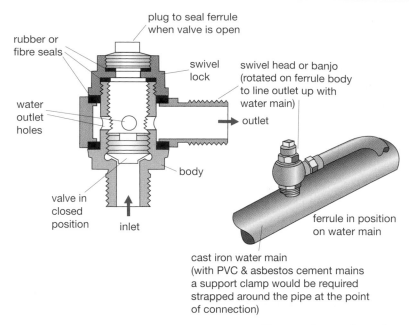

Screwdown ferrule

a threaded spindle which is turned through several revolutions before finally fully opening or shutting the valve, there are variations to the design of spindle these being either rising or non-rising. For examples, see STOPCOCK, BIB TAP, PILLAR TAP, and SUPATAP.

SCREWDRIVER

A tool used to turn screws. A screwdriver consists of a steel blade with a wooden or plastic handle. There are several designs of screwdriver and it is essential to choose one with the correct size blade, thus ensuring full leverage and avoiding damage to the screw head.

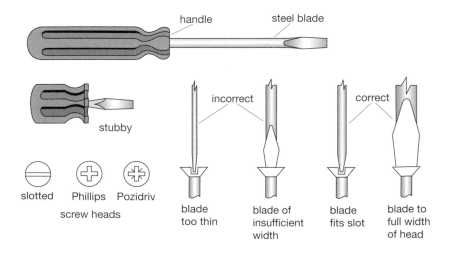

Screwdrivers

SCREWED GLAND JOINT

A design of flexible joint sometimes used on cast iron pipes.

SCRIBING PLATE

A piece of hard sheet metal used to mark out the shape that needs shaving from a lead pipe when forming a wiped solder branch joint. It is used in conjunction with a set of dividers.

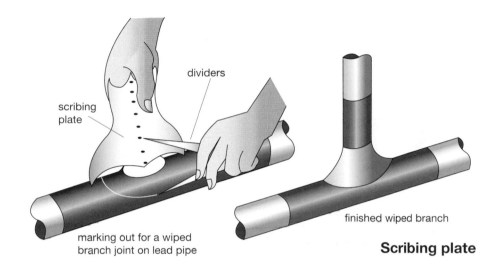

dividers

scribing plate

finished wiped branch

marking out for a wiped branch joint on lead pipe

Scribing plate

SE DUCT (SOUTH-EASTERN DUCT)

A design of flue sometimes used in multi-storey buildings to receive BALANCED FLUE appliances. See also SHARED FLUE.

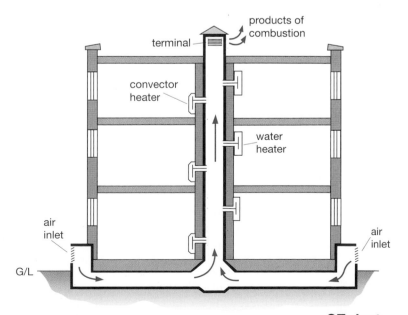

products of combustion

terminal

convector heater

water heater

air inlet

air inlet

G/L

SE duct

SEALED EXPANSION VESSEL (FLEXIBLE MEMBRANE VESSEL)

A gas/air loaded vessel used to take up the expansion of water in hot water pipework systems such as SEALED SYSTEMS OF HEATING. Expansion vessels use an internal flexible diaphragm to keep the water separate from the gas, as the water heats up it compresses the gas. When fitting expansion vessels, into systems, they should be sited either in the cold feed supply pipe or on the return pipe, this would provide cooler water and should increase the life of the rubber diaphragm. The expansion vessel should not be connected to the positive side of a pump otherwise it would be subjected to unnecessary pressures. It is important to ensure that the expansion vessel is correctly sized, water heating from 4°C to 100°C would expand by approximately 4% ($^{1}/_{25}$) in volume.

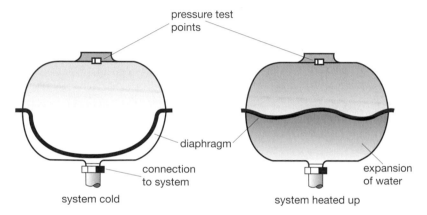

Operation of a sealed expansion vessel

SEALED SYSTEM OF HEATING (CLOSED OR UNVENTED SYSTEM)

A hot water heating system in which the water supplied to the system is fed via a temporary supply pipe which has a DOUBLE CHECKVALVE ASSEMBLY connected to the circuit or some other means to prevent BACK-SIPHONAGE. Some sealed systems are designed with a top-up bottle so that once the system is filled and the mains connection has been removed, which is at a high point, the water in the bottle can make up small amounts of water loss. Once the system is filled with water the inlet supply is shut off thus all the water is entrapped in the circulating pipework. When the boiler is fired the water heats up and expands. Because this water, which is expanding, cannot be taken up in a feed and expansion cistern it generates a pressure on the internal pipework, eventually this pressure compresses the air and nitrogen gas located in a SEALED EXPANSION VESSEL thus taking up the expansion of the water. It must be noted that due to the fact that a pressure is created in this type of system higher water temperatures can be achieved due to the fact that water boils at a higher temperature under pressure (see BOILING POINT). Therefore convector heaters or heating panels are often chosen in preference to panel radiators to prevent scalding. The advantages of sealed systems over more conventional systems include:

(1) Less pipework being necessary on installation.
(2) Pumping water over vent or drawing air into the system is eliminated.
(3) Higher water temperatures can be achieved.

(4) The position of the boiler can be anywhere, even in the roof space, as no header cistern is required.

See also HOT WATER HEATING SYSTEMS and UNVENTED DOMESTIC HOT WATER SUPPLY.

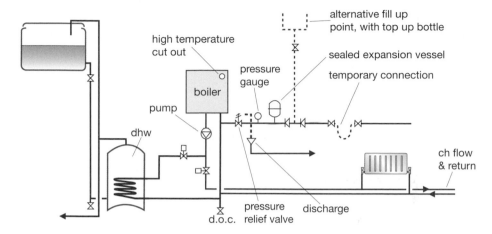

Sealed system of heating

SEAMING PLIERS
A tool used to assist the forming of sheet metallic materials. See PLIERS.

SEAMING TOOL (SEAMING CHISEL)   See GROOVING TOOL.

SEATING
The position in a valve where a washer would come to rest. The seating consists of a hole through which the water supply passes. Examples of seating positions can be seen in the various SCREWDOWN VALVES and BALLVALVES to name just a few. In the case of many ballvalves the seating is interchangeable allowing for variations in water pressure.

SECOND FIX
The final installation of appliances, such as baths, basins or radiators, etc. undertaken following the FIRST FIX operations and the completion of the necessary surface finishes, such as the plastering or the replacement of floor boarding, etc.

SECOND STAGE REGULATOR   See REGULATOR.

SECONDARY AIR
The air supply that supplies a proportion of that needed in a PRE-AERATED FLAME. Refer to this entry for further study.

SECONDARY CIRCUIT
The pipework which circulates from a HOT STORAGE VESSEL to the DRAW OFF POINTS and back again. Its purpose is to permit hot water to circulate around always keeping a hot supply close to the draw off points. Thus when using a hot tap,

it is not necessary to wait for a period allowing the cold water out before getting hot water. Secondary circulation can be designed to work by CONVECTION CURRENTS but in many cases a pump is fitted. It must be noted that if installing a secondary circuit the secondary return must be connected to the top third of the hot storage vessel, if this is not done the hot water will mix with the cooler water in the vessel reducing its temperature. See also DEAD LEG.

SECONDARY CIRCULATION
The water passing around a SECONDARY CIRCUIT.

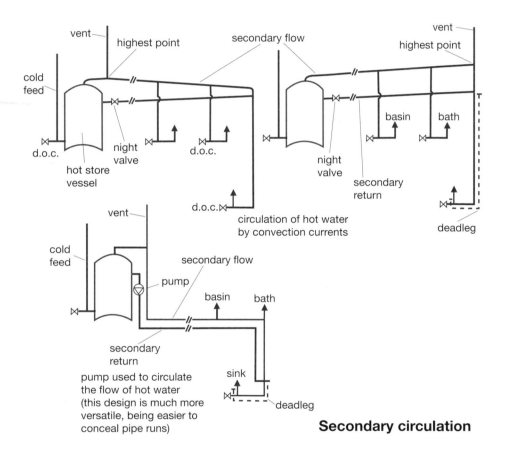

Secondary circulation

SECONDARY FLOW AND RETURN
The main distribution pipework in a SECONDARY CIRCUIT.

SECONDARY FLUE
The part of a conventional flue which removes the products of combustion to the outside environment. It is the length of flue from the connection of the PRIMARY FLUE to the flue terminal. For example, see OPEN FLUE.

SECONDARY GAS METER   See GAS METER.

SECONDARY VENTILATED STACK SYSTEM (VENTILATED STACK SYSTEM/ MODIFIED ONE PIPE SYSTEM)

A design of above ground sanitary pipework in which the main discharge stack is vented via a 50 mm minimum diameter ventilating pipe connecting to the main stack at each or alternate floors. This design of system is normally employed only in buildings of over five storeys where COMPRESSION tends to be the main cause of TRAP SEAL LOSS; the ventilating pipe tends to relieve any build up of pressure within the main discharge stack. When this system is employed, the pipework should follow the same criteria as the PRIMARY VENTILATED STACK SYSTEM to avoid trap seal loss by SIPHONIC ACTION. For an example of this system, see SANITARY PIPEWORK.

SECRET GUTTER

A concealed gutter which runs down the slope of a roof at a position where the roof meets an ABUTMENT or at an internal angle such as at a VALLEY GUTTER. Secret gutters are chosen when as much of the roof surface as possible is desired to be seen and not the valley. Generally when a roof meets an abutment SOAKERS and STEP FLASHINGS are chosen but sometimes, where there is a possibility of water washing back under the slates, a gutter proves the best alternative. The main disadvantage of secret gutters is their danger of becoming blocked with leaves, etc. and therefore the cause of leaks into the building, also if blocked they can be difficult to unblock. For this reason it is recommended that these gutters should not be used in wooded areas.

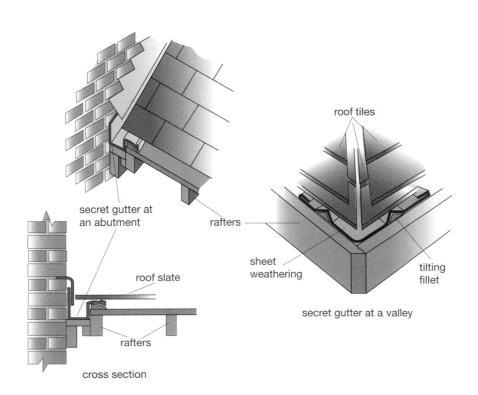

secret gutter at an abutment

rafters

roof tiles

roof slate

sheet weathering

tilting fillet

rafters

cross section

secret gutter at a valley

Secret gutters

SECRET OVERFLOW

An overflow passage forming an integral part or connected to those waste appliances such as wash basins, sinks, baths, etc., which conveys overflowing water into a slotted waste fitting. For example, see WASTE FITTING.

SECRET TACK

A fixing used on sheet lead to secure it to the building. First a strip of lead about 100 mm wide is lead welded or soldered onto the back of the sheet lead leaving a long tail or free end. The sheet lead is offered up to the building surface and the free end of the 100 mm strip is passed through a slot in the structure. Finally the free end is secured to the internal surface, for example, see SHEET FIXINGS.

SECTIONAL BOILER

A type of boiler generally used in larger buildings which consists of a series of sections bolted together on site. Smaller sectional boilers are available but these are normally assembled by the manufacturer.

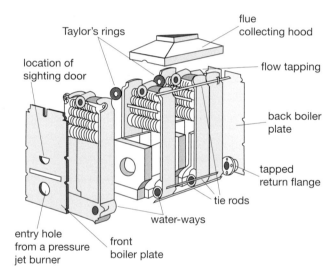

Sectional boiler

SECTIONAL TANK (CISTERN)

A large tank or cistern which is supplied as moulded panels to suit the storage requirements needed and is assembled in position on site. Sectional tanks are essential in buildings requiring very large quantities of water. They are available in some plastics such as reinforced glass fibre, cast iron, or galvanised pressed steel. To give tank stability to the large quantities of water, they hold internal bracing rods are provided.

(See illustration opposite)

SEDBUK EFFICIENCY RATING

The acronym for 'Seasonal Efficiency of Domestic Boilers in the UK'. Basically this is a classification used to compare different boiler efficiencies. SEDBUK is the average annual efficiency achieved by a boiler in a typical domestic situation. Boilers are

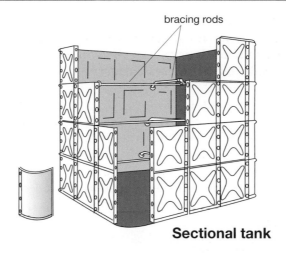

bracing rods

Sectional tank

tested under standard conditions, at 30% and full load. The efficiencies found are used to indicate a typical load profile, which is then classified into one of seven bands. Note that the minimum band under current Building Regulations is band 'D'.

Band	SEDBUK efficiency
A	90–94%
B	86–90%
C	82–86%
D	78–82%
E	74–78%
F	70–74%
G	<70%

SEDIMENT BUCKET (SEDIMENT PAN)
The removable container found in a YARD or GARAGE GULLY for the collection of grit, etc.

SELF-CLEANSING GRADIENT
The GRADIENT to which a horizontal discharge pipe or drain is laid. If the pipe is laid with too steep a fall the water will flow through the pipe very fast and leave the solid matter behind. Equally if the gradient is too shallow the water will flow very slowly and be insufficient to carry along with it any solid matter. To find the required gradient for drainage and discharge pipes, see LOADING UNITS, CHEZY FORMULA or MAGUIRE'S RULE.

SELF-CLEANSING VELOCITY
The VELOCITY or speed at which water flows through a pipe or channel to ensure that no deposits of solid matter are left behind. See SELF-CLEANSING GRADIENT.

SELF-CLOSING TAP   See NON-CONCUSSIVE VALVE.

SELF-GRIP WRENCH   See LEVER WRENCH.

S

SELF-SEALING VALVE (HEPv0 VALVE)
A special waste fitting that has been designed as an alternative to a TRAP. It consists of a collapsible membrane which makes a seal from the house drain, but opens when water is discharged down through the waste pipe.

SELF-SIPHONAGE   See TRAP SEAL LOSS.

SELF-VENTING CYLINDER   See SINGLE FEED CYLINDER.

SEMI-RIGID COUPLING
A steel fitting that incorporates a butyl-rubber seal and is designed to allow for a small percentage of movement. It is therefore ideal for bridges, etc. where distortion is a regular occurrence.

SEMI-RIGID STAINLESS STEEL
A design of stainless steel tube, formed with annular corrugations, that is supplied both in short sections such as the flexible connector to a gas meter (anaconda) or as a coil in various lengths and sizes up to 50 mm. The tube is supplied with factory fitted connections; but some designs use specially designed fittings.

SENSIBLE HEAT   See LATENT HEAT.

SENSOR
A device which is designed to sense or identify heat or light variations, e.g. heat sensor or light sensor. Sensors are used to operate various devices such as fire or burglar alarms, etc. or to control the flow of liquids or gases through pipe lines. See THERMOSTAT for an example of a heat sensor.

SEPARATE SYSTEM OF DRAINAGE
A design of below ground drainage in which FOUL and WASTE WATER are conveyed in one pipe to the sewage treatment works, and all the SURFACE WATER is conveyed in another pipe completely independent of foul water, thus there is no need for water treatment. Connections to surface water drains do not need to be trapped because there should be no unhealthy smells, but with the foul water drain all connections must be trapped. In some cases SOAKAWAYS are permitted to receive the surface water. This system of drainage is more expensive to install than a COMBINED SYSTEM but it is much more economical to operate due to the fact that only the foul and waste water needs to be treated. One of the biggest problems with this system is the danger of cross-connections, e.g. foul water being connected to a surface water drain, thus being discharged to the outfall untreated. If a separate system of drainage is employed, the foul water sewer may require flushing if flows are small in which case a PARTIALLY SEPARATE SYSTEM could be considered. See also BELOW GROUND DRAINAGE.

SEPTIC TANK
A sewage disposal unit sometimes used in rural areas. The house drain is connected to a septic tank with a natural drainage irrigation trench or subsoil drain through which water can drain into the soil. Enough trenching must be constructed to ensure

that flooding of the ground surface does not occur, because the EFFLUENT from septic tanks contains pathogenic bacteria (diseased bacteria), it also must not be allowed to flow directly into streams or underground rivers, etc. The septic tank consists of, usually, a double compartment tank which can be constructed of concrete but more commonly nowadays they are made of reinforced glass fibre. As the liquid sewage flows into the tank, the solids settle out and most are decomposed by anaerobic bacterial action, no chemicals being required. Every 6–12 months the accumulation of sludge and remaining solids must be removed by cleansing contractors. Before the installation of a septic tank it will be necessary to obtain local authority approval and in most cases the siting must be a minimum distance from habitable buildings of at least 15 metres, and preferably sloping away from the building. See also CESSPOOL.

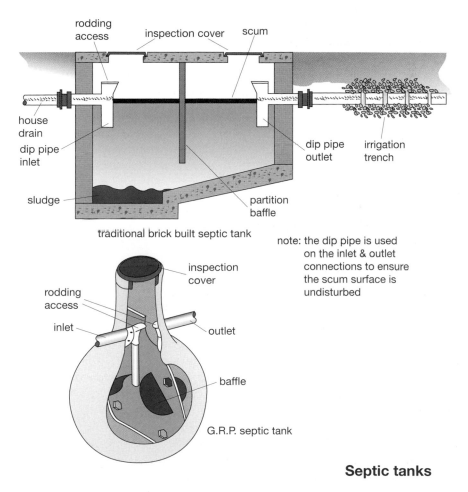

note: the dip pipe is used on the inlet & outlet connections to ensure the scum surface is undisturbed

Septic tanks

SERVICE

A check on a system or appliance to confirm it is operating effectively and with optimum efficiency and will remain so for a projected period without fear of it breaking down.

SERVICE MAIN   See DISTRIBUTION MAIN.

SERVICE PIPE

The pipe which supplies mains water into the premises. This pipe is broken into two sections, the COMMUNICATION PIPE, which is maintained by the water authority and the SUPPLY PIPE which is maintained by the consumer. For example, see COLD WATER SUPPLY.

SERVICE UNION   See SCREWDOWN FERRULE.

SERVICE VALVE (SERVICING VALVE)

An ISOLATION VALVE fitted to an appliance or fitting, or its feed pipe intended to facilitate its servicing or maintenance.

SETTING-IN STICK (SETTING-IN DRESSER)

Tool used to set a crease in the corner of sheet lead prior to BOSSING. A piece of material such as sacking is placed under the lead, the setting in stick is positioned where the crease line is required then the tool is struck on top with a mallet, the lead sinks into the sacking preventing any damage. The setting-in stick also comes in useful when finishing off or setting in the angles where the lead changes directions.

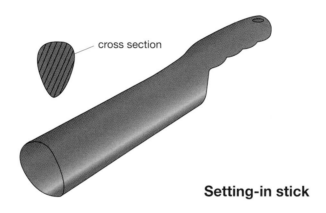

cross section

Setting-in stick

SEWAGE

The domestic water-borne organic refuse and excrement which is disposed of through a system of foul water drains. Sewage may also include TRADE EFFLUENT, or SURFACE and SUBSOIL WATERS.

SEWAGE EJECTORS

A method of pumping sewage at a low level up to a higher level drain. See PNEUMATIC EJECTOR.

SEWER

The main drain which collects and removes the liquid sewage from the individual house drains. The sewer could be private meaning that two or more drain pipes join which belong to separate owners and are laid inside a private boundary and the maintenance cost is shared jointly by the owners, or the sewer could be public in which case it belongs to the local authority.

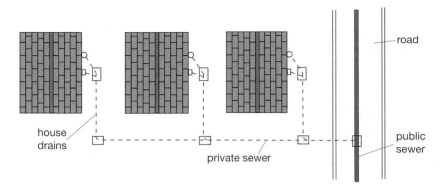

Private and public sewers

SEWER CONNECTION
The join or length of pipe between a public sewer to the last inspection chamber in a house drain or private sewer. See SEWER.

SEWER VENT   See VENT SHAFT.

SEWERAGE (SEWERAGE SYSTEM)
A system of drains and sewers which convey liquid sewage from the various domestic and industrial premises to the water treatment works or other point of discharge.

SHALLOW SEAL TRAP
A trap having a TRAP SEAL of 38 mm. Shallow seal traps are only permitted if the waste appliance discharges over a trapped gully, because, should the trap seal be lost, the GULLY provides a secondary trap between the drain and the appliance.

SHALLOW WELL   See WATER CYCLE.

SHARED FLUE
In multi-storey buildings it is sometimes possible to design a flueing arrangement for appliances in such a way that a main flue is provided to remove the products of combustion from several rooms. There are several basic designs to include the branch or shunt flue system, the SE DUCT system and the 'U' DUCT system. The latter two systems are designed for BALANCED FLUED APPLIANCES and the BRANCH FLUE is designed for OPEN FLUE appliances. Shared flues must be designed so that the products of combustion have a degree of DILUTION AIR mixed with them. All appliances connected to any system of shared flue must be fitted with a suitable FLAME FAILURE DEVICE as a standard point of safety. See also FAN DILUTED FLUE.

SHAVEHOOK
A tool the plumber uses to shave or remove a thin layer of surface lead from sheet or pipe, thus exposing a clean OXIDE-free surface to which it is easy to weld or solder.

(See illustration over.)

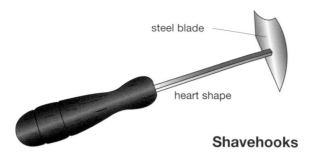

steel blade

heart shape

Shavehooks

SHEET FIXINGS

When covering any roof or wall structure with metallic and non-metallic materials it is essential to secure the sheet weathering material to the building. This is done by incorporating various fixing clips with the various details. Shown in the illustrations are a few methods which are commonly used. One clip too many is better than one clip too few. See also LEAD DOT.

(See illustration opposite)

SHEET WEATHERINGS

The METALLIC or NON-METALLIC materials which are used to cover various forms of building structures and cladding, therefore keeping the weather out.

SHELF SINK

A CERAMIC sink which has a built in ledge through which pillar taps can be mounted, for example, see SINK.

SHERADISING

A corrosion-resistant coating of zinc on the surfaces of small iron and steel objects. The process is carried out by placing the objects to be sheradised into a sealed drum with zinc dust. The drum is rotated and heated to a temperature of about 350°C, being well below the point at which zinc melts. This results in the zinc amalgamating with the iron and gives an adherent coat of pure zinc. See also GALVANISED.

SHETACK

Tool used and referred to as 'Shetack' after the name of the company who manufacturers it, see BASIN SPANNER.

SHINGLES

A thin rectangular piece of timber usually cedar or oak. It is used like a roof slate or tile for covering walls and roofs on buildings such as church spires.

SHOCK ARRESTOR

A device incorporated within a high pressure pipeline to minimise the effects of water hammer. See also AIR VESSEL and HYDROPNEUMATIC ACCUMULATOR.

SHORT HOPPER WC PAN   See LONG AND SHORT HOPPER WC PAN.

SHORT RADIUS BEND   See BEND.

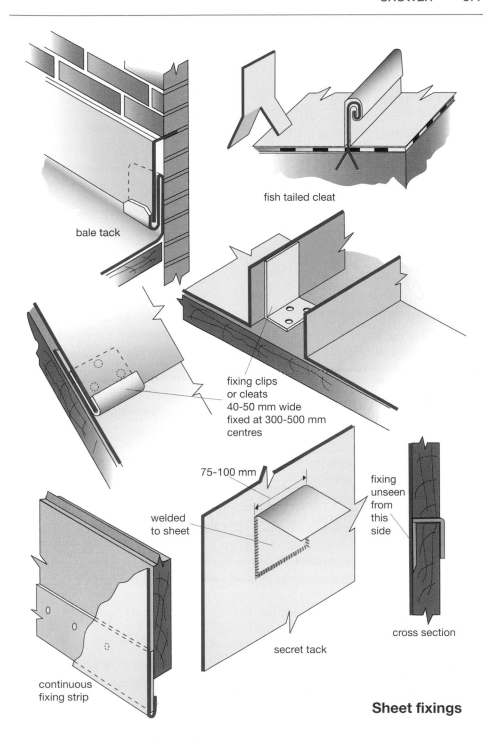

bale tack

fish tailed cleat

fixing clips
or cleats
40-50 mm wide
fixed at 300-500 mm
centres

75-100 mm

welded
to sheet

fixing
unseen
from
this
side

secret tack

cross section

continuous
fixing strip

Sheet fixings

S

SHOWER (SHOWER BATH)

A sanitary appliance consisting of a shower mixing valve and a shower tray.
Showers have many advantages over baths, they occupy less floor space and also
use less water. A shower which might last 5 minutes or so should only use about
35–40 litres, whereas a bath on average could well use over 120 litres.

SHOWER HEAD (SHOWER ROSE)

The part of a shower unit from where the water is discharged. It issues from the shower head as a jet of water usually in the form of a spray.

SHOWER MIXING VALVE

A valve used to control the flow and temperature of water to a shower. There are two basic types either manually or thermostatically controlled. With the manually controlled valve the hot and cold water supply has to pass through a hot or cold port hole respectively. A dial on the valve head can be made to turn, which in effect reduces either the hot or cold port hole size thus restricting its flow. The THERMOSTATICALLY CONTROLLED VALVE is designed to overcome problems such as might be created by a sudden reduction of the cold water supply, causing scalding hot water being produced at the shower head. Thermostatic mixing valves are fitted with a temperature sensing device which would expand due to the heat and should maintain a constant outlet temperature. If the cold supply is taken from a storage cistern it must be run separately from other DRAW OFF POINTS, this should ensure that the shower is not starved of cold water. All shower mixing valves should be supplied with hot and cold water which are at equal pressures. When a shower is supplied with water from a storage cistern it should have a minimum HEAD of water of 1 metre from the underside of the storage cistern to the shower head; if this cannot be achieved a BOOSTER PUMP could be fitted. See also THERMOSTATIC MIXING VALVE.

(See illustration opposite)

SHOWER TRAY (SHOWER RECEIVER)

An acrylic, enamelled cast iron, pressed steel or ceramic receptacle used to catch the water from the shower mixing valve. The size and shape of the tray would depend upon the manufacturer's design but whatever type is used it must be incorporated into a suitable watertight cubicle. Shower trays are usually chased into the wall and where the top edge of the tray meets the wall a fillet of SILICONE RUBBER SEALANT is generally used to make a watertight seal. When installing shower trays on suspended wooden floors it is recommended that a SAFE should be installed underneath because if the tray should move away from the wall breaking the seal the safe would catch any seepages of water. For connections of the waste see WASTE FITTING but it must be borne in mind that access to the TRAP may be required should it block up.

(See illustration on p. 380)

SHRUNK RUBBER JOINT (HEAT SHRUNK JOINT)

A special type of ring seal joint used on some THERMOPLASTICS. A special pipe fitting is purchased which has had a socket formed on one end, during its manufacture, for this socket to be formed the plastic had to be stretched over a large mandrel. The joint is made by first fitting a rubber ring over the spigot of the pipe to which it is to be joined, then the socket of the special jointing pipe is positioned over the spigot with the rubber ring on. Once everything is in place, heat is applied to the socket which in turn shrinks, trying to return to its original diameter, thus it traps the ring tightly onto the spigot. This type of joint proves very useful when joining to pipes of different materials.

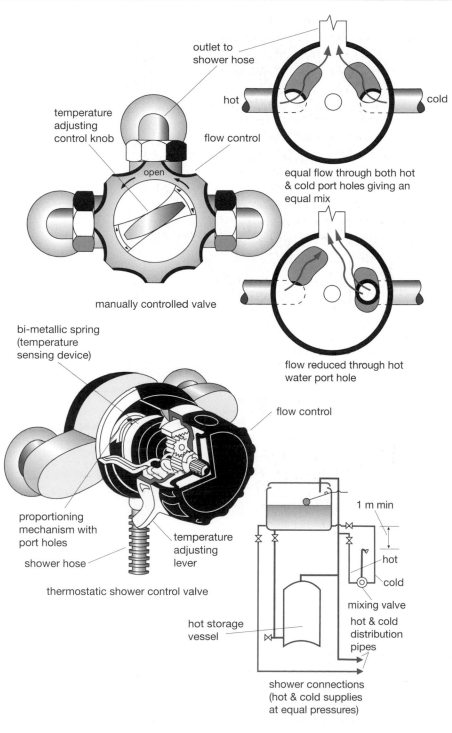

outlet to
shower hose

temperature
adjusting
control knob

flow control

open

hot

cold

equal flow through both hot
& cold port holes giving an
equal mix

manually controlled valve

flow reduced through hot
water port hole

bi-metallic spring
(temperature
sensing device)

flow control

proportioning
mechanism with
port holes

temperature
adjusting
lever

shower hose

thermostatic shower control valve

1 m min

hot

cold

mixing valve

hot storage
vessel

hot & cold
distribution
pipes

shower connections
(hot & cold supplies
at equal pressures)

Shower mixing valves

S

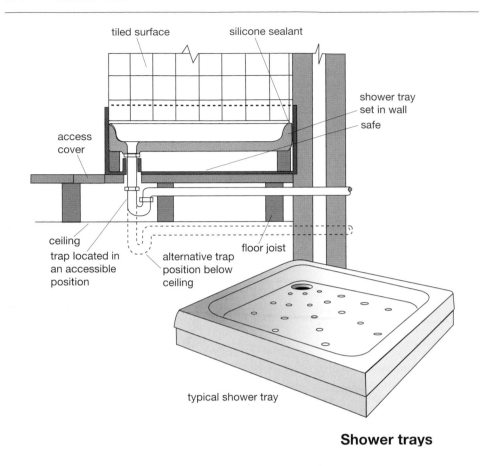

Shower trays

SHUNT FLUE (SHUNT DUCT SYSTEM)   See BRANCH FLUE.

SIDE DRESSER
A SETTING-IN STICK.

SIDE FLASHING   See COVER FLASHING.

SIDE INLET GULLY   See GULLY.

SIDEWALL SPRINKLER
A design of SPRINKLER HEAD which deflects the discharge of water in one direction only.

SIGHT GAUGE
A gauge which is fitted into the side of a storage vessel to indicate the level of the fluid. Sight gauges will be found fixed to various vessels such as oil tanks, AUTOMATIC PNEUMATIC CYLINDERS and BOILER FEED TANKS, etc. See also SIGHT GLASS.

SIGHT GLASS
A special inspection device to enable inspection of the internal surface of a steam heating pipeline to ascertain whether steam or condensate water is present, see TRAPPING SET for example. See also SIGHT GAUGE.

SIGHT RAILS AND BONING ROD

The sight rails and boning rod are tools, often homemade, used to assist in setting out the fall or gradient of a drainage pipe. First the sight rails are positioned at either end of a drainage run at different heights, the difference being that of the required gradient. Then the trench is excavated to the required fall or backfilled as necessary (specified under BEDDING), as shown in the sketch, using two operatives, one to sight their eye between the sight rails and instructing the other to raise or lower the boning rod (also called a traveller) by lifting or lowering the pipe as necessary. See also GRADIENT BOARD.

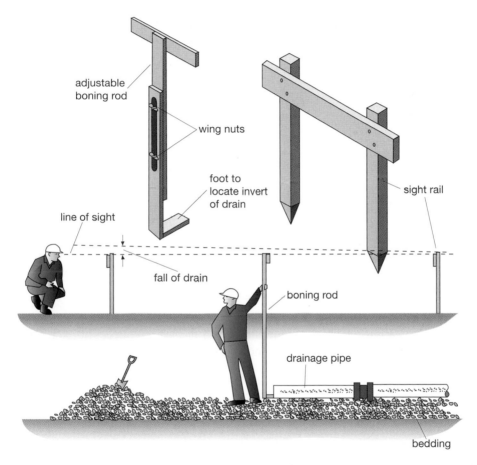

Sight rails and boning rod

SIL-FOS ALLOY   See SILVER SOLDERING.

SILBRALLOY ALLOY   See SILVER SOLDERING.

SILENCING TUBE

A short pipe sometimes found attached to BALLVALVES. Its purpose was to convey the incoming supply of water to a point of discharge below the water line, thus it reduced splashing and the noise created when filling. Unfortunately, due to the fact that these devices caused a potential health hazard should the water be BACK-SIPHONED, they are now no longer permitted to be installed.

SILICA BULB SPRINKLER   See SPRINKLER HEAD.

SILICON

A non-metallic chemical element. Commonly occurs as silicates in rocks and various soils. Pure silicon is a dark-grey solid with a crystalline structure. Silicon is used as a reducing agent in some FLUXES such as BORAX and as an alloying element in brass, bronze and steel.

SILICONE LUBRICANT

A special lubricant available in aerosol cans, tubs or tubes (liquid or paste form), used to assist making push fit joints. It is sprayed or smeared on the pipe end making it very slippery.

SILICONE RUBBER SEALANT

A multipurpose sealant which is used to seal and join many nonporous surfaces. The sealant is usually supplied in tubes and is easily pumped out through a nozzle. On first being discharged from the tube it is very sticky but soon cures quickly on exposure to the air, therefore all jointing must be done within five minutes before it sets to a rubbery finish. Silicone sealant is often used to make watertight the joint between sanitary appliances such as sinks, baths and shower trays where they meet a tiled surface. It is also recommended to join some types of guttering materials such as glass fibre, thus allowing for flexibility in the joint. When using silicone rubber the surfaces to which it will be applied must be dry and when smoothing of any excess material any metal tools used should be slightly dampened to prevent the silicone rubber sticking to them.

SILT TRAP

A GULLY which has a large sediment bucket and is used to collect SURFACE WATER. See also GRIT GULLY, GARAGE GULLY and MASTER GULLY.

SILVER SOLDERING   (SILVER BRAZING)

A method of BRAZING used on light gauge copper tube, it is a type of HARD SOLDERED capillary joint. When silver soldering it is not necessary to use a flux due to the oxy-acetylene flame which is used having a reducing zone. The solders used are either Sil-fos, which consists of 15% silver, 84% copper and about 1% phosphorus; or Silbralloy, which consists of 2% silver, 97% copper and about 1% phosphorus. The greater percentage of silver in Sil-fos solder makes it much more expensive but tends to make for a stronger joint and is used where severe vibration can be expected. The melting temperature depends on the amount of silver/copper in the solder, but the approximate melting range is between 640° and 760°C. To make a simple joint, as with all capillary joints, the joint is first prepared by ensuring that the two mating surfaces are thoroughly clean. The joint is then assembled and heat is applied with an oxy-acetylene blowpipe, a large spreading flame is used to ensure the whole joint is evenly heated all round to about 700°C when the copper will be a cherry red colour, the silver solder is applied, which will rapidly melt, and fill the joint by CAPILLARY ATTRACTION.

SINGLE FEED CYLINDER (self venting cylinder)

A specially designed indirect cylinder for use in a single feed hot water supply. The cylinder differs from the DOUBLE FEED CYLINDER in that the cold feed pipe serving the cylinder also supplies the PRIMARY CIRCUIT. See SINGLE FEED HOT WATER SUPPLY.

SINGLE FEED HOT WATER SUPPLY

A system of INDIRECT HOT WATER SUPPLY where the hot storage vessel used is different from an ordinary indirect cylinder. It is designed in such a way that the primary and secondary waters are kept separate by the use of an air lock situated inside the cylinder. The way in which the primary and secondary waters are kept apart is illustrated in the sketch. Notice how the expansion of the water in the PRIMARY CIRCUIT is taken up by moving the air in the top dome back through its cold feed pipe to the lower dome. The primary circulation system must not be too large (having

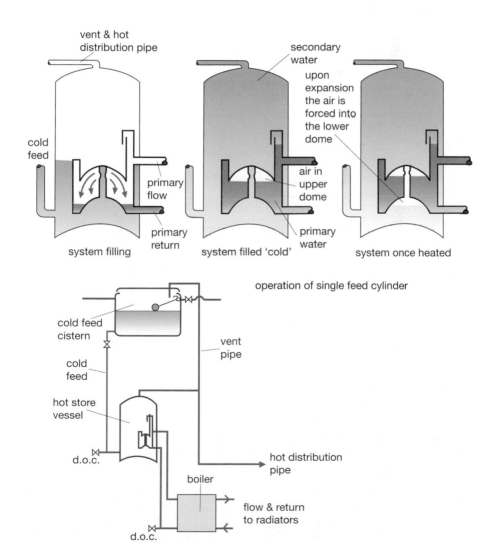

operation of single feed cylinder

indirect hot water system
using a single feed cylinder

Single feed hot water supply

lots of radiators) because the excessive quantity of water that the system will contain would, when expanding, exceed that of the space available in the dome, thus if the air is forced out of this space it would be replaced with water, converting it, in effect, to a direct system of hot water supply, giving rise to corrosion problems. This type of system must not be pressurised (see UNVENTED HOT WATER SUPPLY) because the MAINS pressure would blow out the air lock needed to keep the primary and secondary waters separate.

SINGLE LOCK WELT

An EXPANSION JOINT used on metallic roof coverings for expansion joints which run across the fall of a roof structure or CLADDING (transverse joints) at a pitch of not less than 45°. For roof pitches lower than this a double lock welt should be used. The dimensions for single lock welts are open to debate and in most cases the plumber makes them a size to suit the forming blocks, or special grooving tools, which they have to assist making the joint. Sometimes the size of the welt is specified by the client in which case these specified sizes must be observed. The sizes given in the illustration are sizes that have been found suitable in the past and are used on materials such as copper and aluminium. If a single lock welt is required to be made on sheet lead, generally the sizes given are doubled. See also COPPER SHEET AND ROOF COVERINGS.

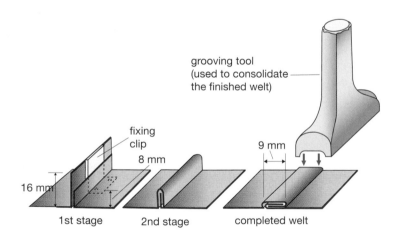

Single lock welt

SINGLE OUTLET MIXER TAP   See MIXER TAP.

SINGLE PIPE CIRCULATION   See ONE PIPE CIRCULATION.

SINGLE POINT HEATER   See INSTANTANEOUS WATER HEATERS.

SINGLE STACK SYSTEM   See PRIMARY VENTILATED STACK SYSTEM.

SINGLE STEP FLASHING   See STEP FLASHING.

SINGLE TRAP SIPHONIC WC PAN   See SIPHONIC WC PAN.

SINK

A sanitary appliance used for domestic, culinary or laboratory purposes. There are many designs of sink, the most common types being made of stainless steel, enamelled pressed steel or ceramic ware. The fixing height of a sink would depend upon various circumstances, in most cases this is determined by the kitchen units used but as a guide in general they are fixed 900–1000 mm from the floor to the front edge. When installing stainless steel sinks it must be noted that due to the 'thinness' of the metal a TOP HAT will be required when securing PILLAR TAPS. See also WASTE FITTING and CLEANER'S SINK.

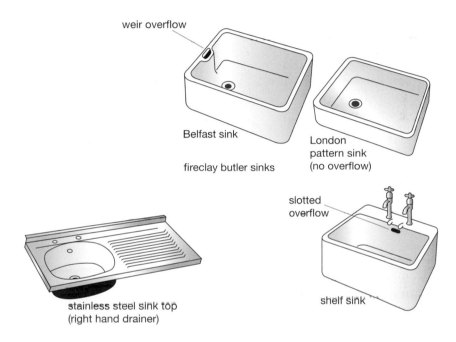

weir overflow

Belfast sink

fireclay butler sinks

London
pattern sink
(no overflow)

slotted
overflow

stainless steel sink top
(right hand drainer)

shelf sink

Sinks

SIPHON TRAP

A RUNNING TRAP.

SIPHONAGE   See SIPHONIC ACTION.

SIPHONIC ACTION

A means of transferring a liquid, from one level up and over the edge of the vessel in which it is contained, to a new lower level without mechanical aid. To siphon a liquid a tube is used to form a letter 'J' upside down, thus producing a long leg and a short leg. The short leg of the tube is inserted into a liquid leaving the long leg to hang down below the surface of the outside of the container (see sketch). The air is now removed from the tube by either sucking it out from the long leg or by pushing it out, forcing water over the CROWN of the bend as in the case of a flushing cistern. Once this air has been removed a PARTIAL VACUUM is created and unless air or liquid replaces this vacuum the tube will flatten, but what in fact happens is that atmospheric pressure exerting a force on the surface of the water pushes down and forces the liquid up the tube and over the crown of the bend down the pipe, and it will continue to do so until air gets into the tube. Why does air not get into the tube from below via the long leg? This is simply explained by the fact that water falling down the tube is at a greater pressure than the air pushing up which is only at atmospheric pressure. Siphonic action can be put to use in plumbing systems as in the case of flushing cisterns and siphonic WC pans. It can also give rise to problems such as the removal by siphonic action of water from a TRAP.

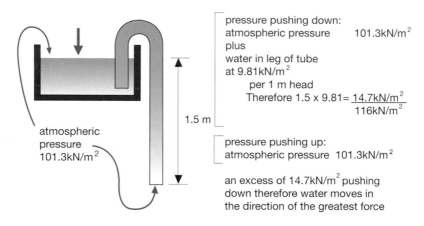

pressure pushing down:
atmospheric pressure   101.3kN/m^2
plus
water in leg of tube
at 9.81kN/m^2
        per 1 m head
        Therefore 1.5 x 9.81= $\underline{14.7\text{kN/m}^2}$
                                        116kN/m^2

1.5 m

pressure pushing up:
atmospheric pressure 101.3kN/m^2

an excess of 14.7kN/m^2 pushing
down therefore water moves in
the direction of the greatest force

atmospheric
pressure
101.3kN/m^2

Principle of siphonic action

SIPHONIC FLUSHING GULLY

A special GULLY in which water from waste appliances or rainwater pipes is allowed to flow into and fill up an enlarged chamber. When the chamber is almost full water discharges over the CROWN of the bend (X), see sketch, and pushes the air out of the pipe between the two traps causing a partial vacuum and lowering the air pressure thus SIPHONIC ACTION is set up and siphons out the collected water. This type of gully is sometimes used where the drains require occasional flushing due to, for example, an insufficient gradient of the pipe being available. See also AUTOMATIC FLUSHING TANK.

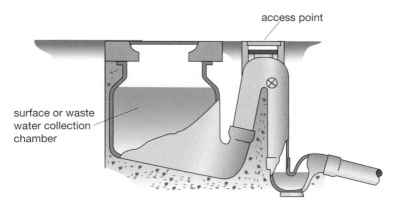

access point

surface or waste
water collection
chamber

Siphonic flushing gully

SIPHONIC WC PAN

A WC pan where the contents of the TRAP are removed by SIPHONIC ACTION which has been induced by the flushing water. Siphonic pans are more silent and positive in operation than WASHDOWN WC PANS, plus most are designed with a larger water surface which is closer to the human body thus helping to prevent splashing and fouling of the bowl. Siphonic pans can be of two general designs, either single trap or double trap. With the single trap type, siphonic action is started when water, trying to exit down the outlet, is restricted by a reduced outlet size, this momentarily fills the trapped outlet with water and as it runs away siphonic action is started. The other design which is much more common is the double trap siphonic pan and most modern designs use a pressure-reducing fitting which is located between the flushing cistern outlet and the WC pan. Older types had a pipe which connected from the space between the two traps and terminated in the flush pipe, this pipe was so designed to cause the flushing water to pull the air from between the traps creating a partial vacuum. Note with the pan which uses a pressure-reducing fitting, as water passes down past the fitting it sucks up the air from between the two traps thus reducing its pressure. Both types cause the water in trap A to fill this partial vacuum which has been created resulting in the commencement of siphonic action. As the siphonic action ceases, the remaining water flushes down the bowl and reseals the traps of the pan. See also PRESSURE-REDUCING FITTING.

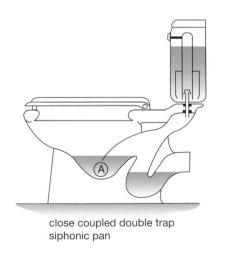

close coupled double trap siphonic pan

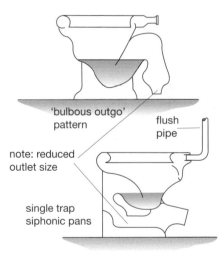

'bulbous outgo' pattern

flush pipe

note: reduced outlet size

single trap siphonic pans

Siphonic WC pans

SITE DATUM (TEMPORARY BENCH MARK)

The term site datum denotes the position from where all measurements are taken on a building site. Before a building is erected the plot of land may be no more than a field. Therefore in order to know the position of the finished floor level of a building, or the position of drain INVERT LEVELS, etc., a position is needed from which it is possible to measure up or down. For example the finished floor level may be 2 metres above the site datum, or the drainage pipe may have to be 1 metre below it at its lowest end. Therefore the plumber has a datum mark transferred across the site by the surveyor to, for example, a post positioned from where the drain is to be run. The plumber can then measure down 1 metre from this point and it is from here the drain run can be set out to give the required incline of the drain. The datum point on a site can be a position marked on a wall or a post which has been carefully levelled from a local Ordnance Survey bench-mark, or, on some contracts, it may simply be taken from an INSPECTION COVER, etc.

SITZ BATH ('SIT UP' BATH)

A special bath designed so that the bather can sit as in a chair and wash. These baths are designed specifically for old or physically handicapped people. Some types are designed with a removable door fitted to assist getting in and out. The door is made watertight by a special arrangement which allows air to be pumped round the door seal compressing it against a rubber rim.

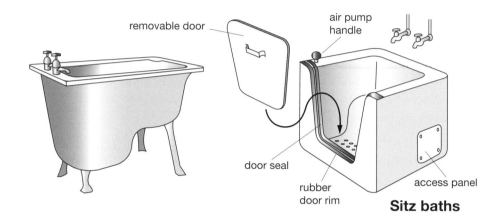

Sitz baths

SIXTEENTH BEND   See BEND.

SKELETON FLASHING

A CONTINUOUS STEP FLASHING.

SKELETON WASTE FITTING

A design of waste fitting used on waste appliances such as sinks or baths, which is clamped between the top and bottom parts of the appliance, the two halves of the fitting are pulled together assisted by a long bolt between the two pieces. The joint is made watertight by a jointing medium being clamped between the surfaces. The term skeleton waste fitting originally denoted the type of waste fitting used in conjunction with lead pipes, the lower portion was lead welded or soldered onto the pipe. Today this is still very much the meaning of the term but some forms of COMBINATION WASTE and overflow fittings are designed in a similar fashion and are therefore called the same. See also WASTE FITTING.

(See illustration opposite)

SKIRTING HEATING

A convector heater fitted at low level around a room. Skirting heating relies on the natural movement of air due to CONVECTION CURRENTS, for example, see CONVECTOR HEATER.

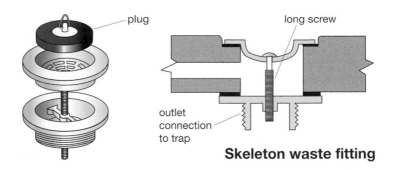

Skeleton waste fitting

SKYLIGHT (ROOF LIGHT)

A sash window fixed into a pitched roof, but more generally the term skylight can mean any window such as a LANTERN LIGHT or NORTH LIGHT or any design of patent glazing which allows light into the building from above. The window may or may not be openable.

SLAB URINAL

A flat IMPERVIOUS slab fixed to a wall in a male toilet to receive urine. See also STALL URINAL and URINAL.

SLATE PIECE

The sheet weathering used to ensure a watertight joint where a small obstacle such as a VENTILATING PIPE passes through the roof. Slate pieces can be made from almost any of the sheet weathering materials although lead tends to be the favourite due to its ease of fabrication and long life. It is also possible to purchase ready-made slates to suit any roof pitch.

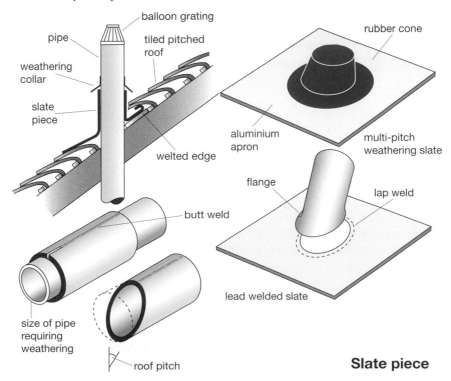

Slate piece

SLEEVE

Odd lengths of tube built into the wall, through which smaller pipes can pass, thus allowing for THERMAL MOVEMENT due to expansion and contraction.

SLIP COLLAR

Larger version of a SLIP COUPLING.

SLIP COUPLING

A straight COUPLING which has no stops in the fitting therefore it will slide along the pipe. The slip coupling proves useful when jointing two pipes in which there is no lengthways movement. The slip coupling is inserted over one pipe and slipped along it, then the two pipes are aligned and the coupling is slid back to the joint, being positioned equally over each pipe. The joint is then made watertight as necessary.

SLIPPER BEND   See BARRON BEND.

SLOP HOPPER (SLOP SINK)

A SOIL APPLIANCE which is bowl shaped with a flushing rim, it is used to receive and discharge the contents of bed pans and urine bottles. The hopper is flushed with water from a FLUSHING CISTERN. In the days of large houses and the lack of WCs throughout the building slop hoppers were often installed and thus were called housemaid's closets, as it was one of the duties of the housemaid to empty the bed pans in the mornings. For example, see HOSPITAL APPLIANCES. See also SLOP SINK.

SLOP SINK

A sanitary appliance designed to serve in the same way as a slop hopper. Slop sinks are in fact generally called slop hoppers, the difference being that the slop sink is often installed in conjunction with a BED PAN SINK and is installed at a working height of about 1 metre to the front lip edge, whereas the slop hopper sits on the floor and is as low as a WC pan. See SLOP HOPPER.

SLOTTED OVERFLOW

A SECRET OVERFLOW found in waste appliances such as wash basins, bidets and some sinks etc.

SLOTTED WASTE FITTING   See WASTE FITTING.

SLOW IGNITION DEVICE

A device found on gas instantaneous water heaters, that is incorporated to make the gas valve open slowly, and allows for a rapid closure. The slow opening ensures a safe, quite non-explosive ignition; whereas its quick closure prevents the overheating of the remaining water in the heater.

SLOW SAND FILTER   See SAND FILTER.

SLUDGE

The name given to any form of soft solid deposits which settle in various vessels containing water. For example, the sludge which accumulates in the bottom of a hot and cold storage vessel or the remaining solid matter left in the water after sewage treatment. See SEPTIC TANK.

SLUDGE VALVE

A valve fitted into the base of various cisterns and tanks in order that sludge, which has accumulated, can be drawn off.

SLUICE (BEDPAN SLUICE)   See BED PAN SINK.

SLUICE VALVE

A FULLWAY GATEVALVE with over a 50 mm diameter bore used for mains supply pipes.

SMALLBORE CENTRAL HEATING

A hot water heating system in which the heating circuit pipes do not exceed 19 mm nominal bore. See also MICRO-BORE.

SMOKE MATCH

A special match which, when ignited, produces a small quantity of smoke that can be used to determine the effectiveness of a flue system from a gas-burning appliance. See SPILLAGE TEST.

SMOKE PELLET

A special pellet which, when ignited, produces large quantities of smoke. This is used to test a flue system for its effectiveness, although it can also be used to check both flue and drainage systems for soundness. See FLUE FLOW TEST.

SMOKE ROCKET   See SMOKE TEST and TRACING EXISTING DRAINS.

SMOKE TEST

A pressure test applied to above and below ground drainage installations to check for leaks. To carry out a smoke test first light the smoke testing machine and allow smoke to fill up the pipework. As and when smoke appears at the open ends of pipes, blank off the exits with DRAIN PLUGS or stoppers and allow the pressure to build up within the pipe. As the pressure builds up the dome on the smoke machine will rise and remain stationary if the drain is sound. Should there be a leak the dome will fall. Smoke testing can have an advantage over air testing because leaks can easily be detected. Another kind of smoke test can be made with a smoke rocket, these are devices which are put into the pipe and give off a dense smoke, they are not for pressure testing and are only used to locate leaks or to trace existing drain runs. See also AIR TEST, WATER TEST and FLUE FLOW TEST.

(See illustration over.)

SMOKE TESTER   See COMBUSTION EFFICIENCY TESTING.

SMUDGE   See PLUMBER'S BLACK.

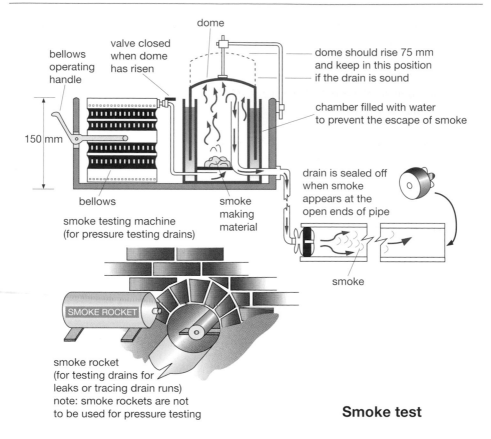

dome

bellows operating handle

valve closed when dome has risen

dome should rise 75 mm and keep in this position if the drain is sound

chamber filled with water to prevent the escape of smoke

150 mm

bellows

smoke testing machine (for pressure testing drains)

smoke making material

drain is sealed off when smoke appears at the open ends of pipe

smoke

SMOKE ROCKET

smoke rocket (for testing drains for leaks or tracing drain runs) note: smoke rockets are not to be used for pressure testing

Smoke test

SNAKENTAINER
A DRAIN AUGER which is attached to an electrically or manually rotated drum, which also houses the snake. The snakentainer can provide a more powerful leverage when turning the spring inside the drain and therefore can be very effective when unblocking a drain. For example, see DRAIN CLEANING.

SNAP ACTION CUTTER   See PIPE CUTTER.

SNATCH WEIGHT
A large metal bead usually brass and slightly smaller in diameter to a bobbin. It is tied onto the end of the rope which passes through the bobbin and is used to assist in pulling the bobbin through a lead pipe. For example, see BOBBINS.

SNIFTING
The PURGING of oxygen and acetylene cylinders, etc. prior to fitting the regulator. The cylinder is purged or sniffed to remove any grit which may be sitting in the valve seating.

SNOW BOARD
A slatted timber framework set on blocks which has been positioned on top of a parallel or tapering VALLEY GUTTER. Its main purpose is to ensure an uninterrupted

flow of water along the gutter should there be heavy snow falls. The snow accumulates on top of the boards and as it melts, it drips through the slats and flows down the gutter. If a snow board is not fitted, it is possible that the blockage caused by the snow could lead to a leak into the building. The timber slatted boards also tend to give protection when people walk across the gutter surface.

SOAKAWAY

A hole sited well away from the building which is filled with brick rubble or stones to prevent the side walls caving in. The purpose of a soakaway is to receive SURFACE WATER and they are often used in country areas where there is a lack of sufficient sewerage pipes. Soakaways should only be used in areas which will allow the water to percolate into the surrounding ground. The size and distance from the building is usually specified by the local authority who must be notified of their use, but in general the following formula is used:

$$C = \frac{AR}{3}$$

C = capacity
A = area to be drained in m^2
R = rainfall in m/hr (in England and Wales the rainfall is usually taken as 0.05 m per hour.)

Example
An area of 125 m^2 is to be drained, find the size of the soakaway required.

$$C = \frac{AR}{3}$$

$$C = \frac{125 \times 0.05}{3} = 2 \text{ m}^3 \text{ (approx)}$$

If possible, the soakaway should be sited on ground which slopes away from the building in case of flooding.

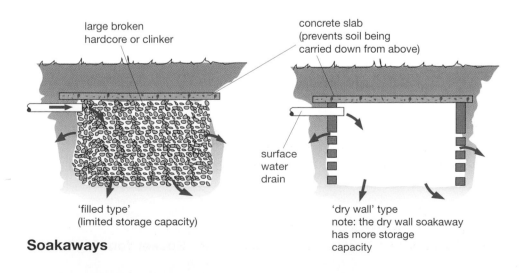

large broken hardcore or clinker

concrete slab (prevents soil being carried down from above)

surface water drain

'filled type' (limited storage capacity)

'dry wall' type note: the dry wall soakaway has more storage capacity

Soakaways

SOAKERED HIP or **VALLEY**   See CLOSE CUT HIP OR VALLEY.

SOAKERS (UNDERS)

Thin sheets of NON-FERROUS METAL bent to form a watertight joint between the roof slates/tiles and the upstanding side of a brickwork abutment, for example, see STEP FLASHING. Soakers are also sometimes used to assist the weatherings of tiles which go round a corner such as a CLOSE CUT HIP. The width of a soaker used to join up to an abutment must be a minimum of 175 mm, this would allow for an upstand of 75 mm against the wall and the remaining 100 mm can lay out across the roof and under the slates. The length of the soaker would vary and depend upon the length of the roof slates. The calculation used to find the length is gauge + lap + 25 mm. The gauge is the distance between the roof slate battens and is calculated as shown below. The lap is the distance one slate overlaps the slate next but one below it and the 25 mm is optional and is purely for fixing purposes.

Example
Length of soaker required for a slated roof with a slate length of 510 mm would be:

$$\text{gauge} = \frac{\text{length of slate} - \text{lap}}{2} = \frac{510 \text{ mm} - 76 \text{ mm}}{2} = 217 \text{ mm}$$

$$\text{soaker} = \text{gauge} + \text{lap} + 25 \text{ mm} = 217 + 76 + 25 = 318 \text{ mm}$$

SOCKET

(1) A pipe fitting with a female thread passing right through it. It is used to join two straight pieces of pipe which have male threads, for example, see LONG SCREW.
(2) The end of a pipe or fitting which has an enlarged bore for the reception of a spigot, e.g. the connection of another pipe end or the outlet from another fitting.

SOCKET-FORMING TOOL (EXPANDING TOOL)

A tool designed to form a raised internal bore to receive another pipe thus facilitating a connection for capillary jointing. There are two basic types of socket-forming tools, first those which are struck with a hammer to drive the socket former onto the pipe, this type of socket former is often called a bumper tool. The second type which

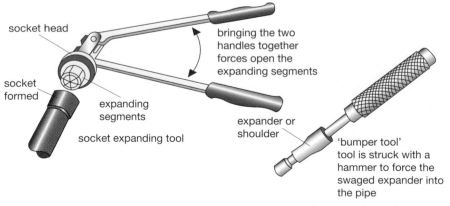

socket head

bringing the two handles together forces open the expanding segments

socket formed

expanding segments

socket expanding tool

expander or shoulder

'bumper tool' tool is struck with a hammer to force the swaged expander into the pipe

Socket-forming tools

is much easier to use, consists of a device with a lever handle and a socket head made up of expanding segments. The socket head is inserted into the pipe end and the lever handles are pulled together, this forces a tapered shaped pin into the head forcing it to become larger. Sometimes it is necessary to ANNEAL the tube end before forming the socket, this makes forming the socket easier and prevents the pipe end splitting.

SODIUM HYPOCHLORITE   See DILUTED CHLORINE.

SOFFIT
A board which is secured to the underside of rafters on an overhanging pitched roof, see BUILDING TERMINOLOGY.

SOFFIT LEVEL (CROWN)
The highest point on the internal surface of a pipe when in the horizontal position.

SOFT SOLDERING
The term soft soldering refers to any soldering process in which the solder used makes a joint which will not withstand too much stress, generally soft soldered joints are made with a flame temperature no higher than about 450°C. See also HARD SOLDERED JOINTS.

SOFT WATER
Water which is free from dissolved salts such as calcium carbonates and sulphates. It tends to be more pleasant to wash in and is much easier to work up a soap lather than when using HARD WATER supplies. Soft waters can be either naturally soft, being slightly ACIDIC, or can be softened by the use of a WATER SOFTENER.

SOIL APPLIANCES
Fittings which are used for the removal of human excreta from a building, they include slop hoppers, urinals and water closets. See also WASTE APPLIANCES.

SOIL STACK
A discharge stack which carries FOUL WATER from various sanitary appliances.

SOIL WATER
The water discharged from SOIL APPLIANCES such as WCs or urinals.

SOLAR COLLECTOR
A device used to collect solar heat. A simple solar collector consists of a thin vessel painted matt black with piping attached flowing to and from a hot storage vessel. It is covered with double or triple glazing and backed by thermal insulation material. The collector is sited at a convenient position to catch the solar heat from the sun, usually at an angle of 40° and facing south. See SOLAR HEATING.

SOLAR ENERGY
Solar energy is RADIANT HEAT from the sun. When the sun is at its maximum intensity, such as on a clear midsummer's day the Earth's surface yields about 0.75 kW of

power for every square metre of ground. This power can be collected in a SOLAR COLLECTOR and used, for example, to heat the water in a building.

SOLAR HEATING

Solar heating utilises the SOLAR ENERGY from the sun and collects the RADIANT HEAT waves in a SOLAR COLLECTOR, usually located on the roof. This heat can be used to heat up the domestic hot water supply. The designs of solar heating systems vary and generally in this country, because of unreliable weather, it would not be the only form of heating the water. A form of back-up heating arrangement would therefore be required. Note in the system shown in the diagram the solar collector is fitted above the cold feed cistern, this causes no problems because the PRIMARY CIRCUIT from the solar collector forms part of a CLOSED SYSTEM. The water in the primary circuit is made to circulate by means of a pump which is switched on automatically should the temperature in the top of the collector be higher than that in the base of the hot storage vessel.

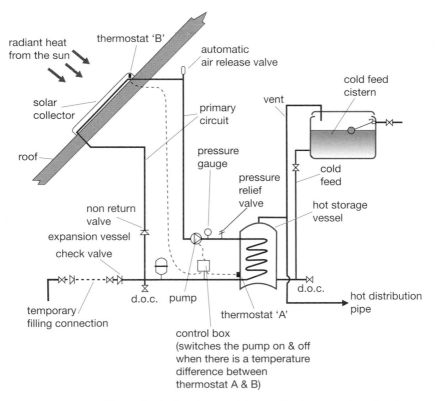

Solar heating to a domestic hot water system

SOLDER

An ALLOY used to bond metals together, see SOLDERING. There are various kinds of solders used for various purposes, they are generally referred to as HARD or SOFT SOLDER. The following chart shows some common soft solders used but see also SILVER SOLDERING, BRAZING and BRONZE WELDING for solders used at much higher temperatures, these being hard solders. See also PLUMBER'S METAL, TINMANS SOLDER and FLUX.

Melting characteristics and uses of soft solders

Composition	Uses	Solid–liquid 'plastic range'
Antimony, tin and lead BS grade F	Capillary joints	183–212°C
Antimony and tin (HT3)	Capillary joints	236–243°C
Copper and tin	Capillary joints	230–335°C
Pure tin	Capillary joints	232–232°C
Silver and tin (HT6)	Capillary joints	221–221°C
Antimony, tin and lead BS grade J	Wiped solder joints	183–255°C
Antimony, tin and lead BS grade D	Wiped solder joints	185–248°C

Note Solders containing lead are prohibited on domestic water services.

SOLDER PAINT (SOLDER PASTE)
A soldering flux/solder which is available, it consists of a finely ground solder mixed with an ACTIVE FLUX to form a paste, which can be smeared on tube ends and inside the fittings to assist making CAPILLARY JOINTS without the separate use of flux and solder. The biggest disadvantage with this paste is the cost, it is very expensive and the flux used is very corrosive.

SOLDER RING CAPILLARY FITTING   See CAPILLARY JOINT.

SOLDERED DOT
A method used to cover and weather the fixings used on vertical sheets of lead, see LEAD DOT.

SOLDERED STRUT SPRINKLER   See SPRINKLER HEAD.

SOLDERING
A process in which ALLOYS are used to join metals below their melting temperature, unlike welding, in which the joint is made by fusing with the parent metal. The basic procedure to carry out in order to solder successfully is as follows:

(1) Thoroughly clean the joint.
(2) Apply FLUX (some soldering processes using an oxy-acetylene flame do not require a flux).
(3) Apply heat (either directly or via a soldering iron).
(4) Feed solder to the joint.
(5) Allow the joint to cool, without movement.
(6) Remove excess flux. Failure to do so could result in corrosion problems.

See HARD SOLDERING and SOFT SOLDERING. See also CAPILLARY JOINT and ZINC SOLDERING.

SOLDERING IRON (COPPER BIT/COPPER BOLT)
A tool with a copper end, steel shank and a wooden handle, it is used when TINNING metal or soldering small jobs. Copper is used due to the fact that it is a good conductor of heat. In most plumbing tasks heat is applied with a blowlamp to the soldering iron but for very small jobs electric soldering irons are available.

(See illustration over.)

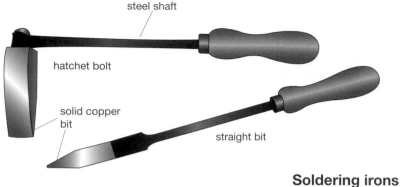

steel shaft

hatchet bolt

solid copper bit

straight bit

Soldering irons

SOLDERING MACHINE   See ELECTRIC RESISTANCE SOLDERING MACHINE.

SOLE BOARD
The name sometimes given to the timber decking of a PARALLEL or TAPERING VALLEY GUTTER.

SOLE BOARD
The section of timber located behind a chimney onto which the BACKGUTTER sits, see BUILDING TERMINOLOGY.

SOLENOID VALVE (MAG VALVE/ELECTROMAGNETIC VALVE)
A solenoid valve is an electrically operated valve used to control the on/off flow of gas or liquid through a pipeline. The solenoid consists of a coil of wire in the form of a cylinder; when an electric current is allowed to flow through the wire it creates a magnetic field and draws a plunger usually of iron, into the solenoid opening the valve. When the power is switched off, the return spring pushes the valve back onto its seating, closing the valve.

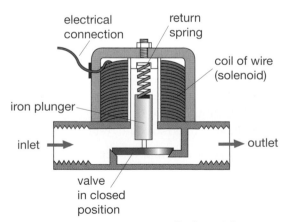

electrical connection

return spring

coil of wire (solenoid)

iron plunger

inlet

outlet

valve in closed position

Solenoid valve

SOLID FUEL
Solid materials, such as wood, various ANTHRACITES, coal or coke, etc., used for burning.

SOLID ROLL   See WOOD CORED ROLL.

SOLUBLE OIL (CUTTING OIL)
A special oil to which water is added to make it thinner and go further. Soluble oils are used in many cutting machines, etc., to give a longer life to the cutting edge of blades. See CUTTING LUBRICANTS.

SOLVENCY (SOLVENT)
The term solvency denotes a liquid's ability to dissolve or disperse other substances. Water is by far the most common solvent and should the water contain dissolved gases such as oxygen, carbon dioxide or sulphur dioxide, etc. its ability to dissolve becomes even greater. See CUPRO-SOLVENT, PLUMBO-SOLVENT and ZINC-SOLVENT.

SOLVENT WELD CEMENT
A special SOLVENT liquid which is available for jointing plastic pipes such as UPVC and ABS. Solvent weld cement is not a glue, it does not simply stick pipes together but when the liquid is applied to the plastic surface it temporarily dissolves it. Therefore when making a solvent weld joint between two surfaces, they must be joined very quickly, thus the two dissolved plastic surfaces can mix and truly bond together. When making a solvent weld joint, the correct solvent jointing cement must be used as specified by the manufacturer of the plastic. To make a joint it is first thoroughly cleaned with the special cleaning fluid supplied, then the pipe end and socket are smeared with the solvent and the two are simply pushed firmly together. The joint will set hard within 10–15 minutes, but needs at least 12 hours to dry out completely. As a note of safety, the fumes from solvent weld cement are toxic therefore it is essential to ensure good ventilation, also the liquid is highly inflammable. See also PLASTIC PIPE AND FITTINGS.

SOLVER FLUE   See VERTEX FLUE.

SOUNDNESS TEST
A term applied to mean the testing of an installation to confirm it does not leak. See TIGHTNESS TESTING, AIR TEST, WATER TEST and HYDRAULIC PRESSURE TEST.

SPACE HEATING
The source of heating for a room (a space), etc. Space heating could consist of a simple gas fire located in the room to be warmed, but more generally the term refers to some form of CENTRAL HEATING.

SPACE NIPPLE   See NIPPLE.

SPADE OFF
Term sometimes used meaning the same as capping off, which is the disconnection of a section of pipework and the making of the pipe end gastight or watertight.

SPANNER
A tool used to turn nuts without slipping off, there are four main types of spanner these being open ended, ring, adjustable and box spanners. Ring and box spanners

have the advantage of gripping the nut all round its head thus there is less risk of it slipping off. The most common spanner found in a plumber's tool kit is the adjustable spanner which is available in a range of sizes, the jaws can be adjusted to suit several sizes. On a point of safety, never extend the handle with a pipe in order to gain more leverage as it could damage the tool and a nasty accident may result. See also WRENCH and BASIN SPANNER.

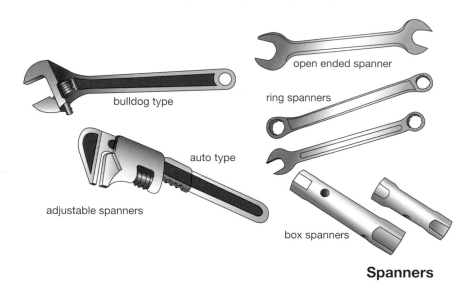

open ended spanner

ring spanners

bulldog type

auto type

adjustable spanners

box spanners

Spanners

SPARGE PIPE
A horizontal FLUSH PIPE having a perforated outlet along its length through which water is sprayed to cleanse a urinal slab.

SPARK IGNITION   See IGNITION DEVICES.

SPECIFIC GRAVITY (RELATIVE DENSITY)
The weight of a substance per volume, compared to an equal volume of water, for example 1 m³ of water weighs 1000 kg whereas 1 m³ of lead weighs about 11 300 kg; it can be seen that the lead is 11.3 times heavier. From this we can calculate the weight of any substance by dividing the density of a substance by the density of water. Water is always shown with a specific gravity number of 'one' therefore any material with a number higher than one will sink in water and those with a number less than one will float.

Specific gravity at atmospheric pressure

Material	Specific gravity	Material	Specific gravity
Water	1	Tin	7.3
Alcohol	0.8	Mild steel	7.7
Linseed oil	0.95	Copper	8.9
Aluminium	2.7	Lead (milled)	11.3
Zinc	7.1	Mercury	13.6
Cast iron	7.2	Gold	19.3

It is worth noting that we can compare materials with each other for example we can see that in fact cast iron weighs less than tin 'per equal volume'.

SPECIFIC HEAT

Specific heat is the amount of heat required to raise 1 kg of material 1°C. The heat required would differ from material to material, for example, it requires 4.186 kJ to heat 1 kg of water 1°C, whereas it would only require 0.385 kJ to heat 1 kg of copper.

Specific heat values

Material	kJ/kg°C
Water	4.186
Aluminium	0.887
Cast iron	0.554
Zinc	0.397
Lead	0.125
Copper	0.385
Mercury	0.125

It must be noted that specific heat values vary as the temperature changes. See also QUANTITY OF HEAT.

SPHERICAL PLUG VALVE

A lever or a screwdriver-operated BALL VALVE.

SPIDER MANIFOLD

The name sometimes given to refer to a pipe manifold, such as used in a MICRO-BORE heating sysytem.

SPIGOT

The end of a pipe or fitting. See also SOCKET.

SPILLAGE

A term used to mean that the PRODUCTS OF COMBUSTION from a boiler or fire are not being drawn up the flue pipe which could be the result of a wrongly positioned flue or inadequate ventilation.

SPILLAGE TEST

A test carried out on open flued gas-burning appliances, designed to confirm that the appliance is not 'spilling' combustion products into the room in which the appliance if installed. To test for spillage, all doors and windows should be closed and after the appliance has been alight for five minutes, a smoke match is held about 3 mm up inside the DOWN DRAUGHT DIVERTER or just below the outer edge of the chimney opening; spillage is indicated by the smoke being blown outwards into the room.

SPILL-OVER LEVEL (OVER-SPILL LEVEL)

The highest possible level of water in an open top vessel such as a cistern, a bath or basin, etc. Above this level water will discharge over the sides of the vessel onto the floor. Note the spill-over level is above the overflow level.

SPLASH BACK

The protection given to the wall surface area at the back of or beside waste appliances such as sinks, baths or wash basins, etc. In most cases the splash back consists of tiling to the wall surface, but plain glass, a mirror or even sheet metal such as stainless steel could be used as an alternative.

SPLASH LAP

The part of the OVER CLOAK on a roll or drip which lays onto the surface of the next sheet. For example, see LEAD SHEET AND ROOF-COVERINGS.

SPLASH STICK

A homemade tool consisting of a piece of mild steel or hardwood about 180 mm long, 40 mm wide and being slightly curved along its length. The splash stick is no longer used but was, in the past (before the use of blowlamps) a tool used to flick molten solder out of a ladle onto an upright or similar WIPED SOLDER JOINT to assist heating the metal and to apply the solder for wiping.

SPLAY BRANCH   See ANGLE BRANCH.

SPLAYED DRIP

A design of DRIP sometimes used on copper or aluminium roofs. Should a splayed drip be chosen it is essential that the drip still maintains a vertical drop of 50 mm. For example, see COPPER SHEET AND ROOF COVERINGS.

SPLIT COLLAR

A COLLAR which is formed in two separate pieces. The joint is made by clamping a rubber or fibre gasket of some sort between two lengths of pipe which are butted up to one another, the collar is positioned around the joint and the two halves are bolted together. For one design of split collar see CAST IRON PIPE AND FITTINGS.

SPLIT PIN (COTTER PIN)

A pin which passes through a hole and is opened up to prevent its removal. A split pin will be found in many designs of BALLVALVE to secure the float lever arm to the body and act as the fulcrum.

SPOON DRESSER   See BENDING STICK.

SPOUT   See NOZZLE.

SPOUTING

An EAVES GUTTER.

SPRAY MIXING TAP

A design of TAP used mostly in public buildings such as factories and offices, designed to cut down on the fuel used to heat the domestic hot water; saving up to 66% of water per person per day at the appliance. The tap is designed to deliver 1.8 to 3 litres per minute, therefore in buildings such as public conveniences and schools, because the water delivery rate does not exceed 3.6 litres per minute, the appliance can be installed without a waste plug, thus reducing vandalism problems. Hot and cold water supplies are connected to the tap and mixed together before being discharged out through a spray outlet. When fitting these taps, as with any single outlet mixer tap, it is essential that the hot and cold supplies are at equal pressures. One point which must be strictly observed in order to obtain satisfactory usage of the tap, is to ensure that the hot water DEAD LEG is as short as possible, otherwise the tap would have to be open for an excessively long period before the hot water, from the hot water storage vessel, reached the outlet. Therefore SECONDARY CIRCULATION is generally required.

SPREADER
A special fitting which is fixed to the lower end of a urinal FLUSHPIPE and is designed to spread the flow of flushing water over the urinal surface. For example, see URINAL.

SPREADER TEE
A tee fitting which is designed to spread the flow of water entering a vessel in different directions therefore dispersing its flow. One example of a spreader tee could be found in a cold feed pipe which enters directly through the bottom of a hot water heating apparatus. For example, see HORIZONTALLY FITTED CYLINDER.

SPRING   See BENDING SPRING and DRAIN AUGER.

SPRING BACK   See ELASTICITY.

SPRING-LOADED SAFETY VALVE   See PRESSURE RELIEF VALVE.

SPRING-LOADED TAP   See NON-CONCUSSIVE VALVE.

SPRINKLER HEAD
The outlet discharge point of water from a SPRINKLER SYSTEM. There are two basic types of sprinkler head, the silica or quartzoid bulb type and the solder strut type. Both types work by holding the valve onto the seating of the water discharge point; should the bulb or solder be removed the valve falls away allowing a stream of water through, which impinges on a deflector and is sprayed like rain. The principle of operation of sprinkler heads is that when the temperature around the sprinkler head rises to a predetermined level the bulb will break, or conversely the solder will melt, allowing the valve to fall and open. The bulb type can have an advantage over the solder strut type where corrosion of the solder is likely, but to overcome this the metal work is sometimes coated with vaseline or a sprinkler head, known as the Duraspeed type, is used in which the solder is protected from corrosion by its design, being enclosed by an insulating metal and a protective film. The deflector from a sprinkler head can be made to spray the water in all directions or can be designed to deflect the spray sideways away from a wall, etc. Sometimes multi-jet sprinkler heads are used, these distribute a flow of water to various open heads or distributors. This type of head is used where the standard type of head would be located at a position unlikely to be heated quickly enough, due to an obstruction, etc. Both the bulb and the solder strut sprinkler heads are designed to operate at various temperatures, in general about 68°C but sometimes it is necessary to fit heads

S

with a higher rating due to the temperature of the building being usually high. The temperature rating is stamped on the side of the solder strut type and for the bulb type the breaking temperature is determined by the colour of the liquid used in the bulb.

Bulb temperature ratings

Colour of liquid	Breaking temperature °C
Orange	57
Red	68
Yellow	79
Green	93
Blue	141
Mauve	182

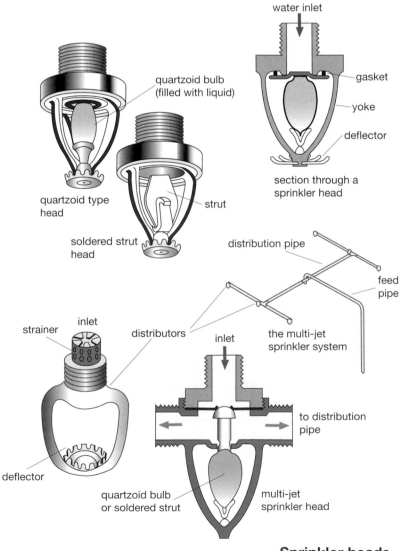

Sprinkler heads

SPRINKLER SYSTEMS

A means of protecting a building against fire damage. A sprinkler system causes an automatic discharge of water to be sprayed usually from SPRINKLER HEADS located near the ceiling. The design shown in the sketch is a typical arrangement of the pipe layout. There are three types of sprinkler installations:

(1) The wet-pipe system, in which the sprinkler system is permanently charged with water.

(2) The dry-pipe system, in which the sprinkler system is charged with compressed air and is used in unheated buildings where the temperature may fall below 0°C therefore the pipes, if charged with water, are liable to freeze.

(3) The alternate system, in which the sprinkler system is filled with water during the summer and works as a wet-pipe system, and filled with compressed air in the winter, working as a dry-pipe system.

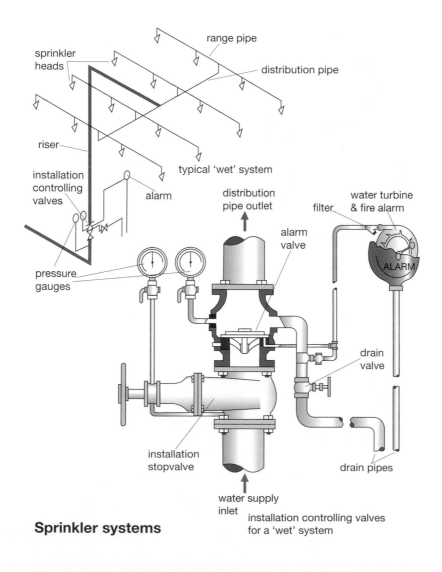

Sprinkler systems

In the wet-pipe system water will discharge immediately the sprinkler head bursts, whereas in the dry-pipe system there is a time delay in the discharge of water from the head, this is due to the compressed air which has to be expelled from the system first. All three sprinkler systems are designed so that an alarm will ring should a sprinkler valve blow open. A valve fitted on top of the main stopvalve is kept closed while the pressure remains static within the system. The bursting of a sprinkler head causes an immediate drop in pressure on top of the valve and causes it to open thus allowing water to flow to the sprinklers and to pass down a pipe which discharges water onto a water turbine activating the alarm. For variations to the design of sprinkler systems see DELUGE SYSTEM, DRENCHER SYSTEM AND MULSIFYRE SYSTEM.

SPUN CAST IRON PIPE
CAST IRON PIPES produced by a water-cooled mould rotating at a high speed. When molten metal is introduced into the mould, via a protruding arm, it is flung out by centrifugal force against the inside of the spinning mould, once the metal has cooled the pipe is withdrawn. This type of casting gives a much more even textured pipe than that produced by SAND CASTING, also spun cast iron has an equal DENSITY and strength throughout its length.

SPUN LEAD   See LEAD WOOL.

SPUR OUTLET
An electrical outlet point taken from the final ring main circuit. A spur outlet is often used to make a permanent switch connection for the electrical supply to items such as a pump or boiler, etc.

SQUARE BEND
A term sometimes used to mean a 90° bend.

SQUATTING PLATE
The raised tread plate of a SQUATTING WC PAN upon which the user stands.

SQUATTING WC PAN (ASIATIC CLOSET/EASTERN CLOSET)
A WC pan with an elongated bowl, the squatting WC is designed so that the user stands or adopts the squatting position. It is rarely found in this country.

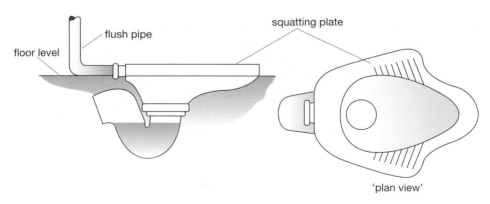

Squatting WC pan

SQUIRREL'S TAIL　See RUNNING ROPE.

STABILITY BRACKET
A bracket that is screwed into the floor or wall and is designed to hook into the lower rear of a cooker to prevent it from tipping forward should the door be opened and leant upon.

STACK　See DISCHARGE STACK.

STAINLESS STEEL
An ALLOY consisting mainly of iron, chromium, nickel and carbon, in varying proportions with the possible additions of manganese, molybdenum, niobium, phosphorus, selenium, silicon, sulphur, titanium or zirconium, depending upon the type of steel required. There are three basic groups of stainless steel:

(1) Ferritic steels, which are non-hardenable (containing 15–30% chromium).
(2) Martensitic steels, which can be hardened by quenching (containing 10–18% chromium).
(3) Austenitic steels, which are hardenable without quenching (containing 16–26% chromium). Austenitic steels usually have the highest resistance to corrosion.

The chromium added to the iron imparts remarkable resistance to corrosion and heat. Stainless steel has a silvery, shiny appearance and is highly resistant to corrosion due to the chromium oxide which forms on its surface. Stainless steel is used for the manufacture of sanitary appliances, but will be found in wire, pipe and sheet form. See also STAINLESS STEEL ROOF COVERINGS.

STAINLESS STEEL PIPE AND FITTINGS
There is only one grade of stainless steel used for plumbing works, having an average wall thickness of 0.7 mm. The method of jointing is the same as for copper pipes and most fittings used on copper can simply be used on stainless steel. The material is available in 3 m and 6 m lengths in sizes ranging from 6–35 mm outside diameter. Stainless steel is harder than copper and will therefore withstand damage to a greater degree. The material can be bent using the same bending machine as for copper although more force is generally required.

STAINLESS STEEL ROOF COVERINGS
Stainless steel has been widely employed for roofing on the continent, particularly in France, since the early 1950s, but it is only since 1974 that the material has been used in this country. One of the main features of stainless steel as a roofing material is its very low scrap value, thus it does not present an attractive proposition to the thief and is likely to remain on the roof for 100 years plus. The methods of laying stainless steel to a structure are the same as those for the zinc roll cap system (see ZINC SHEET AND ROOF COVERINGS) or the LONG STRIP ROOFING SYSTEM. This material is obtained either with a low reflective finish being a silvery grey colour or it can be obtained with a 'solder coating' finish (called a terne coated finish) designed to weather to a dull lead like appearance. Stainless steel and lead are chemically compatible, so a stainless steel roof can be dressed with lead FLASHINGS without fear of corrosion to either metal.

S

STALL URINAL

A soil appliance, installed in a male toilet, which has a curved back and forms a large area to receive urine, it has an integral floor channel and when fixed in lines of more than one CLOAKING PIECES are usually provided for privacy. The stalls are flushed periodically from an AUTOMATIC FLUSHING CISTERN to help keep the appliance clean. For example, see URINAL.

STANDARD WIRE GAUGE   See GAUGE.

STANDING OVERFLOW

An overflow or warning pipe which passes through the base of a cistern and extends vertically inside, therefore avoiding a connection through the side wall of the vessel.

STANDING PRESSURE   See METER PRESSURE and GAS PRESSURE.

STANDING SEAM

An EXPANSION JOINT used on metallic roof coverings for joints which run with the fall of the roof structure or CLADDING. The finished standing seam should be about 25 mm high, as shown opposite, and to achieve this, upstands of the sheet edges of 32 mm and 38 mm would be required, see sketch. Standing seams are not to be recommended for roofs where foot traffic may be expected, as the joint would be ineffective if trodden on and flattened. See also COPPER SHEET AND ROOF COVERINGS.

(See illustration opposite)

STANDING WASTE AND OVERFLOW

An overflow pipe consisting of a vertical pipe with a tapered end to fit a waste fitting. This type of overflow fitting is commonly used in school and hospital kitchens and laboratory sinks because of its advantage of being easily cleaned, therefore being more hygienic.

(See illustration opposite)

STANDPIPE

A vertical pipe terminating with a tap which is installed to extend the water main up above ground level thus enabling water to be drawn off more easily, at a higher level.

STANFORD STOPPER

A clayware stopper used to seal off the rodding eye in an intercepting trap. For example, see INTERCEPTING TRAP. See also CLENCHER STOPPER.

STAT

An abbreviated term often used when referring to a thermostat.

STATIC HEAD

The pressure created by the vertical height of a column of water when at rest. See HEAD, WATER PRESSURE and HYDRAULIC GRADIENT.

STEAM

One form of water (H_2O) in its gaseous state. When water is heated to 100°C a physical change occurs and the liquid water expands some 1600 times and changes

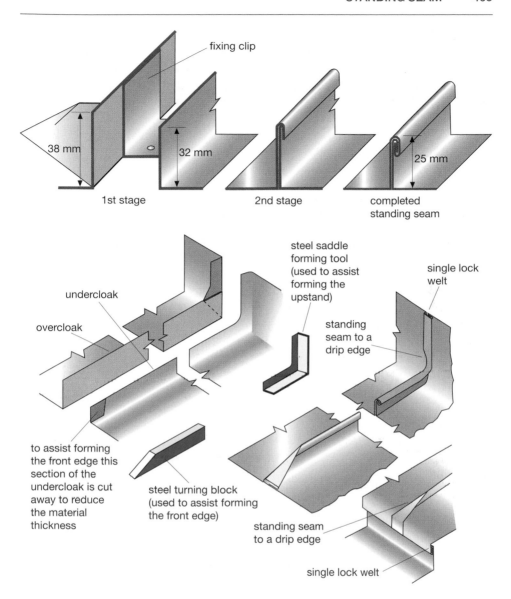

fixing clip

38 mm

32 mm

25 mm

1st stage

2nd stage

completed
standing seam

undercloak

overcloak

steel saddle
forming tool
(used to assist
forming the
upstand)

single lock
welt

standing
seam to a
drip edge

to assist forming
the front edge this
section of the
undercloak is cut
away to reduce
the material
thickness

steel turning block
(used to assist forming
the front edge)

standing seam
to a drip edge

single lock welt

Standing seam

S

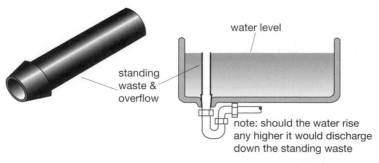

water level

standing
waste &
overflow

note: should the water rise
any higher it would discharge
down the standing waste

Standing waste and overflow

to steam. The production of steam can be used for various applications in industry such as a means of CENTRAL HEATING, or as a sterilising medium in hospitals, etc., or even used as a source of power to drive a steam turbine. See STEAM HEATING. See also LATENT HEAT, VAPOUR and CONDENSATION.

STEAM-HEATED CALORIFIER

A steam-heated calorifier is a device which utilises the heat energy of steam to heat the water for heating and domestic purposes. There are two basic types of calorifier, the storage type, used to heat water for domestic purposes, and the non-storage type designed to heat water for a means of central heating. The storage type works on the same principle as an INDIRECT CYLINDER, passing the steam through a HEAT EXCHANGER. The non-storage type consists of a series of small diameter tubes which pass through a water compartment or battery in the water filled heating system. As water circulates in the heating system it is heated as it passes through the battery and over the steam heated tubes. See sketch.

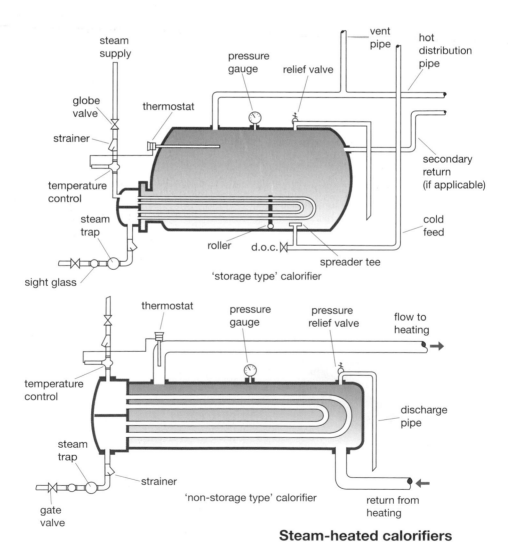

Steam-heated calorifiers

STEAM HEATING

A CENTRAL HEATING system in which the water is heated to produce steam which is circulated, under its own pressure, to the various HEAT EMITTERS. Steam heating systems are not used in the normal domestic dwelling but in large factories and hospitals where the steam can be used for various other purposes such as processing and sterilisation. The principle upon which it is based is that when water being heated reaches 100°C, it begins to boil and will remain at 100°C until all the water boils away. From the time of first boiling to completely boiling away a great amount of heat energy is put into the water (see LATENT HEAT). Therefore before the steam can be changed back to liquid this extra heat supplied must be given up, which is done as the steam circulates around the STEAM MAIN.

In the steam-heated system the boiler produces its own steam pressure to force itself around the steam main, which must be designed in such a way that as the steam slowly condenses back to a liquid it can flow through the pipes and escape from the steam main via a special steam trap and condensate pipe. If water cannot readily escape from the steam main, it could block the flow of steam at low points (called waterlogging) or cause a form of water hammer. Water hammer in steam mains is caused by small droplets of condensate being picked up by the steam which passes through the pipes at high velocities, when the condensate meets an obstruction such as a bend it is crushed against its surface, resulting in a loud bang, and possibly causing damage.

The steam main is taken from the boiler to the highest point of the system, then slowly all the horizontal pipes run with a slight gradient (see table below) to terminate with a dirt pocket and steam trap.

Minimum gradient to horizontal steam mains

Nominal bore of pipe mm	Gradient	Drop per metre run
13–19	1 in 211	4.7 mm
25–32	1 in 288	3.5 mm
38–64	1 in 432	2.3 mm
over 76	1 in 864	1.2 mm

To prevent pockets of water forming in horizontal pipes, which could give rise to waterlogging, when reducing the bore of the pipe eccentric reducers must be used as shown in the diagram, also intermittent drainage points must be provided in pipe runs over 30 m, this allows for the free flow of condensate. All connections through which steam is intended to flow must be made directly into the top of the horizontal steam main and never from the sides or bottom as condensate water could flow to the appliance reducing its efficiency, see sketch. Once the condensate water has been removed from the system it is generally re-routed or pumped back to the boiler for re-use, see CONDENSATE PIPE.

There are two basic designs of steam heating, low pressure and high pressure. A low pressure system is shown in the illustration over and although would rarely be installed today may still be found in older Victorian properties, it operates as follows:

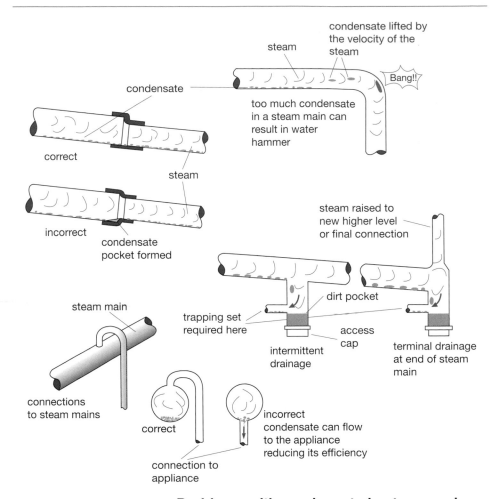

Problems with condensate in steam mains

(1) The system is filled with water until the correct level is achieved as indicated in the sight gauge. Note that there is only water in the boiler and the short section of vertical condensate pipework, the rest of the system is filled with air at atmospheric pressure.

(2) The boiler is fired and when the water reaches 100°C and is subsequently vaporised the steam generates a pressure which tends to compress the air in the system causing it to be expelled through the AIR ELIMINATORS.

(3) The steam flows from the boiler and through the pipework to the heat emitters slowly condensing on route, and falling back to the boiler for reheating.

See also HIGH PRESSURE STEAM HEATING, AIR ELIMINATOR, STEAM TRAP, STEAM-HEATED CALORIFIER and RELAY POINT.

(See illustration opposite)

STEAM MAINS

The pipe through which steam passes to the various HEAT EMITTERS, STEAM-HEATED CALORIFIER or sterilisation unit, etc., see STEAM HEATING.

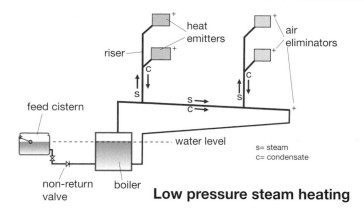

Low pressure steam heating

STEAM TRAP

A special valve which permits the flow of condensate but will prevent the flow of steam. Steam traps are fitted to the outlet side of steam operated appliances to prevent the steam continuing along the pipe going to waste. They are also located at low points in the STEAM MAINS to permit the condensate to escape from the system, which would otherwise cause a blockage to the passage of steam. There are three types of steam trap, the float operated, the thermostatic and the thermodynamic, but there are many variations of these three types of trap.

There are two basic types of float operated or mechanical steam trap. Firstly the ball float trap; as condensate enters the trap body the ball float rises thus opening the outlet valve allowing the condensate to discharge through the outlet pipe. Should steam enter the trap it would first push down on the condensate and slowly the float would lower because the steam would have insufficient density to hold the float up, therefore as the float falls the valve closes. A second design of float trap is the open bucket float trap which works by allowing condensate, which enters the trap, to fill up and overflow into the bucket which, when full, drops allowing the valve to open and also allowing any more condensate entering the trap to discharge through the outlet. Should steam enter the trap it forces down on the water pushing it out of the bucket, therefore with the removed condensate the bucket becomes buoyant again and floats on the condensate surrounding it, thus rising to close the valve. There is a variation to this type of trap in which the bucket is inverted.

The thermostatic or expansion type of steam trap, again to which there are many variations, works by having an expansion element of some sort, e.g. an expanding bellows filled with a VOLATILE FLUID, or a BIMETALLIC STRIP arrangement. Should condensate enter the trap it can continue and discharge through the outlet; should steam enter the trap the element expands due to the heat and seals off the outlet, which will remain sealed until the steam cools and condenses back to a liquid.

For the remaining type of steam trap, see THERMODYNAMIC STEAM TRAP. When fitting any float-operated steam trap in the pipeline, it is often necessary to fit an AIR ELIMINATOR prior to the trap, to allow the air to escape from the system. The condensate water from a steam trap flows along a CONDENSATE PIPE usually back to the boiler house, see STEAM HEATING. When a steam trap is fitted, it has several other controls installed with it to ensure that maintenance and inspection of the trap are possible, for these controls, see TRAPPING SET.

(See illustration over.)

S

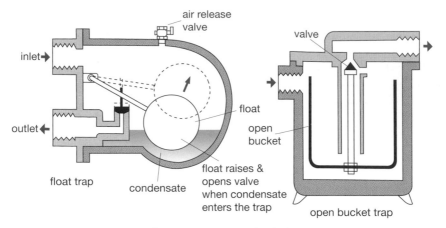

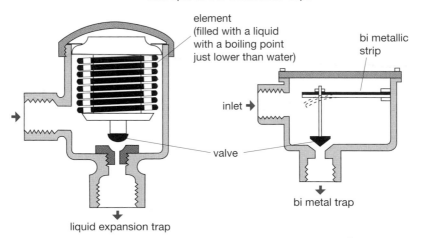

float operated or mechanical traps

thermostatic or expansion traps   **Steam traps**

STEEL
An ALLOY of iron and carbon. The carbon is added to the steel to make it much harder and tougher. Steel is available as a low carbon steel (commonly called mild steel) which has up to 0.2% carbon; medium carbon steel which has 0.2–0.6% carbon; and high carbon steel which has 0.6–0.9% carbon. See also STAINLESS STEEL.

STEEL WOOL   See WIRE WOOL.

STEP AND COVER FLASHING (RAKING)
A means of weathering the joint between a pitched roof as it meets a brickwall ABUTMENT. The step and cover flashing removes the need for SOAKERS but has the disadvantage of having more material on show to the finished job therefore, for reasons of appearance, it is often not chosen. See STEP FLASHING.

STEP FLASHING
The COVER FLASHING which weathers a pitched roof to brickwork. The flashing could be either single steps or a continuous running strip, not exceeding approximately

1.5 m in length, longer lengths are achieved by over lapping further strips. There are two basic ways in which step flashings are designed, either by using the step flashing in conjunction with SOAKERS, or using what is called STEP AND COVER FLASHING. The setting out of step flashing is best described in the sketch. Using a piece of material 150 mm wide allows for a water line of at least 65 mm, in the case of step and cover flashing add to this width 150 mm to lay out across the tiles. Step flashing can be made from any of the standard roofing materials, although lead is generally accepted as the best, due to it being easily dressed into awkward positions.

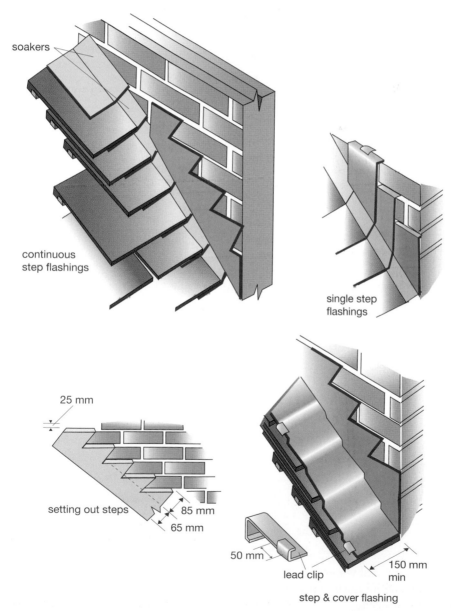

Step flashings

S

STEP IRON

The galvanised malleable iron treads which have been set into the brickwork of deep inspection chambers to assist getting in and out. The step irons should be set into the brickwork at every fourth course of brickwork, or 300 mm in depth apart and they should be staggered to assist climbing. See INSPECTION CHAMBER.

STEP TURNER

Homemade tool used to form the turn-in on STEP FLASHINGS.

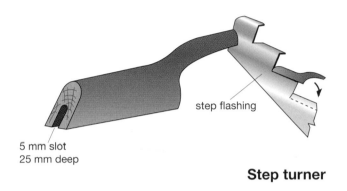

step flashing

5 mm slot
25 mm deep

Step turner

STEPPED DUCT SYSTEM   See WARM AIR CENTRAL HEATING

STERILISATION OF INSTALLATIONS WITHIN BUILDINGS

When a new domestic water supply system is completed generally all that is required is that the pipe work and storage vessels are thoroughly flushed with fresh MAINS supply water. But for buildings other than private dwellings where the domestic SERVICE and DISTRIBUTION PIPES serve only that building the following procedure should be adopted:

(1) The cistern should be cleaned of visible debris and the system flushed with water.
(2) The system should be re-filled.
(3) A measured amount of sodium hypochlorite solution is then added to the water in the cistern to give a free residual chlorine concentration of 50 mg per litre (50 ppm) in the water.
(4) After a period of one hour every DRAW-OFF POINT is opened progressively working away from the cistern, as the smell of chlorine is noticed at each draw-off point the valve is closed (the cistern being kept full of water at all times).
(5) After a second period of one hour a sample of water from the draw-off point furthest from the cistern is taken and the level of free residual chlorine should not be less than 30 mg/l (if it is the process must be repeated).
(6) After a final period of at least 16 hours the system is thoroughly flushed out with clean mains water and the free residual of chlorine concentration should be no greater than that in the water suppliers main.

Where chlorinated water is to be discharged into a sewer or natural water course, the local water authority must be informed.

STICK WELDING   See ARC WELDING.

STIFFENER (CLOSED WELT)

The free unfixed edge of a sheet weathering material which has been turned directly back onto itself by 8–10 mm. The reason for this is to provide a safe edge and prevent the material flapping due to wind pressures. See also STIFFENING BEAD.

STIFFENING BEAD (BEADED EDGE)

A half round bead which is formed on the free or unfixed edges of copper, aluminium and zinc roof coverings to stiffen the edge and prevent the material lifting in strong winds, for example, see COVER FLASHING. See also STIFFENER.

STIFFENING CREASE

A slight crease which has been temporarily put into a piece of sheet lead to assist keeping the internal corner in shape while lead bossing. For example, see BOSSING.

STILLS BOILER

The trade name of a steam or pressure type café boiler. The boiler is generally housed below a counter with the draw off point mounted above. The boiling water can only be drawn off when the water is boiling hot as the steam produced is used to provide the force to push the water out. Note that some heaters also include a steam draw off, designed for heating liquids rapidly.

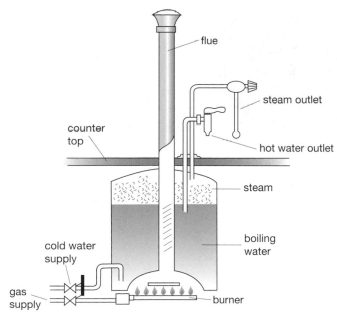

Pressure type café boiler

STILLSON PIPE WRENCH

A pipe wrench used and referred to as 'Stillson' after the name of the company that designed it. See WRENCH.

STOCKS AND DIES

Devices which cut threads onto pipes and other circular sections of metal. Dies are held in stocks which have handles thus enabling them to be rotated around the pipe.

There are two types of stocks and dies, chaser and receder dies. Chaser dies have a stock which can be adjusted to suit a range of pipe sizes, or are also available with an interchangeable drop head. The drop head type is not adjustable but can save time when threading pipes with different pipe diameters. When chaser dies are used they are drawn onto the pipe as they cut the thread and as a result can cause a great strain on the pipe sometimes causing it to be crushed. The second type, the receder die, can overcome this problem. The die stock is designed with a clamping device which screws back and forward on the die stock via a feed screw. To use the receder die stock, first the clamping device is wound forward and then clamped onto the pipe. As the stock is rotated it winds back up this feed screw, thus it is pulled onto the pipe cutting the thread, alleviating any strain. Most dies cut a TAPERED THREAD onto the pipe end for the first 20 mm or so, then when the die is on to its full depth, it starts cutting a PARALLEL THREAD. Dies are designed in this way for two reasons, firstly it is much easier to start the thread and secondly the tapered thread is useful when screwing into the fitting, it slowly tightens into the female thread. See also THREADING MACHINE.

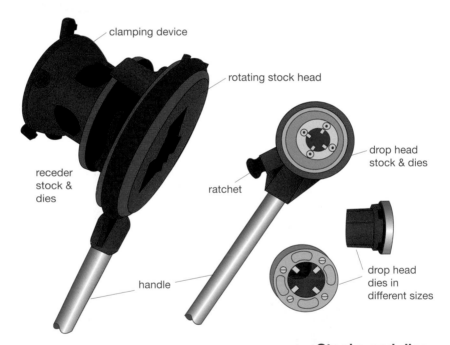

clamping device

rotating stock head

receder stock & dies

drop head stock & dies

ratchet

handle

drop head dies in different sizes

Stocks and dies

STOICHIOMETRIC MIXTURE

All substances react according to certain qualities or relationships with other substances. The term stoichiometric mixture refers to the weight relationships in volumetric analysis. As an example 10.57 volumes of air need to be mixed with 1 volume of natural gas in order to produce complete combustion. From this we can see how important it is to ensure that good ventilation is available when burning this fuel.

STONEWARE   See CERAMIC WARE.

STOPCOCK

A stopcock is of the SCREWDOWN VALVE type and is fitted in a straight run of pipe which is under high water pressure. The inlet and outlet of these valves are made in different designs to suit connections to copper, lead, plastic and low carbon steel pipes. It should be noted that a stopcock must be fitted the correct way round, so that the water passing through the valve would push the jumper up from its seating, for this reason stopcocks are stamped with an arrow indicating the direction of the water flow. In the manufacture of stopcocks the jumper should always be fixed to the spindle. In the past this was not the case, because, should a negative pressure occur on the inlet side of the valve, the jumper would close like a NON-RETURN VALVE, preventing backsiphonage. Under current Water Regulations this method of BACKFLOW prevention is now not permitted.

STOPCOCK KEY (TURNKEY)

A special key used to turn a STOPCOCK fitted underground on or off. Stopcock keys are usually homemade consisting of a piece of mild steel bent to form a U shape and welded onto the end of a piece of low carbon steel pipe or rod. As a quick measure one can be formed by cutting a V shape into the end of a piece of timber. See also FERRULE KEY.

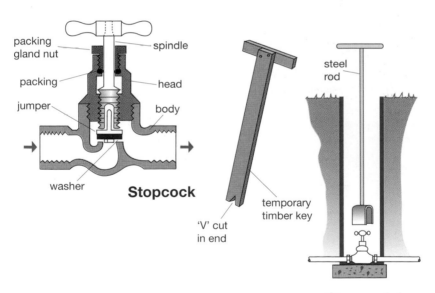

Stopcock

Stopcock key

STOPPER

A plug which is used to block a drainage pipe. See DRAIN PLUG and INFLATABLE STOPPER for drain testing stoppers; or CLENCHER STOPPER and STANFORD STOPPER for rodding eye stoppers.

STOPVALVE (STOP TAP)

A valve fitted in a pipeline to control the flow of water. For example, see STOPCOCK, FULLWAY GATEVALVE and PLUG COCK.

STORAGE CISTERN

An open topped vessel holding a supply of cold water for the various appliances fitted to the INDIRECT SYSTEM OF COLD WATER SUPPLY. The storage cistern should have a minimum capacity of 100 litres. If the cistern is also to act as a feed cistern for the hot water supply as well (being a combined storage and feed cistern) it should have a minimum capacity of 230 litres. Cold DISTRIBUTION PIPES from storage cisterns should be connected so that the lowest point of the water outlet is a minimum of 30 mm above the base of the cistern, this is to prevent sediment passing into the pipework. Connections of feed pipes to hot water apparatus from cisterns should be at least 25 mm above cold distribution pipes, if applicable, this should minimise the risk of scalding should the cistern run dry. Overflow pipes should be a minimum of 19 mm internal diameter and in all cases should be greater than the inlet pipe. Note the overflow/warning pipe must be fitted with a filter or screen to prevent the ingress of insects. See also COUPLING OF STORAGE CISTERNS.

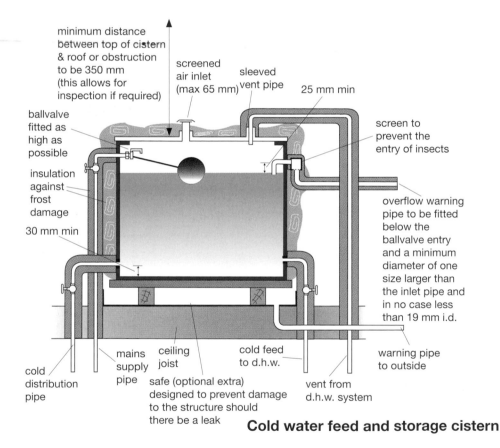

Cold water feed and storage cistern

STORAGE HEATER   See ELECTRIC STORAGE HEATER.

STORAGE WATER HEATER

A type of gas or electric water heater which heats a quantity of water for domestic use. Storage heaters can be of two designs, either LOCALISED or CENTRALISED HOT WATER SUPPLY. See ELECTRIC WATER HEATING and GAS STORAGE HEATERS.

STORED COLD WATER SUPPLY   See INDIRECT SYSTEM OF COLD WATER SUPPLY.

STRAIGHT PEIN HAMMER   See HAMMER.

STRAIGHT THROUGH TRAP
Another name to indicate an 'S trap'. See TRAP.

STRAINER   See FILTER.

STRAP ON BOSS   See BOSS.

STRAP WRENCH
A special kind of wrench used to grip thin walled piping (especially if it has been plated), without marking the tube.

Strap wrench

STRATIFICATION
An arrangement of layers or stratas. The word stratification is often used when describing the temperature difference of water in a hot storage vessel. The hot water accumulates at the top of the vessel and slowly down the vessel the water gets cooler and cooler, this is due to the fact that the hotter water rises, see CONVECTION CURRENTS.

STREET ELBOW   See M AND F ELBOW.

STRIKE BACK   See LIGHT BACK.

STRIP LEAD
Lead sheet which is bought in a much smaller width than the normal standard roll 2.4 m wide. The purpose of buying strip lead is to reduce the work necessary on site cutting small widths from a roll for COVER FLASHINGS, etc. Also in this form it is much easier to handle, the rolls of lead being supplied in weights of up to about 50 kg.

STUB DUCT SYSTEM   See WARM AIR CENTRAL HEATING.

STUB STACK (UNVENTED DISCHARGE STACK)
A discharge stack which is capped off with a RODDING EYE at its top end. Stub stacks are only permitted to be installed where they connect to a ventilated discharge stack

or drain within 6 metres from the base of the stack. If a stub stack is used of no branch waste connections may be made into the stack higher than 2 metres above the INVERT LEVEL of the drain and in the case of a WC pan connection this distance is reduced to 1.3 metres maximum to floor level. See sketch.

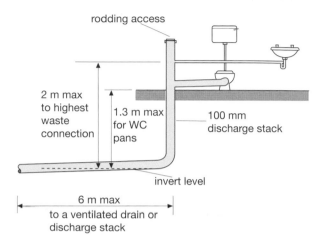

rodding access

2 m max to highest waste connection

1.3 m max for WC pans

100 mm discharge stack

invert level

6 m max to a ventilated drain or discharge stack

Stub stack

STUD (STUDDING)
A length of rod which has a continuous thread throughout its length. It is used by the plumber when making pipe hangers to suspend pipes or ductwork, etc. from ceilings. For example, see PIPE SUPPORT. The term stud is sometimes used to mean a RUNNING NIPPLE.

STUFFING BOX
The gland where the packing is positioned in a SCREWDOWN VALVE. See GLAND NUT and BIB TAP.

SUBSIDIARY FLUE
The section of flue pipe from a PRIMARY FLUE to the main SECONDARY FLUE is known as a subsidiary flue. Examples of a subsidiary flue would be found in the case of a BRANCH or SHUNT FLUE, or where two or more boiler flues are connected to use one main flue pipe such as when two boilers are connected together.

SUBSOIL DRAIN   See LAND DRAIN.

SUBSOIL WATER
Any water found in the soil resulting from the collection of rain.

SUCTION PIPE
The pipe which connects to the inlet connection of a PUMP and runs from a sump hole at a lower level. When the pump is operated the water is sucked up the pipe, due to a partial vacuum being formed in the pipe caused by the action of pumping, therefore in effect atmospheric pressure pushes the water up to the pump. See LIFT PUMP and FORCE PUMP.

SUCTION PLUNGER

Any one of a collection of plungers or force cups designed to assist in the removal of blockages from various sanitary appliances. The operating principle of a suction plunger is explained under the entry on DRAIN CLEARING but basically once any overflow connection, connected to the waste pipe, has been sealed off the plunger is pushed back and forth over the waste outlet of the appliance to remove blockages. See also HYDRAULIC FORCE PUMP.

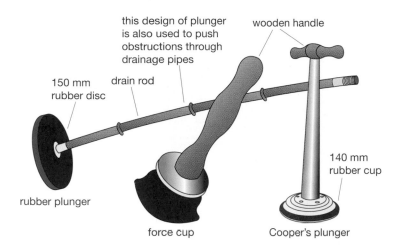

this design of plunger is also used to push obstructions through drainage pipes

wooden handle

150 mm rubber disc

drain rod

140 mm rubber cup

rubber plunger

force cup

Cooper's plunger

Suction plungers

SULPHURIC ACID

A strong mineral acid that is a toxic chemical and very corrosive, yet was once commonly used to pour down drains to assist in unblocking them. This procedure of using neat sulphuric acid in this way is no longer regarded as acceptable, although some drain cleaning solutions do use diluted quantities of this liquid. Ideally other alternative methods of drain cleaning should be sought.

SUMP

The part of a TRAP or GULLY which is below the INVERT LEVEL of the outlet.

SUMP HOLE

A hole sunk below the base of an excavation or below the finished floor level of a structure. Its purpose is to collect unwanted water to facilitate its removal by a LIFT or SUMP PUMP.

SUMP PUMP

A pump designed to lift water from a SUMP HOLE. A CENTRIFUGAL PUMP is generally used which has a submersible section or the pump could be of a submersible design.

(See illustration over.)

SUPATAP

A patent valve, designed so that you can rewasher the tap without turning off the water supply. The tap incorporates an automatic closing valve, thus when the nozzle

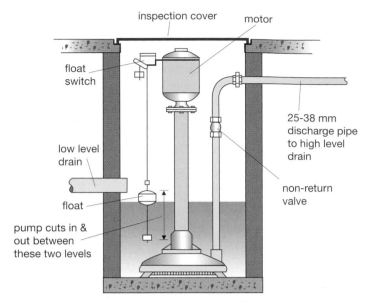

Sump pump

which houses the washer has been removed from the body the automatic valve falls blocking the flow of water. See also BROADSTONE BALLVALVE.

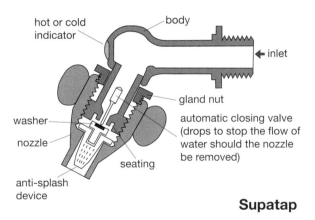

Supatap

SUPERFICIAL EXPANSION   See COEFFICIENT OF THERMAL EXPANSION.

SUPERSLEVE   See CLAYWARE PIPES.

SUPPLEMENTARY BONDING   See CONTINUITY BONDING.

SUPPLEMENTARY STORAGE SYSTEM (CYLINDER TANK SYSTEM)
A domestic hot water system sometimes employed in large buildings in which there are many draw-off points and at a long distance from the hot storage vessel.

The system is designed with one centralised hot storage vessel located at a convenient position and a second or even third auxiliary hot storage vessel, about one fifth the size of the centralised vessel, located at a position furthest away on the SEC-ONDARY CIRCUIT. See sketch. The reason for the second auxiliary vessel is to ensure an adequate flow of water at the draw-off points should several valves be used at once. Running larger sized secondary circulation pipes would also cure the problem, but due to the larger pipe diameters and a greater surface area, through which heat can be lost, designing the system in this way could prove expensive in running costs. Supplementary storage systems are often used in tall buildings to ensure a good flow at the draw-off points on the upper floors and the water is allowed to flow up to the auxiliary tank by CONVECTION CURRENTS, thus reducing the cost of running a pump. A system designed in this fashion often has a valve fitted to prevent PARASITIC CIRCULATION at night.

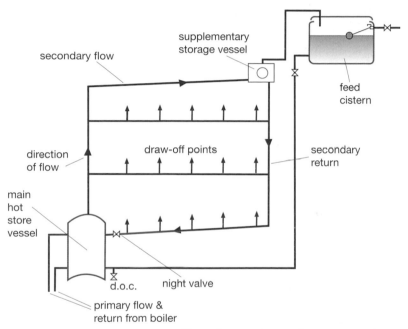

Supplementary storage system

SUPPLY PIPE   See SERVICE PIPE and COLD WATER SUPPLY.

SURCHARGING
Term used to mean that a drain or sewer cannot cope with the volume of water passing down it, thus flooding out of storm overflows or connections at low points will be apparent.

SURFACE TENSION
The tension which exists on the surface of liquids. Surface tension is a strange phenomenon in which it appears that a liquid has an apparent skin, but this, in fact, is not so. Fill a glass to the top with water then slowly add a small drop at a time to see the water rise above the top edge of the vessel, this is surface tension. Surface tension is caused by all the molecules in the liquid pulling together by forces of attraction, none wanting to break from the mass of liquid. See COHESION.

SURFACE WATER
Natural rain water from roofs and the surrounding ground surface.

SURFACE WATER DRAIN
A drain used to convey SURFACE WATER only. See also FOUL WATER DRAIN.

SURGE VESSEL   See AIR VESSEL.

SURGEONS TAP   See ELBOW ACTION TAP.

SUTTONS COLLAR   See FLUSHPIPE CONNECTOR.

SWAGING TOOL (KINGLEY TOOL)
A tool used to form a raised bead on copper pipe, used when making a MANIPULATIVE COMPRESSION JOINT. The purpose of the bead is to replace the need for a COMPRESSION RING or olive. Note this type of joint, which is now obsolete, was only used with special fittings designed for the purpose. See also FLARING TOOL and SOCKET FORMING TOOL.

SWAN NECK   See OFFSET and GOOSE NECK.

SWEPT TEE   See PITCHER TEE.

SWG
Abbreviation for standard wire gauge. See GAUGE.

SWIVEL CONNECTOR   See TAP CONNECTOR.

SYPHONIC ACTION   See SIPHONIC ACTION.

SYSTEM BOILER
A central heating boiler that is designed for a sealed heating system and includes all the necessary components to include the pump; temporary filling connection, sealed expansion vessel, pressure relief valve and high temperature cut-out device. System boilers utilise a separate hot water cylinder, if required for domestic purposes.

T

TACK See LEAD TACK.

TACK WELD
A small short weld with very little PENETRATION which is applied to the two adjoining edges about to be welded. The tack serves two main purposes, firstly, to prevent the weld opening due to distortion, which is caused by expansion as the material is welded along the join, and secondly, to hold the metal to be joined in position.

TAFT JOINT See WIPED SOLDER JOINT and BELL JOINT.

TAGS See WEDGES.

TAIL-PIECE See FERRULE SLEEVE.

TALLOW (TOUCH)
A waxy white fat, obtained from the area about the loins of cattle, horses and sheep. Tallow is used in the plumbing industry mainly as a NON-ACTIVE FLUX for lead soldering, but it is also used to assist in making waxed GROMMETS and to prevent materials sticking unduly. Other uses of tallow are to make materials such as candles, leather dressing, margarine and soap.

TAM PIN (TAN PIN/TAMPION)   See TURN PIN.

TANK
A cube or rectangular shaped vessel not open to the atmosphere, fitted with various connections and possibly an access cover. See also CISTERN.

TANK CUTTER   See HOLE CUTTERS.

TAP
(1) A cutting tool used to cut internal threads inside holes. The tool consists of a male thread which has a groove throughout its length to provide a cutting edge to its thread. Its top end is finished square to facilitate the fixing of a handle or 'tap wrench'.

(See illustration over.)

(2) A valve through which liquid can be drawn-off at a controlled flow rate, see DRAW OFF POINT and COCK.

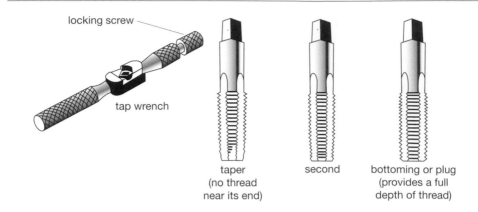

taper
(no thread
near its end)

second

bottoming or plug
(provides a full
depth of thread)

Taps

TAP BOSS
A fitting with a female thread into which a BIB TAP can be screwed, the other end of the fitting having a plain tail for the connection of a pipe.

TAP CONNECTOR (SWIVEL CONNECTOR)
A pipe fitting designed to connect the hot and cold pipe work onto PILLAR TAPS. The tap connector consists of a 'union' nut which is fixed to the fitting due to its design, but can freely move around. Once the fitting has been secured to the pipe the pipe/fitting can be offered to the male thread of the tap and the nut can be screwed on tight, clamping a FIBRE WASHER between the two. For example, see UNION CONNECTOR.

TAP FIXING SET
A selection of plastic and rubber washers used to assist in the fixing of PILLAR TAPS into WASTE APPLIANCES without the need for a jointing medium. It is not essential that a tap fixing set be used when installing pillar taps but if used it should cure any problems such as the tap rotating in the appliance, it can also prevent damage to the appliance due to overtightening. A common tap fixing set used is 'Essex Fix-a-tap' which consists of a rubber washer to sit the tap on in the appliance and a plastic and brass washer which fits to the otherside of the hole.

TAP GAP
The name sometimes given to indicate the AIR GAP between the outlet of a tap to the top, or SPILL-OVER LEVEL of a sanitary appliance.

TAP HOLE STOPPER
A blanking disc to cover up unwanted tap holes in waste appliances.

TAP RESEATING TOOL   See RESEATING TOOL.

TAPER (TAPER PIPE)
A drainage fitting used to increase or decrease the pipe diameter. The taper serves the same purpose as a reducer except that the reduction in the pipe diameter with a

taper is uniform throughout its length and the change in diameter is gradual, unlike a reducer which can be dramatic. There are two different types of tapers, the concentric and the level invert taper. The concentric taper reduces the pipe by connecting the larger pipe to the smaller pipe with the axes of the pipes in the same line. The level invert taper is designed so that the INVERT LEVEL, or the SOFFIT LEVEL, of the pipe run stays at the same plain. See also REDUCER.

TAPERED THREAD

A tapered thread is a male or female thread which gets progressively larger in diameter down the thread, thus as it screws into a fitting, or onto the pipe, it gets tighter. Most stocks and dies cut a tapered thread onto the pipe ends. See also PARALLEL THREAD and LOW CARBON STEEL PIPE AND FITTINGS.

TAPERING VALLEY GUTTER

A common form of timber constructed, metal lined gutter. A tapering valley is located between two pitched roofs or where the parapet wall and a CATCH PIT or CHUTE is located at one end. The gutter is tapered because it has a fall and is higher at the farthest end from the outlet, therefore because the roof slopes away from the gutter the valley has to be wider at its farthest end. See GUTTER.

TAPPINGS

The threaded connections in the sides of various containers such as boilers and hot storage vessels for the connection of pipes or valves.

TARNISH   See PLUMBER'S BLACK.

TARRED YARN (TARRED GASKIN)

YARN coated in a tar-like substance to give it a longer life. Tarred yarn might be used when jointing a sand and cement joint to clayware pipes. It is packed into the socket to prevent the cement mortar falling into the pipe, blocking the flow.

TAYLOR'S RING

A corrugated brass washer which is used to make FLANGE JOINTS. For example, see SECTIONAL BOILER.

TEE

A pipe fitting consisting of three openings, one being a branch connection into the flow of pipe. See BRANCH.

TELL-TALE PIPE

Term sometimes used, particularly in America, to mean a WARNING or OVERFLOW PIPE.

TEMPER

The degree of hardness in a metal. When a metal has been fully ANNEALED its condition would be dead soft temper, conversely a material which has become WORK HARDENED is in a state of hard temper.

TEMPERATURE

A description of heat or coldness measured on any one of several scales. The temperature of a substance determines whether heat shall flow into or out from it, normally heat flows from the hotter to the cooler substance. There are three temperature scales in general use, the Fahrenheit, the Celsius (centigrade) and the Kelvin (absolute) scales. The Kelvin scale cannot have negative values and is measured above absolute zero, which is a hypothetical temperature being the lowest possible temperature which can be achieved and is characterised by a complete absence of heat energy. 0K is equal to −273.15°C, for every 1K rise in temperature 1°C rise in temperature is also experienced, for example 10K equals −263°C or 273K equals 0°C. For conversion between Celsius and Fahrenheit temperature scales see METRIC SYSTEM. See also QUANTITY OF HEAT.

TEMPERATURE CONTROL FOR DOMESTIC HOT WATER SUPPLY

The temperature of the domestic hot water should be no higher than about 60°C in HARD WATER districts and no higher than 65°C in SOFT WATER districts. Therefore some means of controlling the temperature should be supplied. If an IMMERSION HEATER is used to heat the water the task is simple, the thermostat in the heater is adjusted as required. Should the water be heated in a boiler and conveyed to the hot storage vessel via a PRIMARY CIRCUIT the temperature is generally controlled by one of three methods:

(1) The boiler thermostat.
(2) A cylinder thermostat, which operates a MOTORISED VALVE.
(3) A thermostatic control valve.

Often the boiler serves the central heating as well as the domestic hot water, and during the winter the boiler thermostat is generally turned up above that used for domestic purposes therefore the first method often proves unsatisfactory. Control of domestic hot water, from a solid fuel boiler, on the other hand, can pose problems due to the lack of instant control of heat, the boiler being unable to switch on and off quickly, therefore in this case the first method is often chosen as the cylinder may act as a HEAT LEAK for the boiler water. See also THERMOSTATIC CONTROL VALVE.

TEMPERATURE RELIEF VALVE

A valve designed to open should the temperature within a hot water system rise to a temperature in excess of that of the thermostat. Should a fault occur within the system the temperature relief valve acts as a SAFETY VALVE to allow excess hot water to escape from the system, which can only be replaced with cooler water. Temperature relief valves should have the water/steam which discharges from their outlet port directed to a suitable discharge point. The valve shown works by opening should the temperature sensing element, which enters into a tapping in the vessel, expand excessively pushing the valve from its seating. Some designs of valves act as a COMBINED TEMPERATURE AND PRESSURE RELIEF.

(See illustration opposite)

TEMPERING

A process of improving the characteristics of a metal, especially steel. Tempering is carried out by heating the metal to a high temperature and allowing it to cool,

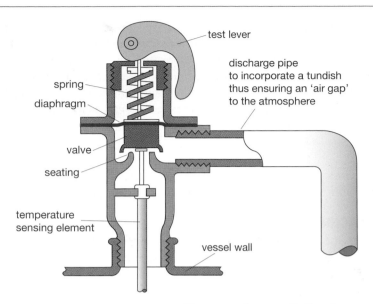

Temperature relief valve

usually by quenching it in oil or water. Should a cold chisel need tempering, or hardening, heat is applied to the chisel about 45 mm from the cutting edge; when the tip of the chisel is at a temperature of about 260–270°C identified by the tip changing to a purple colour, the chisel is at the correct temperature and should then be cooled rapidly by inserting it in a pail of water.

TEMPORARY BONDING WIRE

A special tool, usually homemade, used by the plumber when removing sections of metal pipework from a system, thus ensuring the continuity bond is maintained. A temporary bonding wire consists of two crocodile type clips attached to a length of wire. By ensuring the continuity of bonding is maintained the plumber is protected from electric shock should there be a fault in the electrical system, see CONTINUITY BONDING.

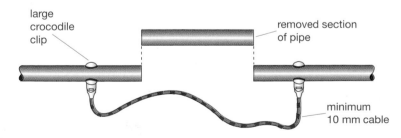

Temporary bonding wire

TEMPORARY HARD WATER

Water containing calcium bicarbonates is called temporary hard because the hardness in the water can be removed by boiling. As rain falls through the atmosphere it absorbs carbon dioxide gas; water containing carbon dioxide dissolves the calcium carbonates in soil, chalk or limestone, for example, and converts it into soluble bicarbonates, these bicarbonates are then carried in suspension within the water. When water is heated to 65–70°C, the water MOLECULES jump about so much that the gases can escape back into the atmosphere, and once the carbon dioxide gas has been removed from the water, the soluble bicarbonates revert back to their original form as calcium carbonate and are precipitated in the form of SCALE, accumulating on the insides of boilers and pipes. See also HARD WATER, PERMANENT HARD WATER, CLARKS PROCESS and BASE EXCHANGE WATER SOFTENER.

TENACITY
Tenacity denotes a material's ability to resist being pulled apart or pulled from its present position. See also TENSILE STRENGTH.

TENSILE STRENGTH
A measure of the TENACITY contained by a material. A sample piece of material is bolted between two clamps in a special tensile strength testing machine, the clamps are then made to pull apart in opposite directions, thus imposing a pulling load on the sample. The sample slowly stretches and eventually breaks. The loading in N/m^2 which finally causes the test piece to break is a measure of the tensile strength of the material.

TENSION
A term used to mean the stretching of a material.

TERMINAL   See FLUE TERMINAL.

TERMINAL FITMENT
Term sometimes used in specification documents or similar to mean a tap, shower control, etc. i.e. the fitting used where the pipe terminates.

TERMINAL GUARD
A wire cage secured around a balanced flue terminal designed to prevent anyone touching the terminal and getting burnt. For example, see BALANCED FLUE appliance.

TERMINAL VELOCITY
The terminal velocity is the maximum steady speed achieved by a substance falling freely through a gas or liquid. If an object was dropped from an aeroplane it would fall and increase in speed until the maximum steady speed of the object is obtained, this speed is known as the terminal velocity. Years ago when tall buildings were first built designers were afraid that the contents falling down a DISCHARGE STACK would hit the base of the stack and cause excessive noises or even damage, therefore they designed the stacks with bends or offsets throughout the length. Today we know this was unnecessary because as water falls through the stack it reaches its terminal velocity within one or two floors and no further VELOCITY is created, therefore discharge stacks should be installed as straight as possible, no matter how tall the building is.

TEST BEND
A special drainage testing fitting sometimes used when carrying out a water test to below ground drainage systems. It is designed so that it can be clamped into the

pipe, plugging off the escape of any water, and looks upwards. On top of the bend can then be fitted an extension piece to extend the height of water test to the recommended 1.5 m, for example, see WATER TEST.

TEST DIAL
The large dial on the front of a gas meter to register the gas flow in ft^3 or m^3. For example, see GAS RATE.

TEST NIPPLE
The test point, designed to facilitate the connection of a rubber hose to the system so that pressure readings can be taken.

TEST PRESSURE
The air or water pressure which is applied internally to any kind of pipework system when it is being tested for leaks. See AIR TEST, WATER TEST and TIGHTNESS TESTING.

TESTING BAG   See INFLATABLE STOPPER.

TESTING FOR SOUNDNESS   See TIGHTNESS TESTING.

TESTING OF DRAINS AND SANITARY PIPEWORK
There are various methods for testing drains and sanitary pipework to check for leaks or obstructions in the pipe line. Whenever possible when testing for leaks the drain should be exposed and the method of testing used could be either by AIR TEST, SMOKE TEST, or WATER TEST. See also TRACING EXISTING DRAINS and INSPECTION OF DRAINS.

THERM   See BRITISH THERMAL UNIT.

THERMAL CAPACITY
A measure of the ability of a material to absorb heat. This is usually identified as SPECIFIC HEAT. For example, the specific heat of water is 4.186 kJ/kg °C, whereas the specific heat of air is 0.993 kJ/kg °C. A consequence of this is that air cannot carry as much heat as water and therefore systems that rely on using air to convey heat must be much larger than those using water.

THERMAL CONDUCTIVITY
The ability of a material to conduct heat. See CONDUCTION.

THERMAL CUT-OFF DEVICE
A specially designed valve which automatically shuts off the flow of gas in the event of a fire. The valve is designed with a special spring loaded gate which is held open only by a metal support which has a low melting temperature. In the event of a fire should the temperature rise to 100–115°C, the metal support melts and the gate is allowed to close due to the force exerted by the spring, see sketch.

(See illustration over.)

T

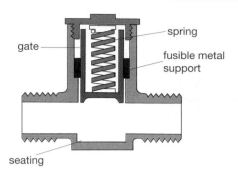

Thermal cut-off device

THERMAL INSULATION
The application of material to pipes and storage vessels to prevent the transmission of heat. Insulation material entraps small air pockets, still air being a bad conductor of heat. Apart from trapping pockets of air a good insulation material should not be flammable and if used externally or in damp conditions sufficient precautions should be made to ensure that the material is kept draughtproof and IMPERVIOUS to moisture. See also FROST PROTECTION.

THERMAL LAG
Term used to identify the time delay in waiting for the heat to arrive at its destination (e.g. the radiators) due to the mass of metal and the sheer volume of water which needs to be heated at the boiler. Modern systems with low water content boilers and better heat exchangers tend to have a greater reduced thermal lag.

THERMAL MOVEMENT
The movement within materials, etc. due to expansion and contraction, see COEFFICIENT OF THERMAL EXPANSION.

THERMAL RADIATION   See RADIATION.

THERMAL STORAGE SYSTEM (WATER-JACKETED TUBE HEATER)
A domestic hot water supply system in which the water is heated instantaneously as it passes through a stored supply of hot water which is being used for central heating purposes. By referral to the illustration it can be seen that cold water enters the hot water cylinder and passes through a heater battery, consisting of a large pipe coil, designed to expose a large area of heating surface contact. As the water flows out of the top end to the taps it will have been heated to a variable temperature depending upon the water velocity and flow rate. The constant 60°C outlet temperature is achieved by the use of an adjustable thermostatic mixing valve which will permit all hot water from the thermal store or a percentage directly from the supply main to cool it down. The advantage of this system is the convenience of having high water pressures and the high RECOVERY TIME to supply the system. The expansion of the domestic hot water is taken up within the vessel by the use of an expansion chamber, or alternatively by the use of a small expansion vessel incorporated within the cold supply pipe.

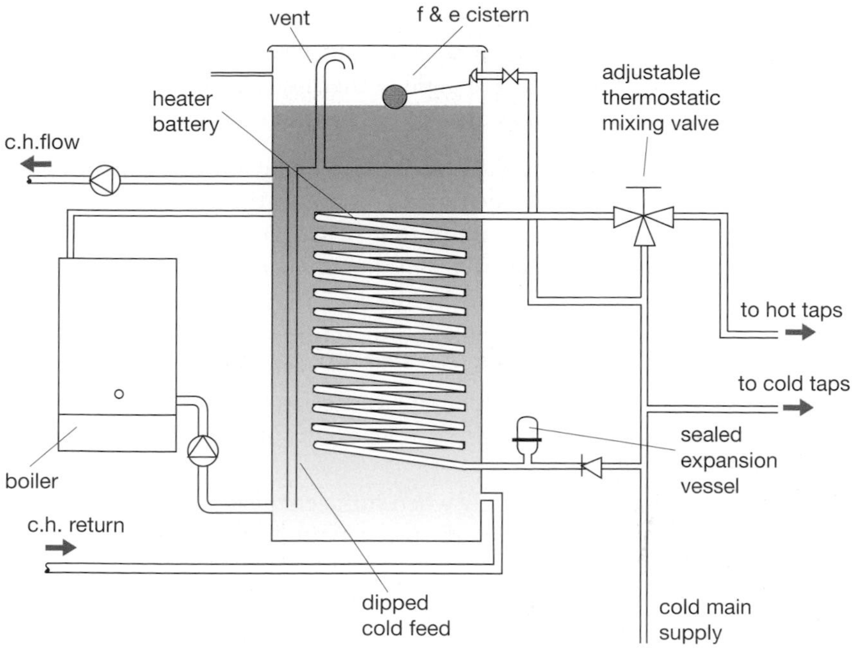

vent   f & e cistern

adjustable
thermostatic
mixing valve

heater
battery

c.h.flow

to hot taps

to cold taps

boiler

sealed
expansion
vessel

c.h. return

dipped
cold feed

cold main
supply

Thermal storage system

THERMAL TRANSMISSION COEFFICIENT   See U VALUE.

THERMISTOR
A special kind of resistor, designed to measure the heat generated at its point of location. These devices are often used to make or break switch lines to various circuits such as HIGH LIMIT THERMOSTATS and ATMOSPHERE SENSING DEVICES.

THERMOCOUPLE
A small heat SENSOR probe which is designed to create a small current of electricity to operate an electromagnet or SOLENOID VALVE. The thermocouple consists of two dissimilar metals, such as constantan (a copper-nickel alloy) and iron, joined at one end, the other end is connected to an electromagnet. When heat is applied to the ends which are jointed a small electric current will be produced, and thus operate the magnet. For example of the use of a thermocouple see THERMOELECTRIC VALVE.

THERMOCOUPLE INTERRUPTER
A special design of THERMOCOUPLE or arrangement in which the current flowing along the inner wire of a thermocouple connecting tube, through which the current is flowing, is made to disconnect. This in effect will break the electrical circuit and cause the THERMOELECTRIC VALVE to close. Thermocouple interrupters are typically used in conjunction with HIGH LIMIT THERMOSTATS and ATMOSPHERE SENSING DEVICES.

THERMODYNAMIC STEAM TRAP

A STEAM TRAP sometimes used in HIGH PRESSURE STEAM HEATING systems. The thermodynamic trap works on the principle of thermodynamics, which is basically the conversion of heat into mechanical energy. The trap works as follows:

(1) Condensate under pressure in the STEAM MAIN can enter the trap and push the disc upwards and allow the condensate to discharge through the outlet pipe.
(2) Condensate at a temperature close to that of steam eventually enters the trap and passes through to the outlet. When the hot condensate reaches the outlet pipe it flashes to steam. This is caused by the fact that the pressure in the outlet pipe is lower than in the steam main and therefore water boils at a much lower temperature (see BOILING POINT).
(3) The flash back to steam in the outlet pipe causes an increase in volume by some 1600 times, therefore creating a pressure in excess of that below the disc thus it is pushed closed.
(4) The disc relifts from its seating only when sufficient heat has been given off from the outlet pipe thus again allowing the pressure in the trap to relift the disc.

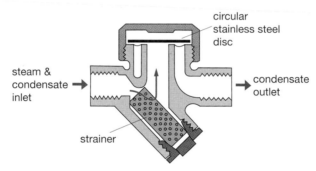

Thermodynamic steam trap

THERMOELECTRIC VALVE (MAGNETICALLY OPERATED VALVE)

A special valve which is a FLAME FAILURE DEVICE designed to prevent gas passing to the combustion chamber of a boiler unless a flame has become established. These valves are normally used to make up one of the component parts of a MULTI-FUNCTION GAS VALVE. The principle upon which a thermoelectric valve works is that when heat is allowed to impinge onto a THERMOCOUPLE an electric current is generated, which is used to operate the SOLENOID VALVE (an electromagnet), thus holding the valve open, should no heat be available to heat the thermocouple, no electric charge will be produced, therefore the valve will not remain open. The valve works as follows:

(1) When no flame is alight and impinging onto the thermocouple the valve remains closed as in sketch A.
(2) The reset button is depressed and held down, this permits gas to flow only to the pilot jet, see sketch B.
(3) The pilot light is ignited and the flame impinges on the thermocouple, thus generating an electrical charge.
(4) After 15–20 seconds the reset button can be released, the solenoid valve will now be held open. As the reset button returns to its original position gas is then allowed to pass onto the main burner pipe, see sketch C.

(5) Should the flame no longer impinge onto the thermocouple the valve will completely close as in A.

Note that when the main flow of gas continues along through the outlet of the valve it has to pass through some other valved control such as another solenoid valve which is opened as and when the thermostat calls for heat.

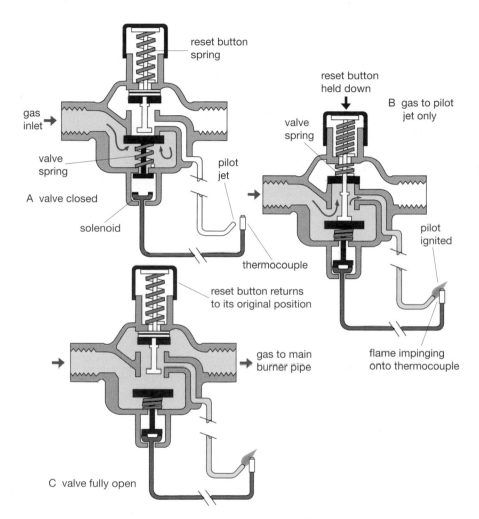

Thermoelectric valve

THERMOMETER

A device used to measure temperature. The thermometer can be calibrated in several scales although the celsius and fahrenheit scales tend to be the most used. A simple thermometer consists of a liquid such as mercury or alcohol coloured with a dye enclosed in a glass tube. Some thermometers work on a similar principle to the BOURDON PRESSURE GAUGE, as the VOLATILE LIQUID in a heat sensor expands it opens out a flattened circular tube, causing it to pull on a toothed cog which in turn rotates the pinion of a pointer, thus with this type of thermometer the temperature is registered on a dial. See also TEMPERATURE.

THERMOPLASTICS

A plastic material which when heated will soften, in this state it can be formed to the shape required and on cooling will harden. If required it could be heated again to form a new shape without any significant change in properties. There are many types of thermoplastics, e.g. acrylics, nylon, polythene, polypropylene and polyvinyl chloride. See also THERMOSETTING PLASTICS.

THERMOSETTING PLASTICS

A plastic material which can initially be heated to soften to form a shape, when it cools the shape will be permanent and no reheating of the material will soften it again. There are many types of thermosetting plastics, e.g. Bakelite (phenoformaldehyde resin), Formica, melamine and polyester. See THERMOPLASTICS.

THERMOSTAT

A device designed to control temperature. There are two basic types of thermostat, the bimetal and the liquid vapour type. A common bimetal type is the rod thermostat which consists of an INVAR rod fixed to the end of a brass tube. As the thermostat rod heats up the brass expands, but the invar does not as it has a low expansion rate, therefore the invar rod is pulled with the brass and slowly closes the gas valve, or in the case of the electric rod thermostat breaks the electrical contact. The liquid or vapour type of thermostat consists of a probe or sensor connected to a capillary tube and bellows as shown in the sketch, as the sensor heats up, the VOLATILE FLUID it contains expands and causes the bellows to become larger and make or break the electric contact, or open or close the gas pipeline.

(See illustration opposite)

THERMOSTATIC CONTROL VALVE

Any one of several valves such as a MIXING VALVE or a radiator valve. One type of thermostatic control valve, the 'Cyltrol' valve, works by having a bellows filled with a volatile fluid, which upon heating up expands and closes the valve. Note with this type of valve the hot water must be flowing in the direction shown in the sketch, this ensures the hot water can surround the bellows keeping it closed. Cyltrol valves are often found fitted on the PRIMARY RETURN pipe to a boiler to control the temperature of water in a hot storage vessel, but if they are used for this purpose they must not be used for FULLY PUMPED SYSTEMS or solid fuel boilers as the flow will be stopped causing a negative pressure to the pump or GRAVITY CIRCULATION. All types of thermostatic control valves can be adjusted to control the temperature of water flowing through the valve, which is basically achieved by turning the temperature adjusting head; this lowers or raises the valve nearer the hot water inlet seating.

(See illustration over.)

THERMOSTATIC MIXING VALVE

A MIXING VALVE which incorporates a special thermostat to ensure that water does not leave the outlet above a predetermined temperature. Thermostatic mixing valves are most commonly used to control the temperature of water to a shower. One type of thermostatic mixing valve, the expanding bellows type shown works as follows:

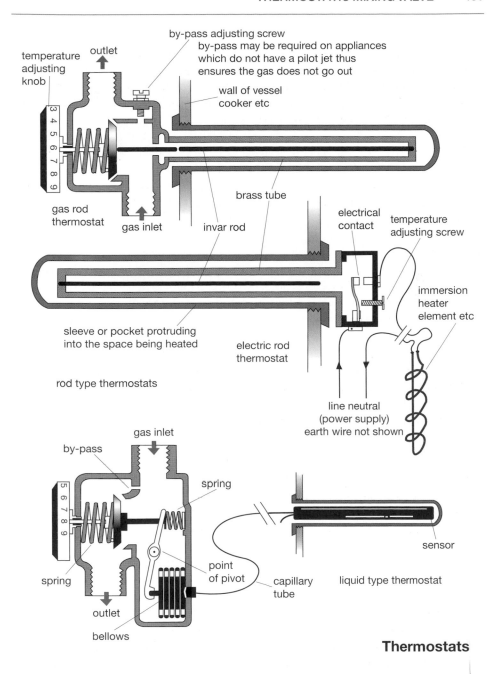

by-pass adjusting screw
by-pass may be required on appliances
which do not have a pilot jet thus
ensures the gas does not go out

temperature adjusting knob

outlet

wall of vessel
cooker etc

gas rod thermostat

gas inlet

invar rod

brass tube

electrical contact

temperature adjusting screw

immersion heater element etc

sleeve or pocket protruding
into the space being heated

electric rod thermostat

line neutral
(power supply)
earth wire not shown

rod type thermostats

by-pass

gas inlet

spring

sensor

spring

point of pivot

capillary tube

liquid type thermostat

outlet

bellows

Thermostats

T

(1) As can be seen in the sketch, both hot and cold water can enter the valve and pass through and across the temperature sensing bellows.

(2) Should the water become too hot the bellows, which is filled with a VOLATILE FLUID, expands and slowly reduces or closes the hot water inlet. As the water cools the bellows it contracts slowly to close the cold inlet.

Another type of thermostatic valve works on the principle of the BIMETALLIC STRIP. Should the water in the valve become too hot the bimetallic strip in the shape of a

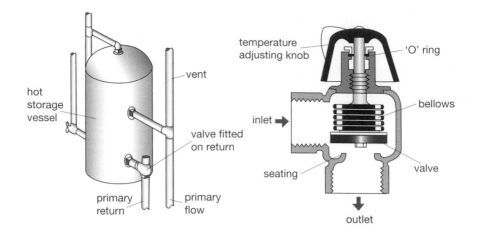

Thermostatic control valve

coil expands turning the head of the valve and reducing the hot or cold port hole size. See also SHOWER MIXING VALVE and THERMOSTATIC CONTROL VALVE.

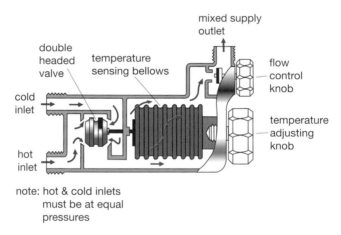

Thermostatic mixing valve

THERMOSTATIC RADIATOR VALVE

A special radiator valve designed to open or close the valve automatically, allowing hot water to the radiator, as the room requires heat. The valve is fitted with a built in heat sensor, or the sensor can be fitted away from the valve and located in a better position; remote sensors prove useful if the radiator valve is often covered with a curtain, etc. The valve works as follows:

(1) Should no heat be affecting the heat sensor the valve will remain open, caused by the pressure regulating spring lifting the valve from its seating.
(2) As the sensor heats up, the volatile liquid expands and is forced into the bellows chamber exerting a pressure on the bellows.
(3) The bellows is slowly compressed and pushes on the pressure pin closing the valve. As the sensor cools the liquid contracts and the pressure regulating spring re-opens the valve.

When installing thermostatic radiator valves into a system it is essential that not all the HEAT EMITTERS are fitted with a means of thermostatic control, as should they all close at once no heat can escape from the boiler and the pump could be pumping against this negative pressure, possibly resulting in unnecessary damage. Should a radiator need to be removed at any time, it is generally necessary to remove the temperature sensing head and secure the pressure pin down with a special manual locking nut, otherwise, should the temperature drop in the room, the valve may open resulting in the discharge of water onto the floor. See also THERMOSTATIC CONTROL VALVE.

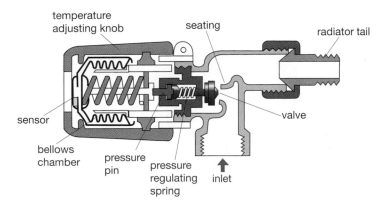

Thermostatic radiator valve

THIMBLE
A copper, brass or lead alloy SOCKET which is used to connect WCs and slop hoppers to discharge pipes made of copper or lead. Due to the fact that lead is no longer used for discharge pipes and stacks the thimble is nowadays rarely fitted. It was connected to the pipe via a wiped solder joint or lead welded; if thimbles are to be connected to copper pipes the joint is usually made by the bronze welding process.

THOMAS BOX FORMULA   See BOX'S FORMULA.

THREADING MACHINE
An electrical machine used to cut threads onto LOW CARBON STEEL PIPES. There are two basic types, those which turn the pipe, and those in which the pipe is held stationary and the stocks and dies are made to rotate around it cutting the thread. The second type of machine has the advantage that pipe can be threaded after bending, whereas only straight lengths of pipe can be threaded with the first type of machine. The STOCKS AND DIES used are the same as those used for the hand held equipment and the dies are adjusted in the same way. Most threading machines incorporate a pipe cutter and DE-BURRING REAMER to assist in the preparation of pipe ends prior to cutting. When the machine is running, a SOLUBLE OIL is discharged over the dies ensuring that they are well lubricated and kept cool, thus giving them a longer life, therefore when using these machines it is essential that this oil is kept well topped up.

(See illustration over.)

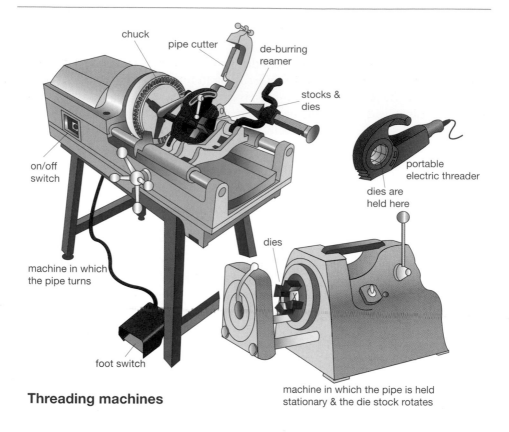

Threading machines

machine in which the pipe is held
stationary & the die stock rotates

THREE PHASE

An alternating current that offers a higher voltage supply than is found in the domestic home. The usual domestic supply is 230 V, whereas the three-phase supply is in the region of 400 V. This high voltage is used in industrial and commercial properties to operate large motors or machinery. Note from the illustration that where a connection is made to a phase conductor and a neutral conductor 230 V will be given. Should a connection be made between two-phase conductors, 400 V is obtained.

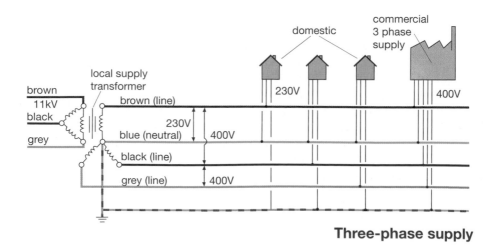

Three-phase supply

THREE PIPE SYSTEM OF CENTRAL HEATING   See TWO PIPE REVERSED RETURN.

THREE PORT VALVE (THREE WAY VALVE)
A DIVERTER VALVE, see MOTORISED VALVE.

THREE QUARTER CHANNEL BEND   See BARRON BEND.

THREE WHEEL PIPE CUTTER   See PIPE CUTTER.

THROAT
The inside part of a bend, having a smaller radius than the outside or back of the bend.

THROATING
The excessive flattening of a pipe bend. It is caused by a bend being pulled too tight and the pipe being compressed in the FORMER, see BENDING MACHINE. See also RIPPLING.

THROUGH FLOW EXPANSION VESSEL
A type of sealed expansion vessel which has both an inlet and outlet. It is designed to reduce the possibility of stagnation occurring within the vessel. See also ANTI-LEGIONELLA VALVE.

THRUST BLOCK
A support located at the back of a bend or tee joint connection of high pressure pipes to prevent undue noise being generated and assist in supporting the pipe. One type of thrust block would include a REST BEND used to support the base of a vertical stack.

THUMB NAIL WELD (HALF MOON WELD)
A finished weld pattern when lead welding. It is generally achieved by using a smaller blowpipe flame, see LEAD WELDING.

THUMB-TURN PIPE WRENCH
A type of footprint pipe wrench in which the jaws are enlarged by the turning of a thumbscrew, see WRENCH.

TIG WELDING   See TUNGSTEN INERT GAS WELDING.

TIDAL GULLY   See ANTI-FLOODING GULLY.

TIGERLOOP   See DE-AERATOR.

TIGHTNESS TESTING
The modern name given to what for many years was referred to as *soundness testing*. Tightness testing refers to the checking of a gas installation for any possible leaks. Several testing methods are employed, depending upon the system size and the gas used for the test purposes. Described here are a few common methods, namely:

(1) the testing of a domestic Natural gas carcass, with no gas meter fitted;
(2) the testing of an existing domestic natural gas installation, with a gas meter fitted;
(3) the testing of a domestic LPG installation.

1. Testing of a domestic natural gas carcass, with no gas meter fitted

(1) All open pipe ends should be capped off except one on to which is fitted a testing tee, comprising a small PLUG COCK and a hand pump.
(2) The manometer, filled with water to register zero, is connected to the test nipple and air is pumped or blown into the system to register a minimum of 20 mbar at the manometer.
(3) The plug cock is shut off and a period of 1 minute is allowed for the air to stabilise.
(4) After the initial minute, a further 2 minutes is allowed in which there should be no further pressure drop at the manometer.

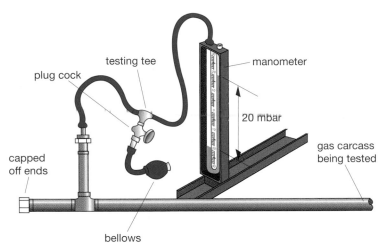

Testing a new installation

2. Testing of an existing domestic natural gas installation
Note either point 3 or 4 below is selected depending upon the system type.

(1) The Emergency Control Valve or Test Valve in the case of systems fitted with medium pressure regulators and all gas appliance pilot lights should be turned off. *Note: ensure the appliance isolation valve remains in the on position.*
(2) Remove the screw from the test nipple, which is located on the outlet from the meter, connect a manometer to this point and zero the gauge.
(3) *This point is only applicable to low-pressure systems and medium-pressure systems in which a test valve has been fitted.* Slowly open the emergency control or test valve until 10 mbar is registered on the manometer, then re-close the valve. Wait a period of 1 minute, during which time there should be no significant rise in gauge pressure (i.e. not more than 0.25 mbar). If this is the case, it indicates that the supply valve does not fully shut off and is referred to as LET-BY. Any drop in pressure may indicate a major leak.
(4) *This point is only applicable to medium-pressure systems in which no test valve has been fitted.* After having released the pressure from the system, including any medium pressure, with zero reading on the manometer the release mechanism on the regulator is held open for 1 minute during which time there should be no rise in pressure at the manometer greater than 0.2 mbar. The release mechanism is allowed to close and the emergency control valve is now turned on, and a

further 1 minute waited. Again no rise at the manometer should be witnessed. Should these tests prove satisfactory, the control valve and regulator can be deemed not to let-by.

(5) Upon a satisfactory let-by test, slowly increase the pressure within the system to 20 mbar and re-close the supply valve. A further period of 1 minute is now waited to allow the gas in the pipe to stabilise. *Note*: any slight rise could be due to high room temperatures. (Note that in the case of a system, such as in (4) above, the test pressure is only 19 mbar.)

(6) After the stabilisation period the manometer is observed and during the following 2 minutes there should be no pressure drop. Note that where the system has appliances connected, a small drop of 4 mbar is permitted where a U6 or G4 diaphragm meter has been installed, or 8 mbar where an E6 ultrasonic meter has been installed, thus allowing for faulty control taps/valves. No such drop is permitted if the isolation control valves to the appliances are closed.

(7) Following a successful test, the manometer is removed and the test point resealed. The gas is turned back on at the supply and the emergency control valve, anaconda, regulator and test nipple are sprayed with leak detection fluid to ensure no bubbles form at any potentially leaking joints. (Note that these areas were not suitably tested during the test.). Providing there is no smell of gas, the system can be deemed sound.

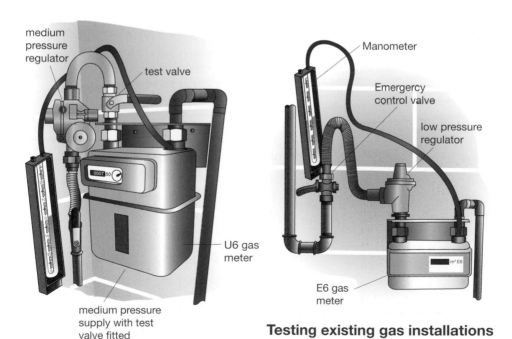

Testing existing gas installations

3. *Testing of a domestic LPG installation*

LPG installations use greater gas pressures than systems using natural gas (i.e. 37 mbar for propane and 28 mbar for butane) therefore a larger manometer is required. Prior to testing the system with gas, the installation should be tested with air in a similar fashion to that described for new installations using natural gas, but at a test pressure of 45 mbar. The main difference being that a 5-minute stabilisation period is required prior to the 2-minute test. Testing the installation with gas is carried out as follows:

(1) Confirm that all appliances are connected to the system with their operating taps closed and connect a manometer to a suitable test point.
(2) Turn on the gas supply and ignite the furthest gas burners to confirm that the system is charged with gas (PURGING if necessary).
(3) Adjust to the design pressure (37 mbar for propane, 28 mbar for butane).
(4) Close the appliance-operating tap and gas supply at the cylinder.
(5) Open an appliance operating tap and burn off the gas to reduce the pressure reading on the manometer to 5 mbar, then close off this tap. Five minutes is allowed to elapse for stabilisation, followed by a further 2 minute test period in which there should be no pressure rise, thus confirming that the supply valve shuts off fully with no let-by.
(6) Recharge the system until governor lock up occurs, closing off the supply valve when achieved. Wait a stabilisation period of 5 minutes then reduce the pressure to 30 mbar for propane and 20 mbar for butane. Wait a further test period of 2 minutes during which no pressure drop should be recorded. However, a small drop is permissible where appliances are connected, typically 0.5 mbar, depending upon the dwelling type.
(7) Finally the manometer is removed and the test nipple screw replaced. With the gas re-established, this joint is sprayed with leak detection fluid to confirm that it is sound.

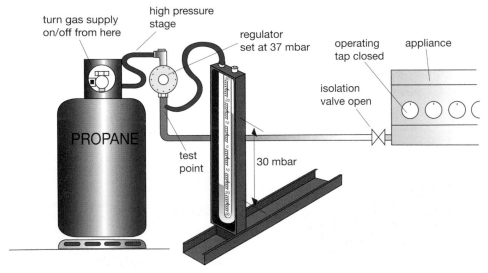

Testing an LPG installation

TILTING FILLET

A triangle shaped piece of timber which is fixed at a point on the under side of the lowest course of tiles on a roof. The tilting fillet is only used on tiled or slated pitched roofs where no fascia board is fitted, such as where a pitched roof discharges directly into a box or VALLEY GUTTER or behind a chimney stack. The purpose of the tilting fillet is to raise the lower edges of the tiles thus preventing wind blowing up under them; the tilting fillet also causes the tiles to be lifted from the surface of any sheet weathering such as lead, this helps prevent water rising up between the tiles and sheet covering by CAPILLARY ATTRACTION. For example of a tilting fillet, see BUILDING TERMINOLOGY.

TIME CLOCK (TIME SWITCH)

A special clock which can be set to automatically switch on and off the power to a boiler or pump, etc. The time clock is a simple device which runs continuously by electricity or by being manually wound once a week. The dial on the clock registers for 24 hours and all that has to be done is to position the start and finish times on the clock face. During the periods set a contact will be made allowing power to be supplied. Time clocks are nowadays commonly installed as a component part of a PROGRAMMER.

TIMESAVER

The trade name of a well-known type of SPLIT COLLAR JOINT used on cast iron pipes.

TIN

A NON-FERROUS METAL hardly used in the plumbing industry except as an alloying material in alloys such as bronze, gunmetal and various solders. The commonest use of tin is as a protective plating to steel cans.

Chemical symbol	Sn
Colour	silvery/bluish-white
Melting point	232°C
Boiling point	2260°C
Coefficient of linear expansion	0.000021/°C
Density	7310 kg/m^3

TIN SNIPS

Sharp cutters used to cut sheet metal. These tools come in a range of sizes and are available with the blades straight or curved. Tin snips are available either left- or right-handed, to assist awkward cuts.

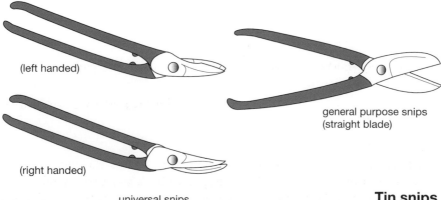

(left handed)

general purpose snips
(straight blade)

(right handed)

universal snips

Tin snips

TINGLE
A LEAD TACK.

TINMANS MALLET   See MALLET.

TINMANS SOLDER (TINSMITHS SOLDER)
Name given to BS grade A solder which has hardly any PLASTIC RANGE (183–185°C) and is therefore commonly used for electrical work. More generally, the term tinmans solder refers to any tin/lead solders which have a large percentage of tin in their make-up, therefore melting at a much lower temperature than solders, such as PLUMBER'S SOLDER which is used for joint wiping. BS grade A has the greatest amount of tin in its make-up, being 65% tin, 0.6% antimony and 34.4% lead.

TINNING (WETTING)
Term used to describe the adherence of a solder to the surface of a metal.

TOGGLE BOLT
A special fixing used to connect appliances, fittings, etc. to thin board such as plaster board or hardboard. There are various types of toggle bolt such as spring loaded or gravity toggles. The toggle bolt proves useful where access to the other side of the board is unavailable and a screw applied directly to the board would simply pull out under a slight load.

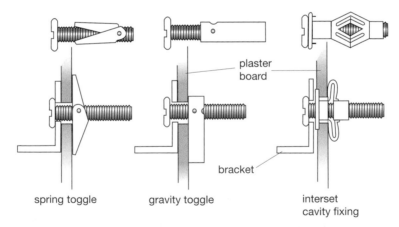

plaster board

bracket

spring toggle   gravity toggle   interset cavity fixing

Toggle bolts

TOILET
A room or compartment in which is installed one or more WC pans or urinals. See also SANITARY ACCOMMODATION and WATER CLOSET.

TONGUE STICK
A BENDING STICK.

TONGUE TEE (LIP TEE)
A specially designed tee fitting having a section protruding into the main flow pipe to divert some of the flow along the branch pipe.

TOP HAT

(1) A purposely designed spacer washer fitted on pillar taps, it is used when the appliance to which they are fitted is made of a very thin material. For an example see PILLAR TAP.

(2) A stopcock cover fitted at ground level to protect debris and dirt falling down the pipe duct to the underground stopvalve.

TOP UP BOTTLE

An open topped vessel, usually an enlarged pipe, which is sometimes fitted at a high point in a sealed system to make up small quantities of water which may be lost from the system. See SEALED SYSTEM OF HEATING.

TORBECK BALLVALVE

The trade name of a design of DIAPHRAGM BALLVALVE, which works on the principle of equilibrium, see EQUILIBRIUM BALLVALVE for example.

TORUS ROLL

The name given to the wood cored roll which is sometimes located at the intersection between the two roof pitches of a MANSARD ROOF, or at the junction of a flat roof and a pitched roof, see sketch

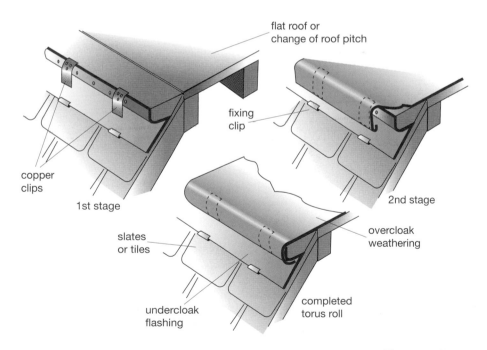

Torus roll

TOTAL HARDNESS

Water which contains both the calcium carbonates of TEMPORARY HARD WATER and the calcium sulphates of PERMANENT HARD WATER.

TOTAL HEAD (TOTAL PRESSURE)

The sum of the POTENTIAL HEAD, the PRESSURE HEAD and the VELOCITY HEAD.

$$\text{Potential} + \text{pressure} + \text{velocity head} = \text{total head}$$

TOTAL HEAT

The total heat contained in a substance, is the sum of the sensible heat, latent heat and super heat. See LATENT HEAT.

TOTAL PRESSURE   See WATER PRESSURE.

TOUCH   See TALLOW.

TOWEL RAIL (TOWEL AIRER)

A specially designed heat emitter which is installed in bathrooms to give a minimum amount of heat, primarily for the warming of bath towels. Often towel rails are designed to be fitted in the primary or secondary circulating pipes, thus providing use in the summer as well as the winter. Towel rails fitted in direct systems of hot water supply or in the SECONDARY CIRCUIT must be made of brass or preferably copper to prevent their corrosion. The arrangements shown in the illustration show how the towel rail can be fitted into the system in the traditional way and work by CONVECTION CURRENTS being set up in the pipework. Alternatively a towel rail could be fitted in the hot water heating system the same way as any radiator.

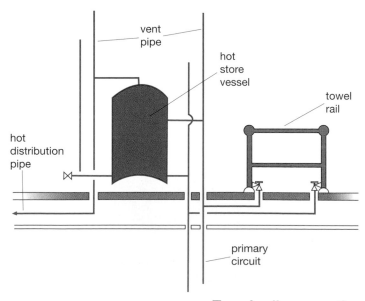

Towel rail connection

TOWEL RAIL BRACKET

Special basin bracket used to hold a wash basin. It is designed so that a towel can be hung over its lower arm, see WASH BASIN.

TOWN GAS

Town gas was the gas once commonly used in this country but has since been super-seded by NATURAL GAS. Town gas is a mixture of COAL GAS and WATER GAS.

Composition of town gas

Constituents	Approximate percentage
Hydrogen	46
Methane	24
Carbon monoxide	15
Nitrogen	9
Propylene	3
Oxygen	1.5
Carbon dioxide	1.5

TOXIC

Something that is poisonous.

TRACING EXISTING DRAINS

There are various methods used to find the run of existing drain pipes from adding a colouring to the water to inserting into the pipe a SMOKE ROCKET which gives off a dense smoke when lit. Whichever method is used, by inspecting various access and rodding points the drain runs can be determined. Sometimes drains are suspected of causing unpleasant odours inside premises; it is possible to determine whether the drain is at fault by adding chemical substances which produce an unmistakable smell when in contact with water. See COLOUR TEST, SCENT TEST and SMOKE TEST.

TRACING TAPE   See HEAT TRACING TAPE.

TRACKING

When using a pipe cutter, instead of cutting the pipe, two or three cuts are made running down the pipe, this is generally caused by the cutting wheels or their hous-ing pins becoming worn.

TRACPIPE

The patented trade name for a range of pre-sleeved stainless steel pipes used for gas installations. The pipes are supplied in coils, the joints being made using special fittings as supplied by the manufacturer.

TRADE EFFLUENT

Any liquid which is produced in the course of industry and removed from the trade premises normally via a system of drainage. Trade effluent does not include sewage from domestic premises.

TRAIN

The name sometimes used to indicate the collection of control valves, e.g. governors, solenoids, etc. fitted to a pipeline to an appliance.

TRANSDUCER   See PRESSURE SWITCH.

T

TRANSFER GRILLE

The name used for the VENTILATION GRILLE positioned in a door or wall between two adjacent rooms, thus allowing the transference of air from one room to another.

TRANSFERENCE OF HEAT

The various ways in which heat is transmitted. See CONDUCTION, CONVECTION CURRENTS and RADIATION.

TRANSFORMER

A device which transfers electric energy from one circuit to another, either reducing the voltage, or boosting it to a higher voltage. Basically transformers work by passing the electric current through a primary coil of wire inside the transformer, this current is then induced onto another coil (the secondary coil). The voltage through the output side is determined by multiplying the primary coil voltage by the ratio of the amount of wire turns in the secondary coil to the amount of turns in the primary coil. In this country all mains power supply is about 240 volts, at this voltage an electric shock could prove fatal therefore when using electrical tools, those with a power rating of 110 volts should be chosen and used with a step down transformer, in the event of a defect only a safe voltage would pass to the operator.

TRANSITIONAL FITTING

A special pipe fitting designed to convert one material to another and thus maintain a sound joint. For example changing from PE to copper or steel. For example, see COMPRESSION JOINT.

TRANSPOSITION OF FORMULAE

Often, when faced with a problem, the formula used to solve it is known, but it is in a form which is of no use in the calculation in hand. If this is the case the formula has to be rearranged to suit the problem. In all formulae there is an equals sign (=) and it is this equals sign which effects the new rearranged formula. Should a sign or symbol be moved from one side of the equal sign to the other its role is reversed, see below.

A plus becomes a minus, conversely a minus becomes a plus:

$$a + b = c \text{ is the same as } a = c - b$$

A multiplication becomes a division, conversely a division becomes a multiplication:

$$a \times b = c \text{ is the same as } a = c \div b \text{ or } a = \frac{c}{b}$$

A squared sign becomes a square root sign, conversely a square root sign becomes a squared sign:

$$a^2 = b \text{ is the same as } a = \sqrt{b}$$

Therefore the simple rule to remember is change the side, change the sign!

Example 1
The volume of a tank is found by multiplying the length by the width by the height ($L \times W \times H = V$). If the volume of the tank is known as well as the length and width, rearrange the formula to find the height.

$$L \times W \times H = V \qquad W \times H = \frac{V}{L} \qquad H = \frac{V}{L \times W}$$

Notice how both the length and width were moved from one side of the equal sign to the other and as a result were divided into the volume, this is simply because before being moved they were multiplied by the height (multiply becomes divide).

Example 2
Shown is a typical example of a formula used to find the capacity of a cylinder:

$$D^2 \times 0.7854 \times H \times 1000 = C$$

If the capacity is known, transpose the formula to find the diameter.

$$D^2 \times 0.7854 \times H \times 1000 = C$$

$$D^2 = \frac{C}{0.7854 \times H \times 1000}$$

$$D = \sqrt{\frac{C}{0.7854 \times H \times 1000}}$$

First the $(0.7854 \times H \times 1000)$ was taken over to the other side of the equal sign and had its role reversed; before being moved all of it was multiplied by the diameter therefore (change the side change the sign) it became a division and is divided into the capacity. Finally, the squared was moved across to become square root.

TRANSVERSE JOINT
Joints in sheet roofwork which run across the fall of a structure, thus the flow of water passes over them.

TRAP
A fitting or part of an appliance designed to retain a body of water thus preventing the passage of foul air. There are many different designs, but in general the trap could be designed with a vertical outlet, an 'S' trap, or a near horizontal outlet, a P trap, or the trap could be fitted in a pipe run in which case it would be called a running trap. The depth of the TRAP SEAL would depend upon the circumstances and usage of the pipe. See also RESEALING TRAP; TRAP SEAL LOSS and SELF-SEALING TRAP.

(See illustration over.)

TRAP INLET
A drainage fitting which is fitted vertically to receive one or more branch connections, designed to connect them all to a TRAP.

TRAP SEAL
The water seal between the inside of a drain and the outside air, which prevents odours entering the building, see TRAP.

TRAP SEAL LOSS
If the TRAP SEAL is lost objectionable smells would enter the building therefore the water seal in the TRAP must be maintained under all circumstances. Trap seal loss can be caused by eight different ways: evaporation, capillary action, momentum, leakage, waving out, compression, induced and self-siphonage. The first five are trap

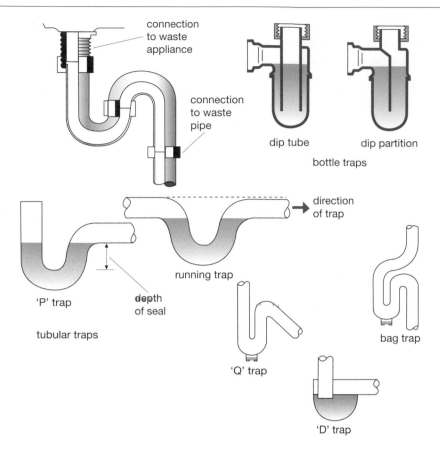

connection
to waste
appliance

connection
to waste
pipe

dip tube   dip partition

bottle traps

direction
of trap

running trap

'P' trap   depth
of seal

tubular traps

'Q' trap

bag trap

'D' trap

unusual & old trap designs

Traps

seal loss due to unforeseen circumstances, but the last three can be avoided by careful design considerations.

(1) Evaporation is caused by the trap not being used for 10–12 weeks in a good hot summer.

(2) Capillary action is another rare occurrence which happens to S traps. A rag or piece of string caught over the bend of the trap absorbs the water and deposits it down the waste discharge pipe.

(3) Momentum is caused by the sudden discharge of water from a bucket, etc. Due to the velocity at which the water is discharging it shoots round the trap bend and goes down the waste pipe, leaving no seal.

(4) Leakage is obviously caused by a fault in the trap or plumbing installation and water on the floor would be apparent.

(5) Waving out is caused by the effects of the wind passing over the top of the ventilation pipe causing pressure fluctuations, thus wave movements in the trap gradually wash over the outlet. If this creates a continued nuisance the problem can sometimes be remedied by extending the ventilating pipe upwards or terminating it with a bend.

(6) Compression only occurs in high rise buildings where the discharge of water down the main discharge stack compresses the air at the base of the stack thus

pushing the water out of the trap back into the appliance. This problem can generally be overcome by ensuring that a LONG RADIUS BEND is installed at the base of the stack and that no connections are made within 450 mm of the INVERT LEVEL of the drain, alternatively a RELIEF VENT should be carried down to connect to the lowest part of the discharge stack. For discharge stacks over three storeys see PRIMARY VENTILATED STACK SYSTEM.

(7) Induced siphonage is caused by the discharge of water from another sanitary appliance connected to the same discharge pipe. As the water falls down the pipe and passes the branch pipe connected to it, it draws air from it, thus creating a partial vacuum and subsequently siphonage of the trap takes place. To overcome this problem, TRAP VENTILATING PIPES could be designed into the system, these would permit air into the discharge pipe preventing the development of a partial vacuum.

(8) Self-siphonage is caused in appliances such as wash basins, designed to be able to discharge their contents of water quickly. As the water discharges, it sets up a plug of water, which as it passes down the pipe creates a partial vacuum, thus causing SIPHONIC ACTION to take place. To overcome this problem of self-siphonage, a larger waste pipe is sometimes used but in most cases a RESEALING TRAP cures the problem.

See also PERFORMANCE TEST.

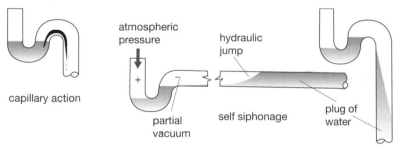

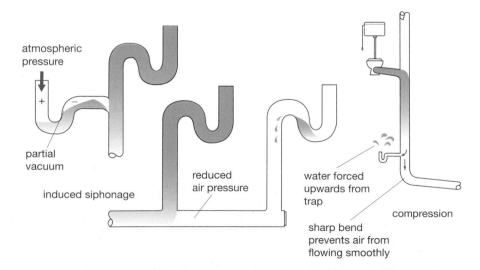

Trap seal loss

TRAP-VENTILATING PIPE (ANTI-SIPHON PIPE)

A VENTILATING PIPE which is extended and connects within 300 mm from the outlet side of a trap. It is designed to prevent SIPHONIC ACTION taking place resulting in TRAP SEAL LOSS. Should the trap ventilating pipe be connected to a main ventilating stack with branch connections it is often referred to as a BRANCH VENTILATING PIPE.

TRAPLESS GULLY (UNTRAPPED GULLY)

A gully used for the collection of SURFACE WATER. It is connected to the surface water drain in a SEPARATE SYSTEM OF DRAINAGE, see GULLY.

TRAPPED GULLY   See GULLY.

TRAPPING BEND

A 90° bend with a long and a short leg, without any sockets at either end, such as a dip pipe in a PETROL INTERCEPTOR or similar situation.

TRAPPING SET

In order to facilitate the maintenance and inspection of a steam trap fitted in a steam heating system several valves and fittings are installed with it. It is this collection of fittings which is known as a trapping set and consists of the following items:

(1) GLOBE VALVE: To prevent the flow of condensate or steam to the trapping set during maintenance.
(2) STRAINER: Designed to filter any grit, etc. from the condensate prior to entering the steam trap.
(3) STEAM TRAP: To allow the passage of condensate from the STEAM MAIN.
(4) SIGHT GLASS: A device to internally inspect the pipeline during operation to see if the steam trap is allowing steam to pass through to the CONDENSATE PIPE.
(5) CHECK VALVE: Designed to prevent the backflow of condensate into the trapping set.
(6) GATEVALVE: To prevent the backflow of condensate to the trapping set during maintenance.

Note that a disconnecting joint, such as a UNION CONNECTOR should be fitted after the globe valve and before the gatevalve to allow the trapping set to be easily removed from the pipeline, see sketch. It is possible to purchase a combined trap and strainer or sight glass and check valve, etc. to reduce the amount of space needed to install all the components within a short space of pipe.

(See illustration opposite)

TREAD PLATE

A hard wearing impervious, non-slip floor finish forming part of a stall or slab urinal. The term is also used to mean the foot rests on either side of a SQUATTING WC PAN.

TRICKLE VENTILATION

An adjustable ventilator usually positioned above a door or window opening in order to maintain a minimal amount of air movement. Note that this kind of ventilation must not be confused with that used for the supply of air for combustion purposes.

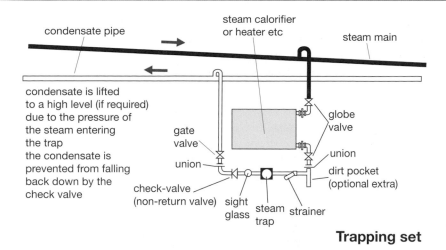

Trapping set

TROUGH CISTERN   See FLUSHING TROUGH.

TROUGH URINAL
A urinal which is in the form of a trough and is hung on the wall at a height of about 610 mm from the floor to the front edge. Trough urinals are generally made from stainless steel or acrylic materials. For example, see URINAL.

TROUGHING
An EAVES GUTTER.

TRUNK MAIN
The large water pipes usually over 1 m in diameter which are used to convey water to and from the various treatment works and pumping stations throughout the country. See also DISTRIBUTION MAIN.

TRUNK VALVE   See ANTI-FLOOD VALVE.

TRV   See THERMOSTATIC RADIATOR VALVE.

TUBE   See PIPE.

TUBULAR TRAP   See TRAP.

TUMBLER COCK
A PLUG COCK.

TUMBLING BAY   See BACK-DROP CONNECTION.

TUNDISH
A funnel-shaped fitting or arrangement of pipe work which is open to the atmosphere. The tundish is used to collect a discharge of water from a pipe or several pipes from, for example, a PRESSURE RELIEF VALVE or an overflow pipe. A tundish may be used for two reasons, firstly, to show that water is discharging through the pipe and,

secondly, to provide an AIR GAP between the appliance and the drain, thus ensuring no BACKFLOW of water is possible.

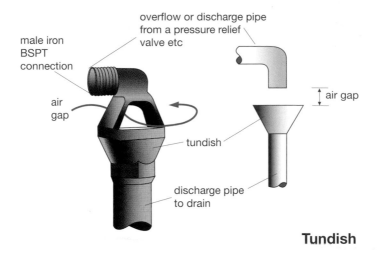

male iron
BSPT
connection

overflow or discharge pipe
from a pressure relief
valve etc

air
gap

air gap

tundish

discharge pipe
to drain

Tundish

TUNGSTEN (WOLFRAM)

A very hard metal which is used to increase the hardness and strength of steel alloys. Tungsten is a white-greyish colour and has the highest melting point of all metals, 3410°C. Tungsten is brittle at room temperatures but can be used in making high speed cutting steels which will easily cut other steels, or in the manufacture of lamp filaments. When tungsten is combined with carbon a substance known as tungsten carbide is produced, this being, with the exception of diamonds, the hardest known material and is used in tools such as MASONRY DRILL bits.

TUNGSTEN INERT GAS WELDING (TIG OR ARGON ARC WELDING)

An arc welding process which uses a gaseous shield of argon to protect the weld zone from atmospheric contamination. TIG welding is often likened to gas welding because the heat used to melt the metal is supplied from an arc and not from a gas flame. The electrical power supply and a flow of argon gas is fed to the welding torch; when it is brought down to the surface an arc is struck between the non-consumable tungsten electrode and the work, this causes heat to be generated, resulting in the melting of the metal to be welded. The filler rod is then fed to the weld pool as necessary, exactly as in oxyacetylene welding. Argon plays no part in the combustion process, being a noble gas, it is simply used to form a shroud around the molten pool to exclude the gases in the atmosphere which might otherwise cause OXIDATION and the formation of nitrides. To prevent the ceramic shield surrounding the electrode cracking under the intense heat, the welding torch is air cooled with currents below 150 amps, while from 150–300 amps the torch should be water cooled, see sketch. TIG welding proves to be very useful when welding metals such as aluminium, copper, magnesium and stainless steel without the use of a FLUX. See also ARC WELDING.

(See illustration opposite)

TURN KEY (TURN OFF KEY)   See STOPCOCK KEY.

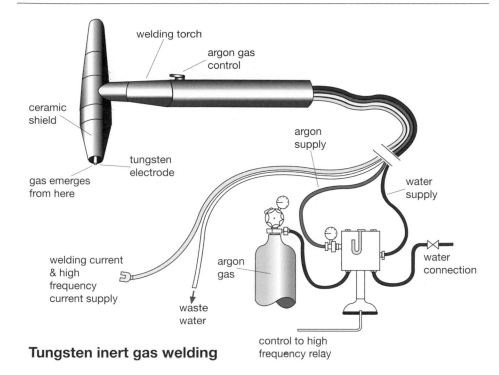

Tungsten inert gas welding

TURN PIN (TAM OR TAN PIN)

A cone shaped, boxwood tool used to bell out the end of lead pipe prior to wiping a soldered joint.

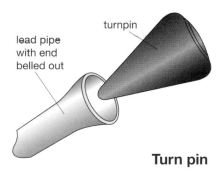

Turn pin

TWIN ELBOW (CURVED TEE)

A specially designed elbow, or tee as it is often called, in which two pipes from opposite directions curve through 90° and join to form a single pipe.

Twin elbow

T

TWIN WALL FLUE

A flue pipe which consists of two concentric pipes. The inner concentric void traps air which acts as an insulator. This type of flue is designed to keep the inner flue warmer, thus reducing the amount of condensation and assisting in creating a suitable pull.

TWINRIB ROOFING

The fore-runner to the D12C method of laying sheet NURALITE. The old twinrib method is no longer carried out but basically consisted of first laying special Nuralite base strips to the roof decking where the joints were to be made, then the sheets of Nuralite were layed onto these strips where a lap joint was made. Finally a cover strip was positioned on top of the join forming a second lap joint, thus the Nuralite sheet was sandwiched between the base and the cover strip. The D12C strip now used does not have the base strip.

TWIST DRILL

A design of drill bit which is used to make holes in various materials such as timber and metal. The tip of a twist drill is usually ground to an angle of about 30° (which results in the point angle being 120°) for general work, but when cutting materials, such as plastic, a much sharper point is required. To drill materials such as brick or stone the twist drill proves ineffective and a masonry drill has to be used having a tungsten carbide tip.

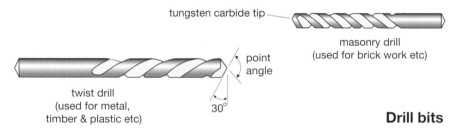

tungsten carbide tip

masonry drill
(used for brick work etc)

point angle

twist drill
(used for metal,
timber & plastic etc)

30°

Drill bits

TWO PIECE WC PAN

A WC pan made with the bowl and trap as two separate pieces. Its purpose is to allow the trap to be swivelled around to discharge in any direction.

TWO PIPE OIL SUPPLY   See OIL STORAGE AND SUPPLY.

TWO PIPE REVERSED RETURN (THREE PIPE SYSTEM)

A special design of the two pipe system of central heating in which the heating circuit to all the HEAT EMITTERS in the system is about the same length. When the cooler water leaves the first heat emitter in the system, it does not simply join the return pipe and travel back to the boiler, as in the two pipe system, instead it travels to the furthest point in the system and upon receiving the return water from the last heat emitter runs back to the boiler return connection. Designing the system in this way ensures the FRICTIONAL RESISTANCE to the water flow is the same to each radiator therefore, although it can prove a little more expensive to install as generally more pipework is required, the BALANCING of the system proves to be a simple task. See also TWO PIPE SYSTEM OF CENTRAL HEATING.

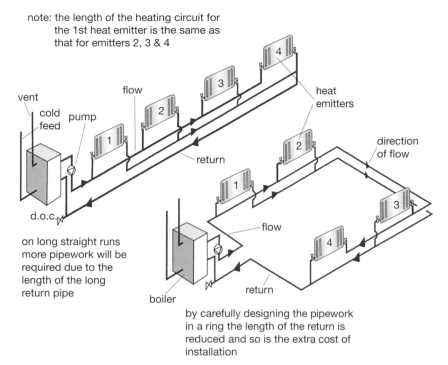

note: the length of the heating circuit for
the 1st heat emitter is the same as
that for emitters 2, 3 & 4

vent

cold
feed   pump   flow   heat
emitters

direction
of flow

return

d.o.c.

on long straight runs
more pipework will be
required due to the
length of the long
return pipe

flow

return

boiler

by carefully designing the pipework
in a ring the length of the return is
reduced and so is the extra cost of
installation

Two pipe reversed return

TWO PIPE SYSTEM OF ABOVE GROUND DRAINAGE (DUAL SYSTEM)

An old-fashioned design of above ground sanitary pipework. The two pipe system
can still be installed although for economic reasons is rarely done so. The two pipe
system consisted of a SOIL WATER discharge stack and a WASTE WATER discharge
stack which ran individually vertically down to the drain. The reason the system
was designed in this way was because waste appliances often had no trap fitted to
them, therefore the TRAPPED GULLY fitted at the base of the waste discharge stack
provided a means of preventing odours from the drain entering the building. The
early designs had a hopper head fitted externally to the building for the connection
of first floor waste appliances, but this arrangement is no longer permitted under the
Building Regulations although may still be found in many older buildings. See also
SANITARY PIPEWORK.

(See illustration over.)

T

TWO PIPE SYSTEM OF CENTRAL HEATING

A design of central heating which has many advantages over the one pipe systems.
The two pipe system, prior to modern designs and standards, often worked by
GRAVITY CIRCULATION, a system design no longer permitted. Today a circulating
pump would be provided to move the water round the pipework. The water which
is pumped around the heating circuit is also pumped through the radiators giving
them a much faster heating up period. With this design of system BALANCING out the
heat to each radiator proves to be reasonably simple, although it is not uncommon
for the first radiator in the system to have its LOCK-SHIELD VALVE just fractionally
opened when balancing; this is due to the FRICTIONAL RESISTANCE to the flow

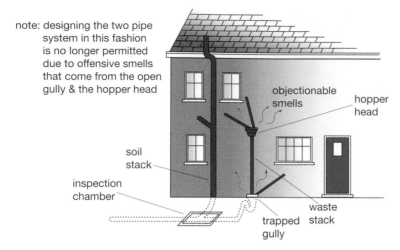

note: designing the two pipe system in this fashion is no longer permitted due to offensive smells that come from the open gully & the hopper head

objectionable smells

hopper head

soil stack

inspection chamber

waste stack

trapped gully

An early design of the two pipe system

through this HEAT EMITTER. Basically there are four designs of two pipe circulation these being, the parallel, the drop, the high-level return and the up-feed systems. Each is shown in the illustrations and the layout chosen usually depends upon the space available to run the pipework. In the case of the high-level return and the drop systems the venting of the radiators is not required as the air can escape from

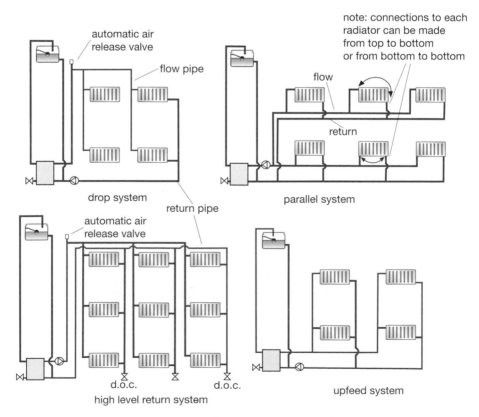

automatic air release valve

flow pipe

drop system

return pipe

note: connections to each radiator can be made from top to bottom or from bottom to bottom

flow

return

parallel system

automatic air release valve

high level return system

d.o.c. d.o.c.

upfeed system

Two pipe central heating systems

the system via the flow pipe. In most cases the design chosen is the parallel system because less pipework is generally on show, being all hidden below the floor. Not shown in the illustration, would be some form of zone control. This would separate the various areas (e.g. upstairs from downstairs) and be controlled via a suitable room thermostat thus complying with the Building Regulations. See also HOT WATER HEATING SYSTEM, ONE PIPE SYSTEM OF CENTRAL HEATING and TWO PIPE REVERSED RETURN.

TWO PORT MOTORISED VALVE   See MOTORISED VALVE.

T

U DUCT

A special flue used in multi-storey buildings into which several BALANCED FLUE appliances can be fitted. The flue duct is in the shape of a U. One leg is to provide air for the combustion of the fuel, the second leg is designed for the connection of the appliances. See also SHARED FLUE.

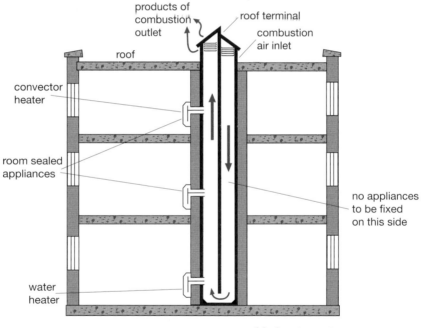

U duct system

U GAUGE See MANOMETER.

U VALUE (THERMAL TRANSMITTANCE COEFFICIENT)

A figure which is determined by experiment and means the rate of heat transfer in watts through one square metre of a structure when the air temperatures at each side of the structure differ by 1 K (or 1°C). The BUILDING REGULATIONS give the maximum permitted U value for a domestic dwelling as 0.16 W/m^2 K for pitched roofs, for exposed walls 0.35 W/m^2 K and floors 0.25 W/m^2 K. This is, therefore, the amount of heat in watts which will pass through each square metre of its surface, in one second, for each degree difference in the temperature of its inside and outside surfaces. The lower the U value, therefore the better the insulation.

UNDER PRESSURE MAINS TAPPING MACHINE

A special machine designed to cut into large diameter distribution mains pipes while the water is still in the pipe and under pressure. Before putting the machine in position, the body of a closed SCREWDOWN FERRULE, with the jointing medium applied to the thread, is fixed into the base of one of the spindles in the headgear of the machine and a combined drill and tap in the other, see sketch. Then the head is lifted onto the pipe and clamped down with a chain sealing off its base with a rubber or leather washer. The ratchet handle is fitted to the spindle with the drill and operated, simultaneously the drill/tap pressure adjusting screw is wound down slowly, this cuts a hole and threads it at the same time into the pipe. Upon completion the ratchet is reversed and the drill is wound up; note water can now flow into the headgear but no further. By a special design the headgear can now be rotated 180° to align the ferrule with the hole just drilled and tapped. The ratchet handle is fitted to the ferrule spindle and operated, thus winding the ferrule into the tapping. When the machine is removed from the pipe only the water in the headgear falls to the ground, the connection made seals the pipe. The other component parts of the ferrule are assembled and the pipe is run to a convenient position for an outside stopcock. This section of pipework is known as the communication pipe. When the connection is completed, the screwdown ferrule is simply opened using a special spanner.

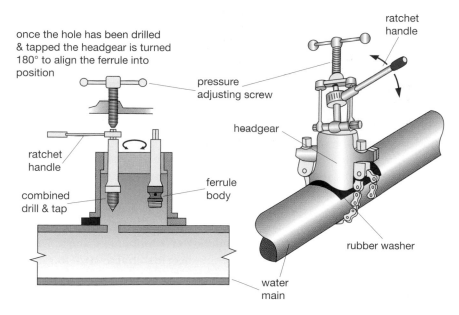

Under pressure tapping machine

U

UNDER PRESSURE SHUT OFF (UPSO–OPSO)   See OVER PRESSURE SHUT OFF.

UNDERCLOAK

In sheet weathering material, the undercloak is the sheet which lies below the next sheet (the overcloak) ensuring a watertight joint, for example, see LEAD SHEET AND ROOF COVERINGS.

UNDERCUT ROLL

A special design of batten roll used on copper and aluminium roofs to allow for large amounts of expansion to the sheet metal. For example, see COPPER SHEET AND ROOF COVERINGS.

UNDERCUTTING

The melting away of the parent metal at the edge of a weld, it is generally caused by welding too fast along the joint without sufficient filler rod being applied, see WELDING DEFECTS.

UNDERFLOOR HEATING   See RADIANT HEATING.

UNDERGROUND DRAINAGE   See BELOW GROUND DRAINAGE.

UNDERHAND JOINT   See WIPED SOLDER JOINT.

UNDERLAY

A material which is used underneath METALLIC ROOF COVERINGS, An underlay is used for several reasons such as:

(1) To prevent condensation from forming under the sheet, which will lead to the ELECTROLYTIC CORROSION of the metal fixings used and the possible rotting of the timber decking. Conversely it also prevents the decking touching the sheet material, which may cause corrosion to the sheet metal itself.
(2) To allow the sheet metal to move freely over the surface of the decking it also hides or smooths out uneven surfaces.
(3) To act as an insulator against heat loss in winter and too much solar heat in the summer.
(4) To act as a sound insulator against rain and wind noises.

The underlay used for metallic roof work is either an INODOROUS FELT and is about 3 mm thick, or a non-woven needle punched polyester geotextile material. When laying, the joints should be butted together. Sometimes as a cheap alternative building paper is used in which the joints should be lapped. Building paper is not as thick as inodorous felt therefore the insulation qualities are greatly reduced and on uneven surfaces ridges appear in the completed work. On no account must ordinary roofing felt, or sarking felt be used as an underlay as it contains bitumen, which will soften due to the heat in summer and cause the sheet material to stick to the surface and thus prevent its free THERMAL MOVEMENT.

UNDERS   See SOAKERS.

UNION CONNECTOR

A pipe fitting consisting of three parts, two of which connect onto the pipe ends, the third piece being a coupling nut (union nut) used to join the two together making a watertight joint. A union connector would be used should a DISCONNECTING JOINT be required, for example, to assist the removal of pipework. Also shown in the illustration is a cap and lining and a plumbers union, these are types of unions used to join to lead pipe. There are many variations of unions for different materials and situations and all are characterised by the large 'union' nut.

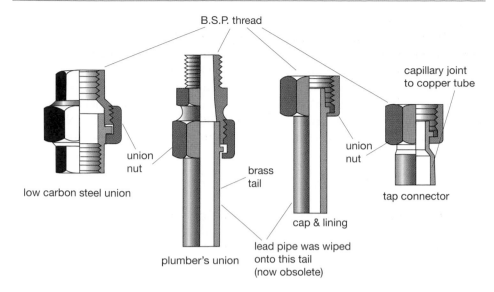

B.S.P. thread

capillary joint
to copper tube

union
nut

brass
tail

union
nut

low carbon steel union

tap connector

cap & lining

lead pipe was wiped
onto this tail
(now obsolete)

plumber's union

Union connectors

UNIT HEATER
A small water or air heat exchanger mounted at high level, on walls or suspended from the ceiling. These appliances are usually found in factories and similar buildings. A fan in the heater blows the warm air down into the space below.

UNIVERSAL SNIPS
A pair of heavy duty cutters, see TIN SNIPS.

UNPLASTICISED POLYVINYL CHLORIDE (UPVC)
UPVC is the same material as PVC except that nothing has been added to the material to make it more resistant to breaking, and, as a result, it can be easily damaged by a heavy blow. UPVC is often mistakenly called just PVC; in fact most of the plastics used for rainwater, soil discharge and drainage pipework are of this unplasticised type, the 'U' simply being dropped. The fixing procedure for this material is the same as for POLYVINYL CHLORIDE.

UNTRAPPED GULLY   See TRAPLESS GULLY.

UNVENTED CIRCUIT
A circuit in a hot water supply or heating system in which no VENT PIPE is connected to the system, the expansion of water is taken up in a SEALED EXPANSION VESSEL. See UNVENTED DOMESTIC HOT WATER SUPPLY and SEALED SYSTEM. See also CIRCUIT.

UNVENTED DOMESTIC HOT WATER SUPPLY (PRESSURISED SYSTEM)
A system of hot water supply where the cold feed is taken directly from the MAINS. The entry described here is applicable to quantities of stored hot water over 15 litres; but see also INSTANTANEOUS WATER HEATERS and LOCALISED HOT WATER SUPPLY for cold mains fed water heaters. Unvented 'bulk storage' systems are relatively new to this country. The system must be purchased as a unit or package and

U

should be installed by firms who are on a register of approved installers, these firms in turn must employ qualified operatives in unvented systems. With the water being taken directly from the mains there is no open vent pipe or storage cistern therefore the expansion of the water, when heated, is taken up in a SEALED EXPANSION VESSEL, usually located on the cold feed pipe to the hot storage vessel or within an air pocket inside the cylinder itself. Should the expansion vessel not function for any reason the water on expanding would be forced out of the pressure relief valve. The design of an unvented supply should be such that adequate precautions are taken to ensure that the stored water does not at any time exceed 100°C in temperature. This is best achieved by the installation of three safety devices these being:

(1) A thermostat to switch off at 60–65°C.
(2) A high temperature non resetting thermal cut out device to shut down at 90°C.
(3) A TEMPERATURE RELIEF VALVE located on the top of the hot storage vessel to open at 95°C.

It is essential to ensure that any hot water discharging from the temperature and pressure relief valves is conveyed to a suitable position where it has no danger of causing harm to anyone in or around the building. The discharge pipe from either of these valves should be the same size pipe as the discharge outlet and be a maximum of 9 m long. The discharge should be via a TUNDISH thus ensuring an adequate AIR BREAK.

When making connections to unvented hot water systems adequate precautions should be taken to avoid BACK-SIPHONAGE. Notice in the systems shown that a PRESSURE REDUCING-VALVE has been fitted on the supply pipe, this is designed to reduce the pressure into the system ensuring it is not too high; for the domestic house the operating pressure would be about 3 to 3.5 bar. To ensure equal pressures at both hot and cold draw-off points it is advisable to take the cold supply to the various sanitary appliances after the pressure reducing valve, failure to do so could result in a higher cold pressure at the taps, alternatively a second pressure reducing valve could be fitted to the cold supply pipe as necessary. Should the hot storage vessel require draining down, once water is flowing from the drain off cock it is essential to open the draw-off taps or the temperature relief valve test lever, this allows air into the hot storage vessel to replace the water and prevents a partial vacuum being formed. In one of the systems shown notice the drain off cock has been fitted to the top of the cylinder, water is removed from a vessel designed in this fashion by siphoning the water out, this is achieved by connecting a hose pipe to the drain off cock and allowing the mains pressure to flow through for a short period, the supply is turned off and the cylinder will empty automatically by SIPHONIC ACTION, again a valve must be opened to let in air. The obvious advantages of this system are the higher pressures obtainable at the draw-off points, the need for less pipework and the time required for installation. The disadvantages are not so obvious. Firstly, such a system can only be installed should the flow rate (litres per second) be sufficient to supply both hot and cold taps at once, bearing in mind several appliances could be running at the same time. Secondly, in HARD WATER districts the build up of SCALE around temperature and pressure relief valves could make these devices ineffective therefore regular checks of the system's safety devices are essential. See also COMBINATION BOILER, VENTED DOMESTIC HOT WATER SUPPLY and CENTRALISED HOT WATER SUPPLY.

1 stop valve
2 drain off cock
3 sealed expansion vessel
4 mains supply pipe
5 checkvalve
6 pressure gauge
7 pressure reducing valve
8 pressure relief valve
9 temperature relief valve
10 motorised valve
11 filter
12 by-pass with lockshield valve
13 circulating pump
14 tundish & discharge pipe
15 high temperature cut-out device

16 hot distribution pipe
17 cold distribution pipe
18 flow to radiators
19 return from radiators
20 temporary fill connection
21 primary flow
22 primary return
23 cold feed
24 vent pipe
25 automatic air release
26 floating baffle plate & air pocket
27 immersion heater

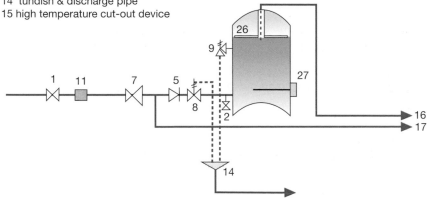

unvented domestic hot water supply
incorporating an air pocket & using
an immersion heater

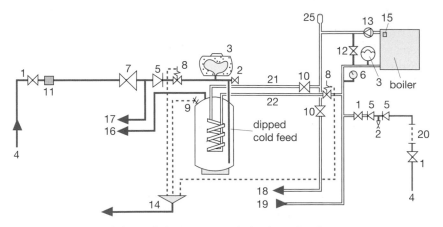

unvented domestic hot water supply & primary/heating system

Unvented domestic hot water supply

UNVENTED HOT WATER HEATING SYSTEM   See SEALED SYSTEM OF HEATING
and HOT WATER HEATING SYSTEM.

UNWHOLESOME WATER
Water that is not fit for drinking.

UPPER FLAMMABILITY LIMITS   See FLAMMABILITY LIMITS.

UPRIGHT WELDING   See VERTICAL WELDING.

UPSO–OPSO   See OVER PRESSURE SHUT OFF DEVICE.

UPSTREAM
A term sometimes used to indicate that part of a water or gas pipeline that is further along from a specific point and closer towards the source or direction to which the fluid has come. Conversely, DOWNSTREAM would be further along the pipe away from the incoming supply.

UPVC
Abbreviation for UNPLASTICISED POLYVINYL CHLORIDE.

URINAL
Sanitary appliance made from stainless steel, ceramic ware and fibreglass. The urinals shown are a cross-section of the various designs. See BOWL URINAL, SLAB URINAL, TROUGH URINAL and STALL URINAL.

(See illustration opposite)

URINAL CONNECTOR (URINAL EXTENSION PIECE)
A pipe fitting consisting of a short pipe with a female thread at one end. The plain spigot end is connected into the inlet of a TRAP and the threaded outlet from a urinal waste fitting can be connected into the female-thread making a watertight seal, see sketch.

URINAL SAMPLING COLLECTOR
A soil appliance found in hospitals and used to collect a small sample of body waste to facilitate its examination, i.e. evidence of blood, tapeworm, etc.; for example, see HOSPITAL APPLIANCES. See also WASHOUT WC PAN, which is basically the same appliance used solely as a water closet.

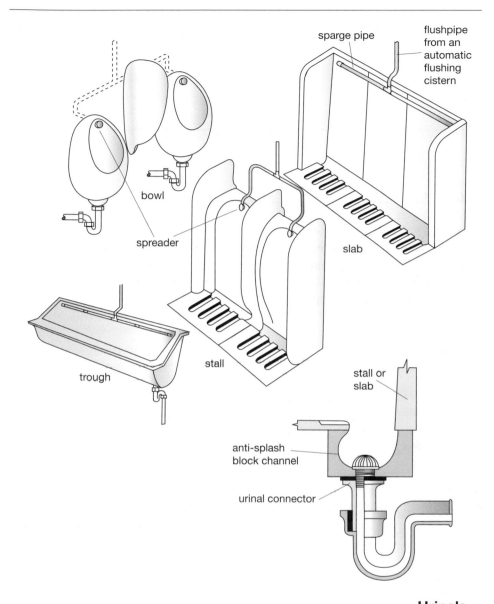

sparge pipe

flushpipe from an automatic flushing cistern

bowl

spreader

slab

trough

stall

stall or slab

anti-splash block channel

urinal connector

Urinals

U

V

VACUUM

A vacuum may be defined as a space devoid of any matter. If no matter is present, then there is also no energy or pressure within the space. Should the gas be expelled from a container a partial vacuum would be formed; it is practically impossible to achieve a complete vacuum as there would always be some trace of a rare gas remaining. Once a partial vacuum has been formed the container used must be able to withstand the pressures imposed on its surfaces due to atmospheric pressure otherwise it will be crushed. See COLLAPSE OF HOT STORAGE CYLINDERS and SIPHONIC ACTION.

VACUUM BREAKER   See ANTI-VACUUM VALVE.

VALLEY GUTTER (CENTRE GUTTER)

A sheet weathering formed between two sloping roof surfaces. A valley gutter may be of an open type and be wide enough for someone to walk through or be less open or closed, this is the secret type (see SECRET GUTTER). The valley gutter formed may take the role of a TAPERING VALLEY if the run-off is at one end. See also GUTTER.

VALVE

A device to regulate, open or close the flow of liquid or gas in a pipeline. The valve could be fitted at outlet points or any intermediate position in the pipe. See also, SCREWDOWN VALVE, FULLWAY GATEVALVE, BALLVALVE and PLUG COCK for examples.

VALVE PLATE   See JUMPER.

VALVED CLOSET

A very early design of WATER CLOSET first invented in the late eighteenth century in which the contents of the pan first fell upon a disc (the valve). When the flush handle was lifted the valve turned allowing the contents to fall to the trap below. A flow of water was then allowed to flush the pan as necessary. When the handle was released it fell, closing the valve to hide the unsightly, smelly pipe. Valved closets were made for some twenty years between 1870 and 1890 when they were superseded by the WASHOUT and WASHDOWN WC PANS, these being cheaper to manufacture.

(See illustration opposite)

VALVED FLUSHING CISTERN

A design of flushing cistern in which a valve opens to allow the permitted volume of water to pass into the flush pipe. This may be either a reduced flush for the purpose

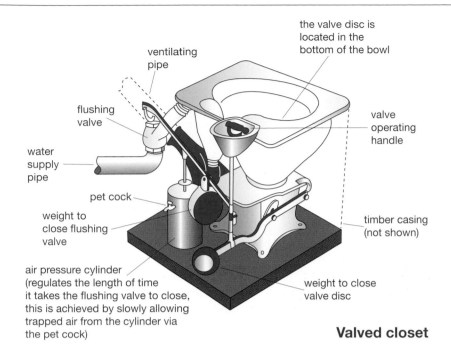

the valve disc is
located in the
bottom of the bowl

ventilating
pipe

flushing
valve

valve
operating
handle

water
supply
pipe

pet cock

weight to
close flushing
valve

timber casing
(not shown)

air pressure cylinder
(regulates the length of time
it takes the flushing valve to close,
this is achieved by slowly allowing
trapped air from the cylinder via
the pet cock)

weight to close
valve disc

Valved closet

of just removing urine from the WC or a full 6-litre flush may be employed where there are solids to be removed. The cistern shown operates should the user press one of the two buttons housed within the cistern lid. One button has a shorter attached rod than the other. When the larger button, with the longer rod is depressed, it lifts the internal mechanism of the valve sufficiently to engage it into a latch and it is then held up by the use of a small float. Water flows from the cistern and the latch only releases as the water level drops taking with it the float. Should the smaller button be pressed, the smaller rod does not lift the valve sufficiently to engage the latch and the valve is as a consequence only raised for a short period whilst the button is held down. The way in which the actual valve lifts is by the use of a linkage cable that when pushed operates a lever to lift the valve from its seating.

Looking at the valve illustrated it should be noted that the overflow passes through a central core, within the valve from the cistern directly into the WC pan.

(*See illustration over.*)

VANITY BASIN (COUNTER TOP BASIN)
A wash basin which has been fixed into a cabinet top or vanity unit. See WASH BASIN.

VAPORISING BURNER
A type of burner used on oil fired boilers. The burner works on the principle of warming the oil to cause it to change into its gaseous state where it can readily mix with the oxygen in the air which is required to support combustion, in this form it can be easily ignited. There are three designs of vaporising burner these being:

(1) The natural draught vaporising burner (natural draught pot burner).
(2) The forced draught vaporising burner (fan-assisted pot burner).
(3) The wallflame or rotary vaporising burner.

V

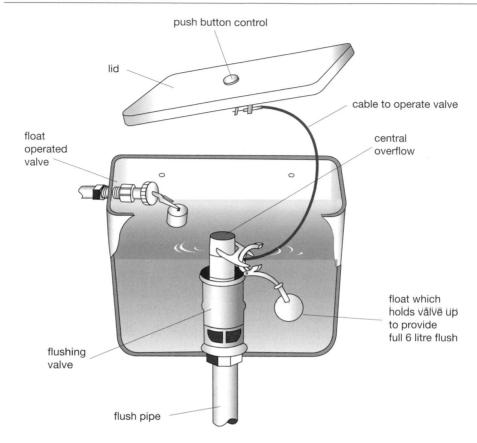

Valved flushing cistern

It should be noted that although the rotary type is classified as a vaporising burner the oil is in fact partly atomised as it leaves the oil outlet point. To ensure that not too much oil is fed to the burner at any time and at any pressure a CONSTANT OIL LEVEL CONTROL is fitted at the bottom of the boiler. This contains a float which allows a small quantity of oil into its reservoir.

Described below is the natural and forced draught vaporising burner, but see also ROTARY VAPORISING BURNER for its explanation. The natural draught 'pot' burner is the simplest type of oil burner and consists of a circular container (the pot) with a series of holes through which air can pass to the COMBUSTION CHAMBER, it works as follows:

(1) When the boiler requires heat, the igniter is activated and glows red hot, then a SOLENOID VALVE opens which allows oil to trickle into the burner from a constant oil level control. Note some burners are ignited manually with a lighted match or taper.
(2) The oil spreads as a thin film over the base of the burner and the vapour from the oil is ignited (it is the vapour which burns, not the liquid fuel).
(3) As the fuel burns, it generates heat which speeds up the vaporising process, thus the flame burns more fiercely and rises up the chamber to obtain sufficient air for combustion, which is admitted through the holes in the sides of the pot.

(4) Eventually the flame burns out at the top of the pot where the waterways of the HEAT EXCHANGER are located.

(5) When the boiler thermostat is satisfied the fuel is cut off to the burner. Note the boiler will continue to burn due to the oil and vapour in the pot, therefore it is sometimes necessary to incorporate some form of 'HEAT LEAK' into the system to prevent overheating.

It is most important that no attempt should be made to relight a warm or hot burner as the ignition of the fuel vapour could blow back with explosive force. The forced draught vaporising burner is identical to the natural draught type except that the pot is enclosed inside a casing into which the air required for combustion is blown as necessary. See also ATOMISING BURNERS and OIL STORAGE AND SUPPLY.

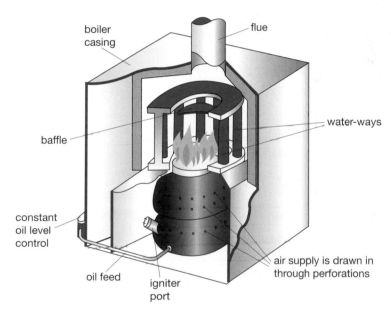

Natural draught vaporising burner

VAPOUR

The gaseous state of a substance, especially if the substance is a solid or liquid at normal room temperatures.

VAPOUR BARRIER (MOISTURE BARRIER)

An IMPERVIOUS membrane which covers various structures and insulating materials to prevent moisture resulting from condensation within a building being absorbed into them. Should these materials become damp it would result in their loss of insulation properties or even their destruction due to rotting.

VAPOUR EXPANSION THERMOSTAT   See THERMOSTAT.

VAPOUR PRESSURE FLAME FAILURE DEVICE   See LIQUID VAPOUR FLAME FAILURE DEVICE.

VELOCITY
The speed at which water or gas is passing through a pipe.

VELOCITY HEAD
The pressure required to set a liquid in motion.

VENT HORN
A socketed projection formed on the outlet of some WC pans and slop hoppers. The vent horn is designed so that a TRAP VENTILATING PIPE can be connected at this point, if necessary, ensuring it connects to the appliance within 300 mm of the trap. See BRANCH VENTILATING PIPE.

VENT HORN STOPPER
A stopper which can be fitted to a VENT HORN, to plug the hole, should it not be required. The vent horn stopper should be fitted ensuring a watertight seal is made.

VENT PIPE (EXHAUST OR EXPANSION PIPE)
An open ended pipe in a hot water system to allow air in and out of the pipework, therefore ensuring the internal surfaces within the system are at the same pressure as that created by the atmosphere. The vent pipe also allows for the safe discharge of any hot water or steam which might be generated due to a fault in the system. The height to which a vent pipe is to rise above the water level in the cold cistern is found by allowing 40 mm for every 1 m head of water in the system, plus an additional 150 mm. For example, if the distance between the base of a boiler and the water level in a cistern is 5 metres the vent pipe should be carried up a minimum distance above the water level of $5 \times 40 + 150 = 350$ mm. See also VENTED HOT WATER SUPPLY.

VENT PIPE (VENT STACK) See VENTILATING PIPE.

VENT SHAFT (SEWER VENT)
A VENTILATING PIPE which is connected to a sewerage system to ensure good ventilation through the pipework. The vent shaft is often recognised as a pipe which stands some 6 m high and is found along the side of the pavement or roadway, its top end is sometimes distinguished as a zig-zag cut, like the shape of a king's crown.

VENTED CIRCUIT
A term used to describe a circuit in a hot water supply or heating system in which the provision of a VENT PIPE ensures that the circuit is permanently open to the atmosphere. See VENTED DOMESTIC HOT WATER SUPPLY and HOT WATER HEATING SYSTEM. See also CIRCUIT.

VENTED DOMESTIC HOT WATER SUPPLY
A method of hot water supply where the cold water to be supplied for heating up is taken from a feed cistern. With this type of system the VENT PIPE and the COLD FEED are two means of ensuring that the system functions under the influence of atmospheric pressure. As the water is heated and expands it can rise up the vent pipe or cold feed pipe to return to the feed cistern. See also UNVENTED DOMESTIC HOT WATER SUPPLY and CENTRALISED HOT WATER SUPPLY.

VENTED HOT WATER HEATING SYSTEM   See HOT WATER HEATING SYSTEM.

VENTILATED STACK SYSTEM   See SECONDARY VENTILATED STACK SYSTEM.

VENTILATING PIPE
A pipe in a sanitary pipework system which helps to keep the inside pressure of the pipe the same as atmospheric pressure thus preventing TRAP SEAL LOSS. It also facilitates the circulation of air within the system preventing bacteriological growth. The ventilating pipe is connected to a discharge pipe at one end, the other end should be open to the atmosphere at its highest point. No water should enter a ventilating pipe, the only exception to this is in the case of WET VENTING. The termination of a ventilating pipe into the atmosphere should be at a position so as not to cause a nuisance or health hazard. It is recommended that if a ventilating pipe is within 3 m of a window opening it should be carried up above the window to a minimum height of 0.9 m. Ventilating pipes should be positioned away from parapets and the corners of buildings where wind pressures are likely and they should be fitted with a balloon to prevent any undue restrictions (such as a bird's nest) to the flow of air through the pipe. See BRANCH VENTILATING PIPE and SANITARY PIPEWORK. See also STUB STACK.

VENTILATION
A term used extensively to refer to the supply of fresh air to, or the removal of air from, a room or compartment. Ventilation is provided for many purposes, including for combustion, cooling, the removal of stale air or smells, the removal of moisture, etc. The VENTILATION GRILLE may be of a type supplied with or without a shutter. See also TRICKLE VENTILATION.

VENTILATION GRILLE
A grille designed to supply air into the building for various reasons, one being to supply air for the combustion of fuel with an OPEN FLUED APPLIANCE. The size of an air grille will be dependent upon the size of an appliance and the fuel being supplied. Where *gas* appliances are being used the supply needs to be 5 cm^2/kW in excess of the first 7 kW net. Where *oil* burning appliances are being supplied this amount would need to be 5.5 cm^2/kW in excess of the first 5 kW. For solid fuel burning appliances and more information on vent sizing see Part J of the Building Regulations.

Example
The free air requirement for a 22 kW net input open flued appliance burning natural gas would be $(17 - 7) \times 5 = 50$ cm^2.

Note One only considers the effective free air requirement and not the actual ventilation grille size. For example, the terracotta grille shown opposite has a free area of $H \times W \times No.$ *of holes* $= 16$ cm^2 (i.e. $8 \times 8 \times 25 = 1600$ mm^2).

The size of the hole should be no smaller than to allow a 5 mm diameter ball to pass through, but equally shall not permit a 10 mm diameter ball to pass. Also note how it is channelled across the cavity. See also COMPARTMENT VENTILATION.

(See illustration over.)

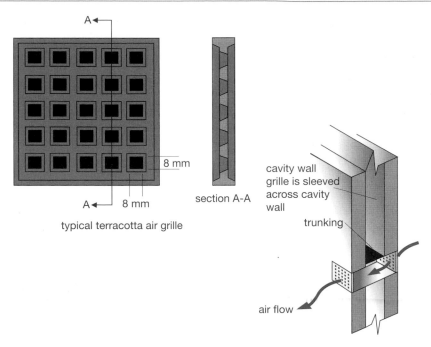

A←—

8 mm

A←— 8 mm

typical terracotta air grille

section A-A

cavity wall
grille is sleeved
across cavity
wall

trunking

air flow

Ventilation grille

VENTILATION OF DRAINS AND DISCHARGE PIPES

It is essential that air is allowed to freely enter discharge and drainage pipes to help maintain an equilibrium of pressure within the pipe and the outside atmosphere; for example should the pressure be greater outside the pipe than inside, the TRAP SEALS of sanitary appliances and gullies would be lost. By allowing a current of fresh air to flow through the whole system any foul matter adhering to the insides of pipes should soon dry and be washed away; ventilating the pipes also prevents any build up of foul and possibly dangerous gases. To ensure good ventilation is maintained, at low level air is permitted to enter the drain via holes in inspection chamber covers, situated in the road or via sewer VENT SHAFTS. Should an INTERCEPTING TRAP be used the air enters the drain via FRESH AIR INLETS located by the sewer connection. The air flows through the drainage pipe and up the discharge stack due to CONVECTION CURRENTS and exits via a VENTILATING PIPE.

VENTURI TUBE

A device which creates a pressure difference in the pipeline through which water flows. Water flowing through a tube is made to pass through a reduced diameter and once through it reverts back to its original bore. This causes the water passing through the restricted hole size to increase in VELOCITY; this increased velocity occurs at the point where the tube is reduced in diameter, however it can only take place at the expense of some pressure energy, therefore there is a consequent loss in pressure at this point. The pressure difference caused when water is flowing can be put to use to operate various devices. A typical example of a Venturi would be found in an INSTANTANEOUS GAS WATER HEATER, gas will only flow to the main burners when water is flowing through the water heater and its HEAT EXCHANGER and works as follows:

(1) When the tap is turned off no water is flowing through the venturi, therefore the water pressure on both sides of the diaphragm are equal thus the gas valve remains closed, see sketch.

(2) Should the tap be opened, water flows through the Venturi and the pressure becomes reduced on the gas valve side of the diaphragm, therefore the greater pressure on the other side pushes on the diaphragm and consequently the gas valve, causing it to open.

(3) Gas flows to the main burner and is ignited by the pilot light. Should the tap be closed, the pressure becomes equalised on both sides of the diaphragm and the gas valve closes.

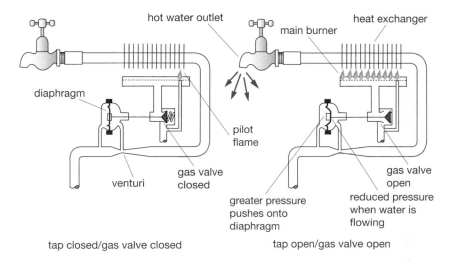

Venturi tube

VERDIGRIS
The green coating which forms on copper roofs, and pipes. Verdigris forms on copper due to a reaction brought about by bringing the copper into contact with an acidic material (such as flux). Unlike PATINA the green OXIDE flakes away and the corrosive reaction continues, slowly destroying the metal.

VERGE
The edge or end of a pitched roof which overhangs a GABLE end.

VERTEX FLUE
A design of gas flue system that takes the air used for combustion from the roof void. The term 'Vertex flue' refers to the manufacturer's name. Another manufacturer uses the name 'Sol-ver flue' for their system. Note that with these systems the room in which the appliance is situated should be free from the effects of cold draughts caused by VENTILATION GRILLES and the appliance is in effect ROOM SEALED. The combustion air is drawn from the roof space, which will need to have an additional air supply provided, as recommended by the manufacturer.

(See illustration over.)

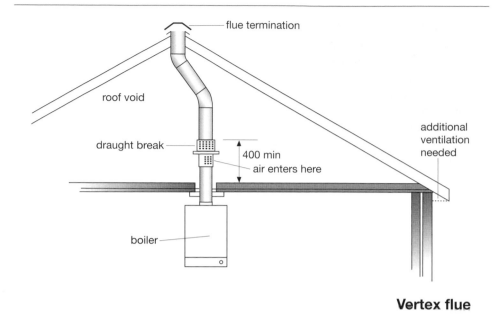

Vertex flue

VERTICAL

A line which is perpendicular to the earth, i.e. going up and down.

VERTICAL WELDING

For vertical lead welding, see LEAD WELDING. The welding process described here is suitable for aluminium, copper and low carbon steel. To successfully weld vertically the two adjoining edges are butted together leaving a suitable gap between them. First a small hole (the ONION) is made with the blowpipe and the molten metal will drop to form the first fusion of the weld. The hole is maintained as the weld proceeds and as the filler rod is introduced to the joint it is allowed to flow down to form a characteristic reinforcing bead. Only when the weld is completed is the hole finally filled with molten metal.

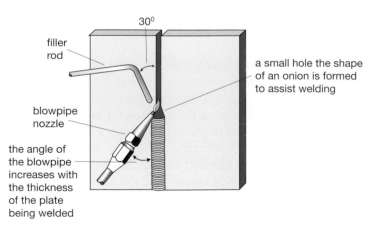

Vertical welding

VIKING JOHNSON FITTING

The trade name for a compression type coupling joint used on low carbon steel pipe. The Viking Johnson fitting incorporates a rubber COMPRESSION RING. See COMPRESSION JOINT.

VISCOSITY

The speed at which a liquid will flow. Water will flow from a container much more quickly than the same quantity of paint or oil. The viscosity of a liquid is indicated by measuring the period of time it takes to pour a standard amount of liquid from a container. For example, domestic fuel oil, used for ATOMISING BURNERS, takes 28–35 seconds to pass through a hole of a specified size (see OIL STORAGE AND SUPPLY), conversely a thick gloss paint may take several minutes.

VITIATED AIR

Air which lacks oxygen, see AIR.

VITIATION SENSING DEVICE   See ATMOSPHERE SENSING DEVICE.

VITREOUS CHINA

A special mixture of clays including china and ball clay, felspar and flint which is used to produce a clay which is completely IMPERVIOUS to moisture. Vitreous china is very light in weight and is used extensively in the manufacture of sanitary appliances. See also CERAMIC WARE.

VITREOUS ENAMEL

A glass-like coating applied to cast iron and pressed steel SANITARY APPLIANCES in order to give protection from corrosion and to provide an easy clean cosmetic finish.

VOLATILE FLUID

Any liquid which expands rapidly and changes to a vapour at a relatively low temperature. Volatile liquids such as mercury and alcohol are often used in thermostats to open and close the gas supply or switch on and off the power supply, see THERMOSTAT.

VOLT

A unit of electrical force which produces an electrical current.

$$\text{WATTS divided by AMPERES} = \text{volts}$$

VOLUME

The space that occupies the void inside an enclosure. To find the volume of a space, the cross-sectional area (e.g. length × breath) is multiplied by the height of the structure. See METRIC SYSTEM.

V

WALK THROUGH BATH   See FOOT BATH.

WALL HUNG WC PAN   (CORBEL WC PAN)
A WC pan suspended clear of the floor, used in situations where hygiene is particularly important. It is supported on a concealed bracket called a chair.

WALL-MOUNTED BOILER
An independent boiler which is hung on the wall, i.e. it is not standing on the floor.

WALLFLAME BOILER   See ROTARY VAPORISING BURNER.

WALLPLATE
A piece of timber laid onto a wall in order to spread the load of the floor, ceiling or roof joist that terminate and rest on the wall structure.

WARM AIR CENTRAL HEATING SYSTEM
A form of CENTRAL HEATING in which warm air is circulated to all the rooms to be heated. There are two basic systems, natural circulation systems and forced circulation systems.

   With the natural circulation system (which is hardly used in this country) the warm air rises up through the building via its system of ductwork by natural CONVECTION CURRENTS. This system is very quiet in operation but can be easily affected by cold draughts which may enter the room from outside. With this system the boiler used to heat the air must be situated in a central position below the rooms to be heated, such as in a basement.

   With the forced circulation system warm air is blown through the system of ductwork assisted by a fan. The air in this system is either heated directly in a special boiler, in which the air circulates around the combustion chamber, or is heated indirectly, in which the air passes over a water filled heat exchanger or battery. The direct system tends to be faster in its heating up period due to the fact that the air is heated directly and that no heat has been lost from the flow and return pipe to the warm air heat exchanger. However, with this system, since a flue is required, it is not always possible to site the boiler in a desired central position, whereas with the indirect system the boiler can be sited away from the warm air heat exchanger, see sketch. The indirect system is generally more expensive to install but has the advantage of also heating water for domestic purposes, the boiler used being the same as that in a water filled system of central heating, although special boilers can be purchased for the direct system which will allow for water to circulate through

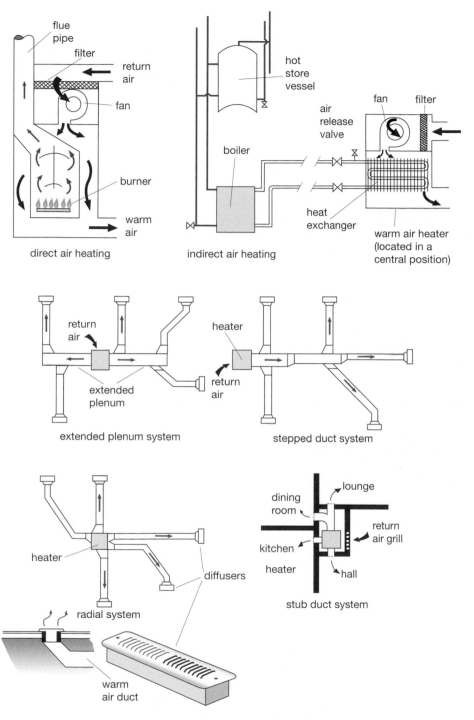

flue pipe

filter

return air

fan

burner

warm air

direct air heating

hot store vessel

air release valve

fan   filter

boiler

heat exchanger

warm air heater (located in a central position)

indirect air heating

return air

extended plenum

extended plenum system

heater

return air

stepped duct system

heater

diffusers

radial system

warm air duct

dining room

lounge

kitchen

return air grill

heater

hall

stub duct system

Warm air central heating system

the boiler for domestic hot water purposes. Once the air has been heated it is passed through a system of ductwork as shown in the illustrations. The delivery air temperature at the room REGISTER should not exceed 60°C. The heater is situated in a central position if possible, this ensures the length to the individual registers/diffusers is reduced to a minimum. Notice where more than one diffuser is supplied from a section of duckwork, its size is increased to account for the required air flow rate. The diffusers should ideally be located in the floor around the perimeter of the room and below window openings for maximum efficiency, although wall or even ceiling positions are used to distribute the flow of warm air where space is resricted. One design of system, the 'stub duct' is designed to reduce the installation costs of ductwork and is sited in the centre of the building where only short ducts pass through the inside walls.

Air from the heated rooms should be returned to the heater for re-heating. If possible the return air grill should be positioned opposite the warm air inlet diffuser, for example should the warm air inlet be at low level on one side of the room the return air outlet should be located on the opposite wall at high level. No return air grills are positioned in bathrooms, kitchens or WC compartments due to the large amounts of condensation and possible odours which can be drawn into the system. The return air is either passed back to the heater by a system of ductwork, or the return air from the rooms is drawn back to the hall and eventually through a duct in the heater. Should the second method be chosen, allowing the return air to flow to the hall the air is simply allowed to flow through the grills in the internal walls, note some thought must be given to their siting in bedrooms, etc., for privacy. To ensure good comfortable room conditions, up to 25% fresh air is often mixed with the return air; the heater air inlet manifold being fitted with a DAMPER arrangement to regulate the proportions of fresh and return air. Often in large industrial type buildings only fresh air is used from outside the building and no return air used at all.

WARNING PIPE
An overflow pipe fixed in such a position that if water is flowing through it, it would readily be seen whether it is inside or outside a building.

WASH BASIN (LAVATORY BASIN)
A sanitary appliance used to assist washing the upper parts of the body. There are several types of wash basin, for example, the vanity type which is installed into, or forms the top of a vanity unit; the pedestal type which is supported off the floor by a column type pedestal, or the wash basin which sits on brackets designed for the purpose, the most common type being a towel rail bracket. The fixing height for a wash basin varies depending upon circumstances but in general they are fixed at a height of 760–800 mm to the front lip edge. For the waste connections to wash basins see WASTE FITTING.

(See illustration opposite)

WASH TUB
A deep, heavy, white glazed fireclay sink which is sometimes found for uses such as washing and rinsing clothes. Wash tubs are rarely found used in England and Wales but proved popular in Scotland. Since the introduction of automatic washing machines wash tubs are no longer installed in domestic dwellings.

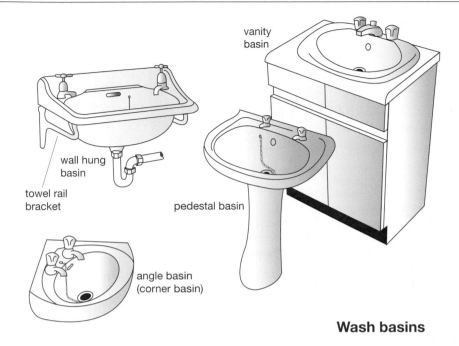

Wash basins

WASHDOWN WC PAN

A WC pan which removes the contents of the trap by the momentum of the flushing water.

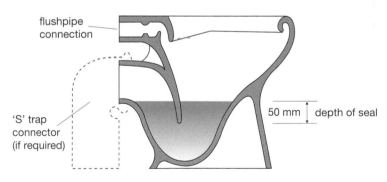

Washdown WC pan

WASHING FOUNTAIN (ABLUTION FOUNTAIN)

A circular shaped hand washing appliance used in industrial premises to allow a maximum number of people to wash their hands at the same time. The water is supplied to the appliance through a central pedestal upon which the circular wash basin sits, the waste is discharged via a similar route. To turn on the supply of water the foot pedal is depressed, this opens the THERMOSTATIC MIXER VALVE to allow water to discharge from the top of the appliance in the form of a spray. When the foot is removed from the pedal, the water is automatically shut off.

(See illustration over.)

W

soap dish

spray of
water

access
panel

foot pedal

Washing fountain

WASHING MACHINE CONNECTIONS

Two connections have to be made to washing machines, the supply and the waste. The machine may be designed for hot and cold supply or cold water only. The cold water only machines tend to take longer to complete their washing cycle because the machine has to heat up the water. The supply connections to the washing machine must comply with the WATER SUPPLY REGULATIONS and have a means of BACKFLOW prevention. It will be found that most modern washing machines incorporate an AIR GAP or PIPE INTERRUPTER in their design. The waste connection is usually via a 20 mm rubber hose located at the back of the machine, this must be connected to a discharge pipe via a TRAP. There are several methods which can be adopted to make this connection but the two most common tend to be, to connect the 20 mm rubber hose directly into a special sink trap with a washing machine branch connection or, if the machine is too far from the sink for the hose to reach, a 40 mm P trap is connected to the waste discharge system and an upstand is left into which the hose

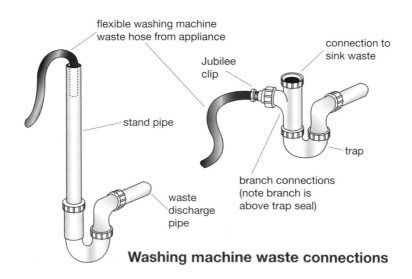

flexible washing machine
waste hose from appliance

connection to
sink waste

Jubilee
clip

stand pipe

trap

waste
discharge
pipe

branch connections
(note branch is
above trap seal)

Washing machine waste connections

is inserted. The rubber hose is left to fit loosely into the 40 mm upstand pipe, this ensures the prevention of siphonage of water from the pump or waste water back into the machine.

WASHING MACHINE TAP
The chromium plated quarter turn BALLVALVES which are often fitted to the hot and cold supplies to a washing machine or similar appliance. These valves are fitted with a plastic head coloured blue (cold) or red (hot) to assist identification of the supply.

WASHING MACHINE TRAP
A 40 mm trap which can be fitted to a sink and has a branch connection for a rubber hose from a washing machine. See WASHING MACHINE CONNECTIONS.

WASHING TROUGH (ABLUTION TROUGH)
A stainless steel or ceramic oblong, trough-shaped wash basin used in industrial premises for the purpose of allowing several people to wash their hands at once. These troughs are often fixed back to back in the centre of a room therefore saving wall space and in this form are sometimes referred to as an island arrangement. Spray mixing taps are usually supplied to washing troughs to conserve water. See also WASHING FOUNTAIN.

WASHOUT PIPE
A pipe which is sometimes connected to the base of large cold storage cisterns of over 2300 litres, for the purpose of draining down and clearing out of any SLUDGE, etc. Cisterns below this capacity are usually emptied by means of the cold DISTRIBU-TION PIPE, the remaining water and sludge is scooped and mopped out. The wash-out pipe is not essential but if fitted should be a minimum diameter of 38 mm and discharged at a convenient position, the sludge and discoloured water often leave stains therefore it is unwise to terminate the washout pipe onto a roof covering which is on show. For example of a washout pipe see COUPLING OF STORAGE CISTERNS.

WASHOUT WC PAN
An early design of WC pan, no longer made for use in this country. The washout pan differs from the modern WASHDOWN WC PAN in that the contents are deposited into a shallow water-filled bowl which is washed out as the pan is flushed, see sketch. See also URINAL SAMPLING COLLECTOR.

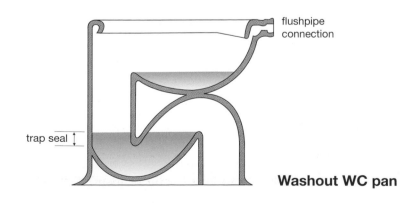

Washout WC pan

WASTE APPLIANCES (ABLUTIONARY APPLIANCES)

Waste appliances are those required for washing purposes such as wash basins, baths, bidets, showers and sinks. See also SOIL APPLIANCES.

WASTE DISPOSAL UNIT

A mechanical device operated and fixed beneath sinks to macerate kitchen refuse into small fragments so that they can be flushed into the drainage system without causing blockages. A waste disposal unit cannot be fitted into a sink with the standard waste hole size as it would not be large enough to house the unit, therefore to fit these devices a special sink is often required. When installing a waste disposal unit a tubular trap must be fitted to its outlet because the units do not have their own integral trap fitted. When using these units it must be remembered that water must be flowing down the waste pipe, otherwise the machine and pipe would soon become blocked up. Should the machine become jammed a special tool is provided to turn the cutting head to loosen the obstruction and on many units there is a cut off switch located on the machine, this is provided to prevent the motor from becoming burnt out should it be jammed. Upon freeing the cutting head the cut off switch must then be pressed to reset it.

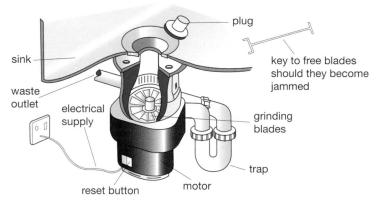

Waste disposal unit

WASTE FITTING

A fitting used to make a watertight seal between an appliance and the discharge pipe. Appliances such as baths and sinks often use a combination waste and overflow fitting, this does away with the need for a separate overflow pipe. The connections shown are for uses in various types of WASTE APPLIANCES.

(See illustration opposite)

WASTE PIPE

A pipe for carrying dirty water from a waste appliance. See also DISCHARGE PIPE.

WASTE PLUG

A rubber or plastic bung which is inserted into the outlet of a waste fitting to make the appliance watertight at its base, thus allowing the container to hold a quantity

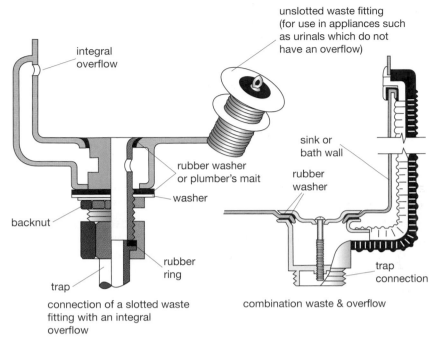

integral
overflow

unslotted waste fitting
(for use in appliances such
as urinals which do not
have an overflow)

rubber washer
or plumber's mait

sink or
bath wall

rubber
washer

washer

backnut

rubber
ring

trap

trap
connection

connection of a slotted waste
fitting with an integral
overflow

combination waste & overflow

Waste fittings

of water. It is a requirement that every waste appliance be provided with a readily accessible plug or some other device capable of closing the outlet; the only exceptions to this are for shower trays or where the water delivered to the appliance does not exceed 3.6 litres a minute. See also ANTI-THEFT PLUG.

WASTE STACK
A discharge stack which receives the water from waste appliances only. The minimum diameter for a waste stack should not be less than 50 mm. See TWO PIPE SYSTEM OF ABOVE GROUND DRAINAGE.

WASTE WATER
The water discharged from waste appliances such as baths, wash basins and sinks, etc.; it is not contaminated by SOIL WATER or TRADE EFFLUENT.

WATER
Pure water is a transparent, tasteless liquid. Water is a chemical compound composed of 2 parts of hydrogen and 1 part of oxygen (H_2O) and exists as a vapour, liquid or solid depending upon the pressure and temperature. It is formed when hydrogen is burned, oxygen being required to support combustion. To revert water back to its original form of gases a current of electricity is passed through it. Water is essential to all forms of life and is generally classified as being either hard or soft, this classification depends on the parts of calcium it contains per million parts of water. See also pH VALUE.

Classification of water

Parts calcium per million parts of water	Classified	
0–50	soft	acidic
50–100	moderately soft	↑
100–150	slightly hard	
150–200	moderately hard	↕
200–300	hard	↓
over 300	very hard	alkaline

WATER AUTHORITY (WATER UNDERTAKER)

The body responsible for providing, by law, a supply of POTABLE WATER. To ensure that the water once treated does not become polluted in any way they are required to ensure that all work installed complies with the WATER SUPPLY REGULATIONS.

WATER CLOSET (WC)

The original definition of the term water closet meant 'a small room' which housed a flushing toilet pan. The term today is used to identify the WC SUITE itself, which consists of a toilet pan and a connected flushing cistern.

The water closet was first invented in the sixteenth century by Sir John Harrington, a member of Queen Elizabeth's court, but because a drainage system was required little was heard of them until 1775 when Alexander Cumming took out a patent for a valved closet. In 1870 T.W. Twyford produced the first WASHOUT closets and later the WASHDOWN closets which are still used today. See also LONG AND SHORT HOPPER WC PANS.

WATER CONDITIONERS

Devices which are used to overcome the problems of SCALE in the pipework. Water conditioners do not soften water they just stabilise the calcium salts which are held in suspension. There are two basic types, those that use chemicals and those that pass water through an electronic or magnetic field. The calcium salts, if viewed through a microscope, would look very much star shaped and it is in this form that they can bind together. The chemical water conditioners use polyphosphonate or metaphosphate crystals, these dissolve into the cold water and form on and in the star shaped salts turning them circular in shape. These polyphosphonate crystals are placed into the STORAGE CISTERN or into specially designed containers, fitted into the pipeline. Periodically the crystals must be replaced. The electronic and magnetic water conditioners are devices fitted in the pipeline which pass a low current of a few milliamps of electricity across the flow of water, this tends to alter the structure of the hard salts making them round or square shaped. As can be seen both methods employed make the hard salts round in shape. In this form they do not stick together but should pass through the system. Electronic and magnetic water conditioners should be installed as close as possible to the incoming main supply. Some types of electronic conditioners are plugged into the mains electricity supply whereas others rely on the current produced by electrolytic corrosion.

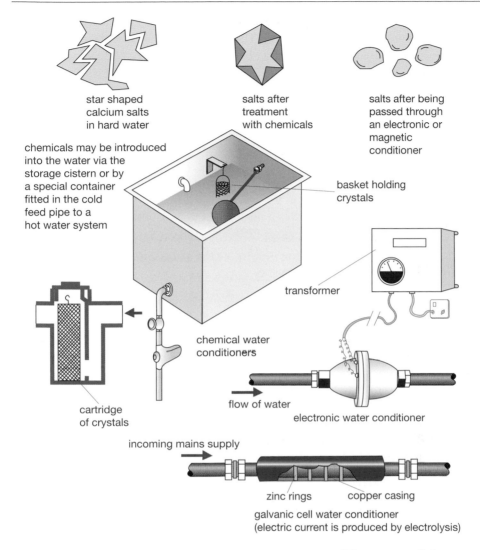

star shaped
calcium salts
in hard water

salts after
treatment
with chemicals

salts after being
passed through
an electronic or
magnetic
conditioner

chemicals may be introduced
into the water via the
storage cistern or by
a special container
fitted in the cold
feed pipe to a
hot water system

basket holding
crystals

transformer

chemical water
conditioners

cartridge
of crystals

flow of water

electronic water conditioner

incoming mains supply

zinc rings   copper casing

galvanic cell water conditioner
(electric current is produced by electrolysis)

Water conditioners

WATER CYCLE

Term used to mean the cycle of water, from the evaporation of the water from seas and rivers, etc. to the formation of clouds and its falling as rain back to the ground to replenish rivers and lakes. When water falls to the ground much of it is collected from streams and rivers by the local water authority and is stored in a reservoir until required for use as a COLD WATER SUPPLY. Water can also be taken from wells dug in the ground to collect water which is below the WATER TABLE. Wells are classified as either deep or shallow according to the water bearing strata from which the water is taken. A shallow well is one dug less deep than the first IMPERVIOUS strata and the water is taken from the subsoil and therefore should be regarded with suspicion. Deep wells are sunk to the water bearing ground deeper than the first impervious strata thus the water has had a chance to be cleaned due to the natural filtration that takes place. There is a third type of well, known as an ARTESIAN WELL, this is a bore hole which is drilled through the impervious strata and allows water out of the ground without the need for pumping. The water rises to the surface by HYDRO-STATIC pressure caused by the well outlet being lower than the source of water.

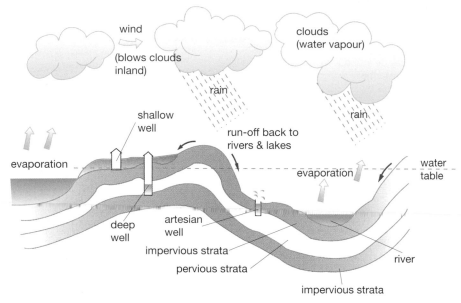

The water cycle

WATER GAS
Decomposed steam which consists of a mixture of hydrogen and carbon monoxide and is produced by passing steam over red hot coke. For one of its uses see TOWN GAS.

WATER GAUGE   See MANOMETER.

WATER GOVERNOR   See PRESSURE REDUCING VALVE.

WATER HAMMER (CONCUSSION/REVERBERATION)
A hammering noise which occurs in high pressure water pipes, the noise is caused by surges of pressure. There are two basic noise types. Firstly, a noise which consists of a sudden loud bang and is very often caused by a loose stopcock JUMPER which quickly flips shut onto the SEATING, or it could be caused by pipes which have not been fixed correctly and flap about, this noise is caused by any sudden back surges of pressure, which could be caused by the rapid closing of a tap. To overcome this problem the fixing of the pipes or any loose jumpers should solve any banging noise created. The second type of noise is often called oscillation or ballvalve murmur which consists of a series of bangs or rumbles generated in the pipeline. The noise is created by a BALLVALVE quickly opening or closing which is caused by ripples or waves which form on the surface of the water in a storage cistern. It is these ripples that cause the ballfloat to bounce up and down, opening and closing the ballvalve. To overcome this problem often a larger ballfloat is used or a damping plate fitted to

the float or ballvalve lever arm, alternatively BAFFLE vanes are fitted in the cistern to prevent waves forming, failing that an EQUILIBRIUM BALLVALVE could be fitted.

To overcome any noises transmitted in water pipes, a PRESSURE REDUCING VALVE could be fitted which should solve the worst problems, but a method often employed is to shut down slightly the incoming stopvalve, this does not reduce the pressure but does reduce the velocity of the water. In larger buildings to minimise the effects of water hammer HYDROPNEUMATIC ACCUMULATORS are sometimes fitted, these take up any surges in pressure that may take place. See also AIR VESSEL and NOISES IN COLD WATER PIPEWORK. For water hammer caused in steam mains, see STEAM HEATING.

WATER JACKETED TUBE HEATER   See THERMAL STORAGE SYSTEM.

WATER LEVEL
(1) A tool used for setting out levels across a room, etc. It consists of a rubber hose terminating at each end with two glass tubes. The hose is filled with water and once all the air bubbles have been expelled from the tube, the water level in the two glass tubes will be the same. The water level is very simple to use; first when the water is at rest a mark is made on each glass tube, identifying when the water in each tube is level with the other. One tube is now held or fixed with its mark on the glass at the position from where the level is to be taken, the other end of the tube is positioned across the room, etc. to where the level is to be transferred, this glass tube is then lifted or lowered as necessary and when the water level corresponds with the line on the glass the correct height has been located and a mark is struck on the wall.
(2) The top surface of water, when it has come to rest in a vessel, for example the water level, or height of water may be 200 mm, 250 mm or 475 mm, etc. above the base of the vessel.

WATER METER
A device which records on a counter or dial the flow of water which passes through it. Meters are used by all water authorities to measure the amount of water being used by the consumer. However, if no meter is supplied, a flat rate is charged for the water irrelevant of the amount of water consumed. The workings of the counter or dial are operated by the flowing water causing it to spin a vane or slide a piston up and down. All meters should have SERVICING VALVES fitted either side to allow for maintenance and, in certain buildings where the water must not be turned off, it will be necessary to design a meter bypass into the pipework.

WATER PRESSURE
The pressure in any plumbing system which can be created by two basic means. Either a pump could be connected to a pipeline, or the pressure could be caused by the weight of the water itself. The higher a column or HEAD of water, the more pressure would be exerted at its lowest point, therefore it is essential to install feed and storage cisterns as high as possible, giving a good pressure at the DRAW OFF POINTS. Water is made up of molecules and they all are influenced by the gravitational pull of the Earth and are pulled directly downwards. If it were possible to imagine the molecules pushing down on one another, and see them, it would be a simple task to

count the amount of molecules acting on a given point, see sketch. In reality this is not possible therefore this pressure has to be calculated. A container 1 m × 1 m × 1 m would contain 1000 litres of water and weigh 1000 kg. The force produced by a 1 kg mass would be 9.81 newtons (see PRESSURE) therefore the 1000 kg of water would exert 1000 × 9.81 = 9810 newtons per square metre (9.81 kN/m²). If a container 1 metre high exerts a pressure of 9.81 kN/m² a container 4 metres high would simply exert 4 times as much pressure.

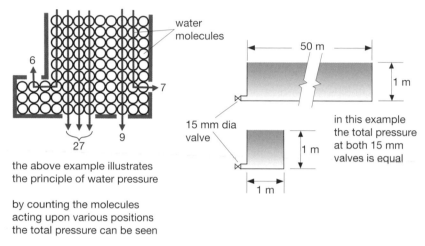

water molecules

6

7

27   9

the above example illustrates the principle of water pressure

by counting the molecules acting upon various positions the total pressure can be seen

50 m

1 m

15 mm dia valve

1 m

1 m

in this example the total pressure at both 15 mm valves is equal

Water pressure

Example
Find the pressure exerted at the base of a container 4 m high with a base 1 m long and 1 m wide.

$$4 \times 9.81 = 39.24 \text{ kN/m}^2$$

The pressure found so far is the total pressure acting upon an area 1 m by 1 m and is known as the intensity of pressure.

$$\text{Intensity of pressure} = \frac{\text{Force}}{\text{Area}} = \frac{\text{kN}}{\text{m}^2}$$

Should the total pressure need to be found for an area larger or smaller than 1 m² then first the intensity of pressure is found and it is simply multiplied by the area which is being acted upon.

Example
Find the total pressure at the base of a boiler which measures 500 mm long, by 300 mm wide, fitted 2.5 metres below the water level in a storage cistern.

$$\text{Intensity of pressure} = \text{head } (2.5) \times 9.81 = 24.525 \text{ kN/m}^2$$

Total pressure = Intensity of pressure × area acted upon
   = 24.525 × area of boiler base (area = length × width)
   = 24.525 × 0.5 m × 0.3 m = 3.6788 kN

By TRANSPOSITION OF FORMULA the height can be found to which the water would rise in a water main.

Example

The mains water pressure is 400 kN/m^2 (4 bar). Ignoring any FRICTIONAL RESIST-ANCES find the height to which the water will rise in a vertical pipe.

$$\text{Intensity of pressure} = \text{head} \times 9.81$$

Transposition of formula

$$\text{Head} = \frac{\text{Intensity of pressure}}{9.81}$$

$$\frac{400 \text{ kN/m}^2}{9.81} = 40.77 \text{ metres}$$

WATER REGULATIONS ADVISORY SCHEME (WRAS)

The independent organisation that advises on the Water Regulations.

WATER SEAL   See TRAP SEAL.

WATER SOFTENER   See BASE EXCHANGE WATER SOFTENER.

WATER SUPPLY REGULATIONS

July 1st 1999 saw for the first time Water Supply Regulations introduced into the UK. Prior to these Regulations there were Water Byelaws, which had variations depending upon the local water supplies. With the Regulations, however, no such variations exist. The document comprises 14 individual Regulations and has been divided into three parts as follows:

- *Part 1* identifies issues such as interpretation and provides definitions of the five fluid categories.
- *Part 2* deals with the bulk of the requirements and sets out what is expected of those who are installing the various water fittings and installations.
- *Part 3* deals with the enforcement of the Regulations and the penalties imposed where contravention occurs.

WATER TABLE

The water level in the subsoil or surrounding ground. The water table is the level below which the ground is saturated with water.

WATER TEST   (HYDRAULIC TEST)

A pressure test applied to below ground drainage installations to check for leaks. Existing drains should not be tested in this way as it is considered too severe a test. To carry out a water test first close up any open ends with a DRAIN PLUG or INFLATABLE STOPPER then fill the drain with water so that it has a minimum test pressure at the highest invert point of 1.5 metres, this pressure can be achieved by using a vertical tube connected to a drain plug, as shown, or by using a special test bend. With the 1.5 metre head achieved allow 2 hours to pass in order to allow for the water absorption by the pipes and joints. Then top up the head pressure to 1.5 metres and there should be no drop greater than 0.05 litres/metre of pipe run over a 30 minute test period (the drop indicated is for a 100 mm pipe). It is advisable to test work in stages as the work progresses, keeping the pipework exposed, therefore any leaks can easily be found. When testing trapped gullies, it is essential to remove the air which would be trapped in the CROWN of the TRAP, this is achieved as shown in the illustration. Note inspection chambers should be tested independently from pipe runs. See also AIR TEST and SMOKE TEST. See also HYDRAULIC PRESSURE TEST for a water test which is carried out to hot and cold water supplies.

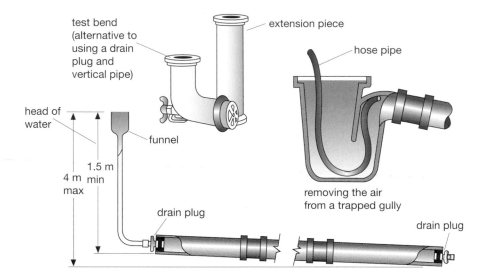

Water test

WATER TOWER
Where a service reservoir cannot be located at a suitable and high enough elevation it is sometimes necessary to install large water reservoirs or tanks at high level. These elevated reservoirs are called water towers and are often seen around the country. See RESERVOIR.

WATER TREATMENT   See CHLORINATION, SAND FILTER and STERILISATION OF INSTALLATIONS WITHIN BUILDINGS.

WATER UNDERTAKER   See WATER AUTHORITY.

WATER WASTE PREVENTER   See FLUSHING CISTERN.

WATERLINE
A line which is marked inside a cistern to identify when the correct quantity of water is contained in the vessel. The supply valve should be adjusted to shut off when the water reaches this line.

WATERLOGGING
A situation which can arise in a STEAM MAIN, in which steam cannot readily pass through the pipe due to a blockage of condensate water. See STEAM HEATING.

WATERWAY

The section through which water flows, see SECTIONAL BOILER for example. The waterway can be defined as the cross-sectional area at any point along a pipe, channel or fitting through which water passes.

WATT

A unit of power: 1 watt produces one joule in one second, see QUANTITY OF HEAT. Many electrical appliances are rated by their consumption of this amount of power (energy).

$$AMPERES \times VOLTS = watts.$$

See also KILOWATT.

WAVING OUT   See TRAP SEAL LOSS.

WC

Abbreviation for WATER CLOSET.

WC BRANCH (WC JUNCTION)   See SANITARY BRANCH.

WC CONNECTING PIPE   See SANITARY CONNECTOR.

WC FLUSHING CISTERN   See FLUSHING CISTERN.

WC MACERATOR PUMP (MUTRATOR PUMP)

A special packaged unit consisting of a macerator and a pump which can be installed behind a WC pan to collect its contents and macerate up any solid matter to allow it to be pumped vertically up to 4 metres or horizontally up to 50 metres and discharged into a small 19 mm nominal bore discharge pipe. Macerator pumps are only permitted to be installed if there is also access to a WC discharging directly into a gravity system of drainage, the reason for this is that should there be an electrical failure, the machine is put out of action. The holding tank to these machines needs to be ventilated to allow for its gravity filling and to facilitate emptying, therefore to prevent unnecessary odours emptying into the room it is recommended that the vent be extended and terminated externally to a safe position.

(See illustration over.)

WC PAN

A soil appliance used for the collection and removal of human solid and liquid excrement. The method of removal of the contents is normally by the flush of water from a flushing cistern as with the WASHDOWN WC PAN or by a flush of water assisted by siphonic action as with a SIPHONIC WC PAN. WC pans can, in general, be of one of three types, either WALL HUNG, pedestal or SQUATTING, the most common being of the pedestal type. See also WC SUITE.

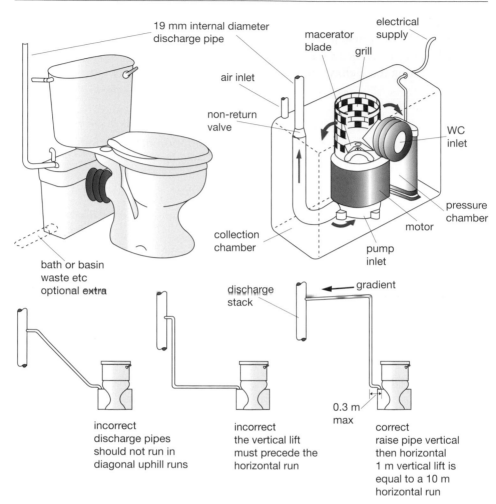

19 mm internal diameter
discharge pipe

air inlet

non-return
valve

collection
chamber

bath or basin
waste etc
optional extra

macerator
blade

grill

electrical
supply

WC
inlet

pressure
chamber

motor

pump
inlet

discharge
stack

gradient

0.3 m
max

incorrect
discharge pipes
should not run in
diagonal uphill runs

incorrect
the vertical lift
must precede the
horizontal run

correct
raise pipe vertical
then horizontal
1 m vertical lift is
equal to a 10 m
horizontal run

WC macerator pump

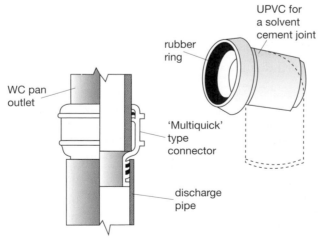

WC pan
outlet

rubber
ring

UPVC for
a solvent
cement joint

'Multiquick'
type
connector

discharge
pipe

WC pan connectors

WC SEAT

There are two basic types of WC seat, the ring type and the open front type, both are designed to make sitting on the WC pan more comfortable. WC seats are generally made of plastic, in a full range of colours, although it is also possible to purchase them made of wood, usually mahogany or oak. The open front type is designed for use in schools and factories, etc. to prevent careless splashes of urine fouling the front surface of the seat. As an alternative to an open front seat, sometimes WC pans with inset pads are used, these solve the problems of rough usage of the appliance, such as slamming down the seat resulting in damage to the pan or seat.

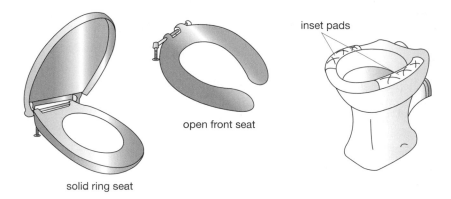

inset pads

open front seat

solid ring seat

WC seats

WC SUITE

A sanitary appliance consisting of a WC PAN, with seat, a FLUSHING CISTERN and, if required, a flushpipe. WC suites are fitted with cisterns at either high level, low level or close coupled. Close coupled WC suites have the flushing cistern bolted directly onto the WC pan. The method of jointing the outlet of the WC pan which leads to the drain/discharge pipe will depend upon the material of which the discharge pipe is made. The most common joint to be made is one which allows the pan to be removed and fixed with ease – such as the WC pan connectors shown opposite. This could be of a type which is manufactured by the makers of plastic pipe, consisting of a rubber collar which fits their own make of pipe. Or a much more

note: high level installations may require a restrictor
at the flushpipe inlet to the pan to prevent splashing
over the rim of the bowl

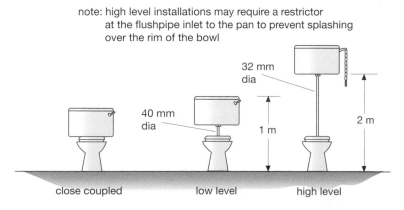

32 mm
dia

40 mm
dia

1 m

2 m

close coupled   low level   high level

WC suite

W

versatile joint is one which is marketed under the trade name of 'Multiquick', this type of joint is made in an assortment of sizes and fits nearly all materials. Solid joints could be made using sand and cement but are not to be recommended as it does not allow for any movement of the pan, also with this method the pan cannot be removed with ease. For the connections to the flush pipe see FLUSHPIPE CONNECTOR.

WEDGES (TAGS/BATS)
Strips of sheet material about 20 mm wide which have been folded over to produce a thick section. They are used to secure a cover flashing into the brick course prior to being filled with cement mortar. Wedges are made from the same material as the metallic cover flashing, thus preventing electrolytic corrosion. For example of a wedge see COVER FLASHING.

WEEP PIPE
A small diameter pipe through which a small quantity of gas can pass from the top side of the diaphragm in a gas valve to the gas thermostat and along to the combustion chamber in a boiler. For example, see RELAY VALVE.

WEIR OVERFLOW
An overflow forming an integral part of some ceramic sinks and wash basins. The overflowing water discharges down the inlet which is in the form of a weir. It is designed in this way so that it is capable of being easily cleaned through most of its length. For example, see SINK.

WELDING
Welding can be defined as the coalescence, or joining of metals using heat, with or without the application of pressure, and with or without the use of a filler rod. The heat required to weld successfully may be supplied by either electricity or a gas flame. Before modern methods of welding, metals were joined by heating them in a forge to welding temperature and hammering or pressing the two metals together. This method of welding now mainly remains as a historical process. All metals to be welded require their own specific rules which must be observed, see ALUMINIUM WELDING, BRONZE WELDING, CAST IRON WELDING, COPPER WELDING, LEAD WELDING, LOW CARBON STEEL WELDING. See also ARC WELDING, WELDING EQUIPMENT, WELDING SAFETY, OXYACETYLENE FLAME AND AUTOGENOUS WELDING.

WELDING BLOWPIPE
There are several types of blowpipe used for various welding processes. The plumber generally uses either the BOC model 'O' or DH blowpipe as shown in the illustration. The model 'O' blowpipe is generally restricted to metals with a low melting temperature such as lead. The cutting blowpipe shown is not used for welding but used when a stream of oxygen is required to cut low carbon steel, see CUTTING BLOWPIPE. When using the welding blowpipe it is essential to use the correct size of welding tip for the job in hand, this ensures the correct amount of heat is produced at the weld surface and helps prevent any FLASHBACK of the flame into the blowpipe. Notice that the blowpipe has one right-handed thread for the connection of the oxygen hose and one left-handed thread for the connection of the acetylene hose. See also WELDING EQUIPMENT.

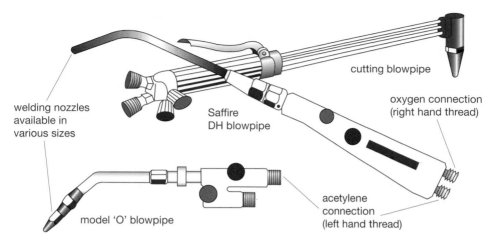

Welding blowpipes

WELDING DEFECTS

A finished weld should have sufficient penetration right down to the ROOT of the weld to ensure the joint will withstand stresses and not fall apart; equally excessive penetration can pose problems especially in low carbon steel where it may lead to blockages, etc. Excessive or insufficient penetration is often the result of the wrong size of welding tip, with the flame being too large or too small. One problem which is often experienced is the undercutting, or thinning, of the material at the weld edge, this is generally caused by welding too fast along the joint without sufficient FILLER ROD or holding the blowpipe welding tip at the wrong angle. The opposite to undercutting is overlapping, generally caused by an excessive build-up of the introduced filler rod. A welding defect which can sometimes be found to a finished bronze welded joint is a series of blow holes, this is caused by the flame being incorrectly adjusted, see BRONZE WELDING.

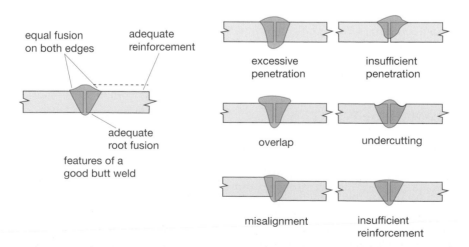

Welding defects

WELDING EQUIPMENT (GAS)

Welding equipment consists of the gases to be used which are supplied in high tensile steel cylinders. The gases most commonly used are oxygen, which is supplied in a black cylinder with a right handed thread for the connection of the regulator; and acetylene, which is supplied in a maroon coloured cylinder with a left handed thread for the connection of the regulator; note the actual valve spindle, to turn the gas on and off, fitted on both cylinders has a right handed thread, and is operated using a special cylinder key, or spanner. Connected to the top of each cylinder is the regulator, followed by a flashback arrestor. It is to these flashback arrestors that the specially made rubber hoses are connected; the acetylene hose is red and the oxygen hose is blue or black. To the other end of the hoses is fitted the required blowpipe with the correct welding tip. See REGULATOR, FLASHBACK ARRESTOR and WELDING BLOWPIPE. When welding equipment is being used one final piece of equipment which must not be forgotten, if the cylinders are to be moved around the site with ease, is a cylinder trolley, which is designed to hold the cylinders secure and upright. See WELDING SAFETY.

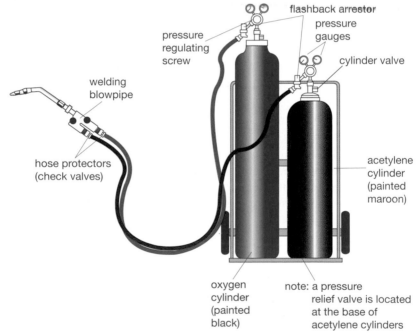

Oxy-acetylene welding equipment

WELDING FLAME   See OXY-ACETYLENE FLAME.

WELDING SAFETY

When carrying out any welding processes several precautions should be taken to ensure the safety of the operator and those around. All personnel concerned with welding should be safety-conscious at all times and as a minimum requirement the following checklist should be observed:

(1) Always wear protective clothing, especially the correct eye goggles or shields.
(2) Ensure good ventilation when welding at all times.

(3) Place any signs or shields as necessary, to give protection and warning to any-one of the welding process taking place.

(4) Always have fire fighting apparatus at hand in case of fire.

(5) Always repair or replace perished or leaking hoses with the correct fittings; do not use odd bits of tubing to join the hoses. Piping or fittings made of copper on no account should be used, as an explosive compound would be produced. Acetylene should never be allowed to come into contact with an alloy containing more than 70% copper.

(6) Store the gas cylinders in a fireproof room. Oxygen should, if possible, not be stored with combustible gases such as acetylene; full and empty cylinders should be kept apart.

(7) Acetylene cylinders should always be stored and used upright to prevent any leakage of the ACETONE.

(8) Oxygen cylinders must not be allowed to fall, as should the valve be broken off the high pressure in the cylinder could cause the cylinder to shoot off like a torpedo, causing extensive damage, for this reason, oxygen cylinders should always be secured. Oil or grease will ignite violently in the presence of oxygen, therefore cylinders should be kept clear of such materials.

(9) Allow an adequate flow of fuel gas to discharge from the blowpipe before lighting up.

(10) In the event of a serious FLASHBACK or BACKFIRE, plunge the blowpipe in a pail of water to cool it, leaving the oxygen running to prevent water entering the blowpipe.

(11) FLASHBACK ARRESTORS should be fitted on the blowpipe to prevent any flash-back into the hoses, also flashback arrestors should be fitted to the regulators to prevent a flashback occurring within the cylinder itself.

(12) Do not weld or cut tanks or vessels which have contained inflammable materials such as paint or petrol in the past without making sure that the vessel has no remaining traces of the substance.

If an acetylene cylinder becomes overheated due to excessive flashback, due to the use of faulty equipment, the cylinder must be shut down and the regulator removed. The cylinder should be taken to the open air at once and immersed in cool water or hosed down. The valve must be re-opened fully allowing the gas out and the cylinder continually kept cool until empty. As this may take several hours further advice should be sought from the supplier of the gas.

WELDING TIP (WELDING NOZZLE)
The position where gas is discharged from a welding blowpipe. The welding tip used depends upon the job in hand, for some examples see LEAD WELDING and LOW CARBON STEEL WELDING. The tips used for a BOC model 'O' blowpipe range from 1–5 and are different from those used for a DH model blowpipe. See also WELDING BLOWPIPE.

WELL   See WATER CYCLE.

WELT
A method sometimes employed to join two sheets of metal together. See SINGLE LOCK and DOUBLE LOCK WELT.

WET PIPE SPRINKLER SYSTEM

A sprinkler system which is permanently charged with water, unlike the DRY PIPE SPRINKLER SYSTEM. This type of system should not be used where there is a possibility of the temperature falling below 0°C causing freezing to the pipework and putting it out of action and/or causing damage. See SPRINKLER SYSTEM.

WET RISER

A 100–150 mm water pipe with one or two 64 mm fire brigade hydrants fitted at each floor, installed vertically in tall buildings over 60 metres in height (20 storeys). The wet riser is used in tall buildings because the mains water pressure would be insufficient to rise to such great heights to assist fighting fires. For buildings less than 60 metres, a dry riser is generally used. The wet riser is charged with water under pressure by a pump capable of delivering 23 litres per second. The water supply to the pumps should be supplied via a BREAK CISTERN of not less than 11500 litres, although some fire authorities require the break cistern to be much larger. Pumping equipment is normally required at about every 20th floor to enable the pumps to be able to push the water to higher levels. Pumping equipment should always be installed in duplicate in case of failure, also an alternative electrical supply or method of pumping should be available in case of power cuts. The water pressure supplied to

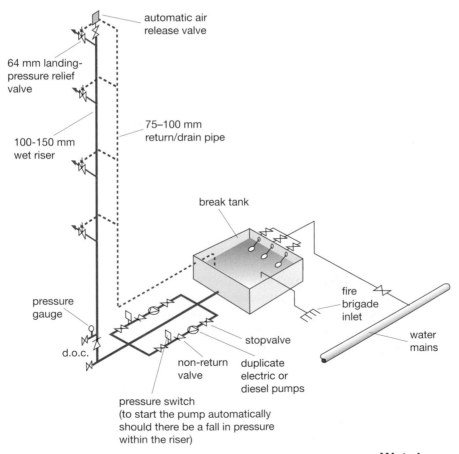

automatic air release valve

64 mm landing-pressure relief valve

100-150 mm wet riser

75–100 mm return/drain pipe

break tank

fire brigade inlet

water mains

pressure gauge

d.o.c.

stopvalve

non-return valve

duplicate electric or diesel pumps

pressure switch (to start the pump automatically should there be a fall in pressure within the riser)

Wet riser

the hydrants on each floor should be a minimum of 410 kN/m² but equally should not exceed 690 kN/m² otherwise damage could be done to the hosepipe when it is connected to, the riser. As a safety precaution a PRESSURE RELIEF VALVE is fitted to the connection of each hydrant; should this valve blow open the excess water returns to a 75–100 mm return pipe and is discharged back to the break cistern. See also DRY RISER.

WET VENTING

A VENTILATING PIPE into which water is allowed to enter to allow it to flush any rust particles, etc. which may settle at the base of a main ventilating pipe. See also RUST POCKET.

WETTED PERIMETER

The distance around the inside perimeter or surface of a pipe or channel which is wet due to contact with water.

WETTING   See TINNING.

WHEEL HEAD

A valve with a round head which can be turned to fully open or close the valve. See also LOCKSHIELD HEAD and RADIATOR VALVE.

WHEEL PIPE CUTTER   See PIPE CUTTER.

WHIRLPOOLS AND SPAS

A selection of baths which are designed to relax the body. There are many variations of these baths which have been specially designed to pump air or circulate water rapidly through jets into the large volume of water causing a mass of bubbles to form which give a stimulating effect to the body.

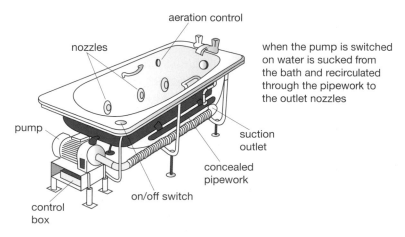

Whirlpool bath

WIPED SOLDER JOINT

The diagram shows a collection of wiped solder joints which were for many years carried out by the plumber. Today however, lead is prohibited for use on water supply pipes. Occasionally the plumber may be called upon to join the lead pipe but some form of compression or pushfit joint will be used. However, in rare cases, an old lead waste discharge stack may need to be connected to, in this case the wiped soldered joint is the only option. Wiped solder joints have been carried out for many years and before the introduction of blowlamps the solder used was heated in a melting pot and when ready was poured onto horizontal pipes or splashed onto the vertical pipes using a ladle and SPLASH STICK, then the joint was moulded as necessary. When the blowpipe was introduced, the pot and ladle method tended to die out.

Wiping a joint with a blowlamp is much easier, especially to vertical joints or in awkward positions. Wiping joints with a blowlamp means heating solder only as and when it is required and having heat available to keep the solder in a nice plastic range. The solder used for wiped joints is either BS grades D, J, or L each of which have a long plastic range. As can be seen there are many designs of finished joint and a lot of practice is required to achieve a good symmetrical finish. Possibly the simplest joint to achieve is the taft joint which is carried out as follows:

(1) Rasp the ends of the pipe square.
(2) Rasp down the spigot end to a feather edge.
(3) Open the socket end with a TURN PIN.
(4) Key the pipe ends with a wire brush and apply chalk followed by a coating of PLUMBER'S BLACK.
(5) Scribe around the spigot end with a pair of dividers to the length of the joint.
(6) Clean, with a SHAVE HOOK, the spigot end up to the scribe mark and the internal edge of the belled out socket.
(7) Apply TALLOW to the cleaned surfaces; heat the two ends and 'tin' the joint (the application of a thin coating of solder).
(8) Assemble the two ends firmly together and secure with FIXING POINTS.
(9) Apply heat to the joint and PLUMBER'S MATAL as necessary, enough to mould the solder, with a WIPING CLOTH, to the rough shape required.
(10) Finally run the wiping cloth around the joint in one continuous sweep to finish it as shown.

All the other joints are carried out in a similar way. The underhand joint shown can also be carried out using a different technique which is known as a ROLLED JOINT.

Wiped solder joints today are generally limited to waste pipe connections.

(See illustration opposite)

WIPING CLOTH

A cloth which is used to assist making wiped solder joints. Wiping cloths are made of fustian, often known by plumbers as moleskin, which is a strong heavy cotton fabric. Wiping cloths can be purchased ready-made or a sheet of fustian can be cut and folded as required by the plumber, they should be a minimum of 15–20 mm wider than the joint length. If the cloths are to be made, it should be noted that fustian has a grain and when folding the cloth to achieve the thickness required (a minimum of 6 layers) to prevent the heat from the solder burning you, the grain should run lengthways with the cloth, for example, see ROLLED JOINT. Sometimes a piece of stiff card is inserted with the folds to assist keeping the underhand wiped

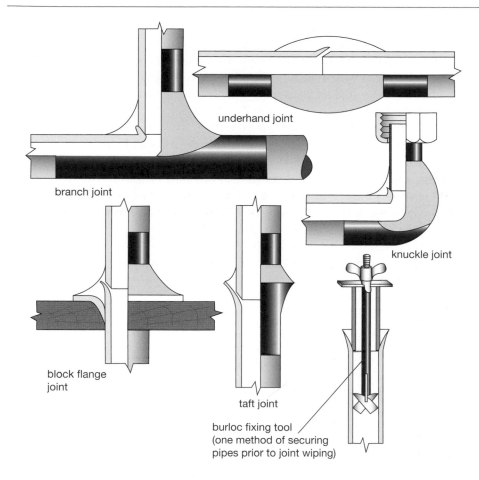

underhand joint

branch joint

knuckle joint

block flange joint

taft joint

burloc fixing tool
(one method of securing
pipes prior to joint wiping)

Wiped solder joints

joints uniform in shape. Before using a new cloth TALLOW should be rubbed well into it, making it supple and giving it protection from the burning solder.

WIRE BALLOON   See BALLOON.

WIRE WOOL (STEEL WOOL)
Strands of steel fibre sold in bundles and used to clean metals as necessary, usually for soldering purposes.

WOBBE INDEX
The heat given off when a gas is burnt at a constant pressure and flow. It is expressed as:

$$\frac{\text{gross calorific value of the gas}}{\sqrt{\text{relative density of the gas}}}$$

For example, see FAMILY OF GASES.

WOOD-CORED ROLL (SOLID ROLL)

An EXPANSION JOINT used on metallic roof coverings running with the fall of a roof. The term wood'cored roll generally refers to rolls used on LEAD SHEET AND ROOF COVERINGS but more widely the term wood-cored roll could be used to mean any of the BATTEN ROLLS as used for copper and zinc.

WORK HARDENED

All metals are formed of many crystals, when the metal has had continuous BOSSING or hammering these crystals become pushed together and crammed, restricting their movement, and resulting in their distortion; in this state the metal is not so DUCTILE and will seem very hard. One should soften the metal if it becomes work hardened, by ANNEALING, as continuous working of the metal could lead to a fracture.

WORKING HEAD

The water pressure created by the HEAD of water to which a pipe, vessel, pump, etc. is designed to work.

WORKING PRESSURE

The normal pressure to which a pipe, fitting or other such vessel is liable to be subjected under ordinary working conditions. See also GAS PRESSURE and METER PRESSURE.

WORKING PROPERTIES   See PHYSICAL AND WORKING PROPERTIES.

WORSHIPFUL COMPANY OF PLUMBERS

One of the ancient livery companies of the City of London, dating back to the fourteenth century and originating in a craft guild. The Grant of Arms to which this company is honoured depicts St Michael the Archangel, the patron saint of plumbing.

WRAS   See WATER REGULATIONS ADVISORY SCHEME.

WRENCH (PIPE WRENCH)

A tool used to grip or turn pipes and other round objects in the absence of a nut shaped grip. There are several types and they are available in a range of sizes. When using any pipe wrench one must look to see that the teeth are not clogged up with jointing compound, etc. and that any turning action would permit the teeth to bite into the pipe, otherwise they will not grip. The lever wrench shown is designed to grip or clamp material as well as turn it. It should be noted that in America the term wrench is also used to identify a spanner. The handle should never be extended with a pipe in order to gain more leverage as it could prove dangerous and could damage the tool. See also SPANNER, BASIN SPANNER and STRAP WRENCH.

(See illustration opposite)

WRIST ACTION TAP

A tap which is identical to an ELBOW ACTION TAP except that it has a shorter operating handle.

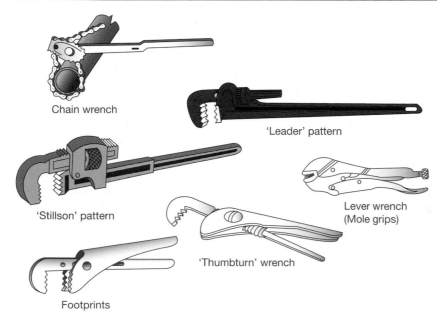

Chain wrench

'Leader' pattern

'Stillson' pattern

Lever wrench
(Mole grips)

'Thumbturn' wrench

Footprints

Pipe wrenches

WROUGHT IRON

A metal consisting of almost pure iron usually containing less than 1% carbon. It appeared in the construction of buildings in the nineteenth century where strength and tension made it superior to cast iron. However, since the introduction of steel, its use in the twentieth century has been mainly decorative.

WWP

Abbreviation for water waste preventer. See FLUSHING CISTERN.

W

Y BRANCH (Y JUNCTION)
A branch in the shape of a letter Y, see BRANCH for example. See also BREECH FITTING.

YARD GULLY (MUD GULLY)
A large GULLY used to collect the SURFACE WATER from paved surfaces. Provided with a grating and large sediment bucket which should periodically be cleaned out. See GARAGE GULLY.

YARN (CAULKING YARN/GASKIN)
Course vegetable fibres which have been spun together to form a rope, it is mostly used to assist making caulked cast iron pipe joints, see CAULKED JOINTS. See also TARRED YARN.

YARNING IRON (YARNING TOOL)
A tool used to compact the YARN down into the socket of a cast iron pipe when making a CAULKED JOINT.

Yarning iron

YIELD POINT
The point at which a metal is strained beyond its elastic limit and any change in shape is permanent. If a test piece is subjected to a pressure or load, the force will tend to alter the shape of the metal. Due to the amount of elasticity within the test piece, caused by the power of attraction of the molecules, the metal will resist distortion, up to a certain point; should the force be removed below this critical point, the metal acts like a piece of elastic and returns to its original shape. If the applied force is greater than the force of the molecular attraction, the metal will yield to the greater force, or become distorted, and the distortion will be permanent. This critical point of distortion is the yield point. See also ELASTICITY.

YOKE VENT   See CROSS-VENT.

Z

ZEOLITES

Minerals used in the base exchange water softener. There are two types of zeolites, the natural and the synthetic. The open framework structure of zeolites consists of linked rings of silicate tetrahedra (4 oxygen atoms surrounding 1 central silicon atom) that provide various channels through which water MOLECULES and various organic molecules may pass, but the calcium carbonates and calcium sulphates in HARD WATERS are sieved, preventing their passage. Natural zeolites are often produced by treating glauconite (or greensand) which is only found in abundance in sea floor areas that are isolated from land-derived sediment. Synthetic zeolites are of a gel-type, known as hydrated sodium aluminosilicates. The exchange value of synthetic zeolites can be almost double that of natural zeolites if excessive amounts of salt are used in its regeneration process. See BASE EXCHANGE WATER SOFTENER.

ZINC

A NON-FERROUS METAL used extensively for GALVANISING to give a protective coat to the FERROUS METAL underneath, in its natural state it is found most commonly in sheet form. See also, ZINC ALLOY and ZINC SHEET AND ROOF COVERINGS.

Chemical symbol	Zn
Colour	greyish-white
Melting point	419°C
Boiling point	907°C
Coefficient of linear expansion	0.000029/°C
Density	7130 kg/M^3

ZINC ALLOY

Due to the fact that 'commercial quality' zinc is very hard to work, zinc ALLOYS are marketed under the names 'Metizinc' and 'Metiflash'. These alloys are very easily bent and formed without the hard hammering required with commercial quality zinc. Metizinc is an alloy of zinc, titanium and copper. The titanium is added to give strength and the copper is added to improve the DUCTILITY of the metal. Metiflash is an alloy of zinc and lead.

ZINC CHLORIDE (KILLED SPIRITS)

An ACTIVE FLUX in liquid form which is used when SOFT SOLDERING materials, such as zinc, copper, brass or iron. The FLUX is simply made by the plumber by dissolving small pieces of zinc in hydrochloric acid (spirits of salts) until all reaction between the two ceases.

Z

ZINC EAVES GUTTERS

Zinc EAVES GUTTERS are rarely fixed today and in most cases any zinc guttering is carried out only as a means of repair work. Zinc guttering and fittings, etc. are not purchased but made on site to suit the requirements. All the sections formed are usually lapped by at least 50 mm in the direction of the fall and soldered together, alternatively the sections are butted together and made watertight by soldering a coverstrip over the two edges to be joined. To give the gutter strength and increased rigidity a beaded edge is formed at the front edge, also gutter bracing rods or stays are fixed at every 380 mm, see sketch. The fixing of the zinc gutter to the building is made through every alternate bracing rod with a long GALVANISED screw, or via an external bracket fixed at every 600 mm. It is also possible to form rainwater pipes and fittings with zinc by rolling a piece around a suitably sized steel pipe and forming a welt or soldered joint as necessary, but if required zinc pipe and fittings can be purchased.

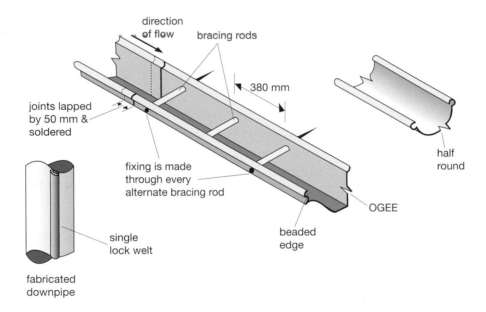

Zinc eaves gutters

ZINC SHEET AND ROOF COVERINGS

Sheet zinc is available in commercial quality, which is very hard to work or as a ZINC ALLOY, which has similar characteristics to aluminium. Sheet zinc has a grain along its length and it will be found that the sheet can be more easily bent across the grain than with it; therefore especially with commercial quality zinc, it is advisable to make all bends with round edged tools, in order to give the bend a slight radius. Zinc sheet can be obtained in almost any size, although the standard size usually specified is 2438 mm × 914 mm; the zinc alloys are also obtainable in rolls of various widths from 150–1000 mm. Sheet zinc comes in a range of thicknesses and the thickness used would depend on circumstances and cost. Sometimes sheet zinc is specified by a zinc gauge which must not be confused with the Imperial STANDARD WIRE GAUGE (see p. 408).

The working properties of zinc are not as good as materials such as aluminium, copper and lead therefore a special design of roofwork is employed called the roll cap system, although it is possible to join the sheets using the STANDING SEAM method. In the roll cap system for joints which run across the flow of water (transverse joints) the method of jointing is either by drips or for roofs with a pitch greater than 1 in 8 (7°) SINGLE LOCK WELTS may be used with a welt size of 45 mm. The joints which run with the fall of the roof are made with a special design of batten roll as shown, the capping piece being held down by a specially formed clip. The drips may be formed in two ways, either as a welt or a beaded edge. The welted edge is usually the easiest to form, note the welt is not turned down as with copper drips. Large roofs are divided up into smaller areas called bays, the size of the bay is usually determined by the size of the sheets obtained. Once the structure has been covered the upstanding edges are weathered with a COVER FLASHING with which the lower edge is beaded to give better resistance to wind pressures. Stainless steel, introduced into this country as a roofing material in the early 1980s, is laid onto a structure using this roll cap design of covering. See also LONG STRIP ROOFING.

Zinc sheet applicable to roof work

mm	Nearest zinc gauge	Nearest swg	Approx weight kg	Uses
0.65	12	23	10.5	soakers
0.7	13	22	11.5	soakers
0.8	14	21	12.5	gutters and roll cap
0.9	15	20	14.5	gutters and roll cap
1.0	16	19	16.0	corrugated sheet without decking

Note The weight given is approximate for sheets 2438×914 mm.

(See illustration over.)

ZINC SOLDERING

Zinc can easily be soldered if necessary, but should, whenever possible, be avoided due to corrosion problems. The FLUX used should be zinc ammonium chloride; which is produced by mixing about 20% ammonium chloride (salammoniac) with 80% zinc chloride (killed spirits). Zinc ammonium chloride is an INACTIVE FLUX and should minimise corrosion due to acid attack. Alternatively zinc chloride can be used directly to the zinc if the metal cannot be suitably cleaned prior to soldering. The solder used should be BS grades F or K which contain 50/50% or 60/40% tin-lead respectively. The soldering iron used to make the joint should not be overheated, nor applied to the zinc longer than is necessary, due to the low melting temperature of zinc. Upon completion of any soldering all flux should be thoroughly washed off.

ZINC SOLVENT

The ability of water to dissolve zinc. See SOLVENCY.

ZONE CONTROL

A situation where a building has been divided, so to speak, into different areas for control of the areas for the purpose of fire protection or heating control. For

Z

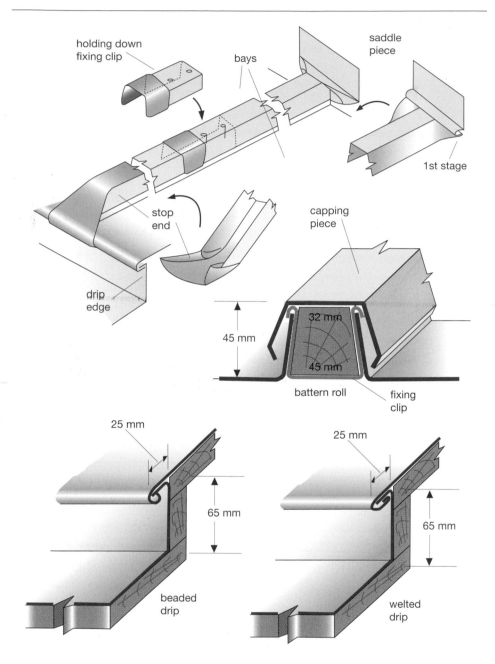

holding down fixing clip

bays

saddle piece

1st stage

stop end

capping piece

drip edge

45 mm

32 mm

45 mm

battern roll

fixing clip

25 mm

65 mm

beaded drip

25 mm

65 mm

welted drip

Allowance for expansion on zinc sheet roof coverings

example, under current Building Regulations where a dwelling has heating, it will need to be designed so that different temperature zones are provided to ensure that unnecessary heat is not lost. Generally no zone or sub zone should exceed 150 m².

ZONE VALVE   See MOTORISED VALVE.